Hermann König • Vitali Milman

Operator Relations
Characterizing Derivatives

 Birkhäuser

Hermann König
Mathematisches Seminar
Universität Kiel
Kiel, Germany

Vitali Milman
School of Mathematical Sciences
University of Tel Aviv
Tel Aviv, Israel

ISBN 978-3-030-13096-1   ISBN 978-3-030-00241-1   (eBook)
https://doi.org/10.1007/978-3-030-00241-1

Mathematics Subject Classification (2010): 39B42, 47A62, 26A24, 47B38, 47J05, 26B05, 39B22

This book is published under the imprint Birkhäuser, www.birkhauser-science.com by the registered company Springer Nature Switzerland AG.
The registered company address is: Gewerbestrasse 11, 6330 Cham, Switzerland

Contents

Chapter 1

Introduction

The purpose of this book is to explain some recent results in analysis, which in general terms may be described as follows: Major constructions or operations in analysis are often characterized in a natural and unique way by some very elementary properties, relations or equations which they satisfy.

A simple example on the real line would be the exponential function mapping sums to products.

To describe the basic theme of the book, let us consider the classical *Fourier transform* $\mathcal{F}$ on $\mathbb{R}^n$ given by

$$\mathcal{F}(f)(x) = \int_{\mathbb{R}^n} \exp(-2\pi i \langle x, y \rangle) \, f(y) \, dy$$

on the Schwartz space $\mathcal{S}(\mathbb{R}^n)$ of "rapidly" decreasing smooth functions $f : \mathbb{R}^n \to \mathbb{C}$. As well-known, $\mathcal{F}$ maps bijectively the Schwartz space onto itself and exchanges products with convolutions. The interesting fact is that these properties (almost) characterize the Fourier transform. As shown by Artstein-Avidan, Faifman and Milman [AFM], any bijective transformation $T : \mathcal{S}(\mathbb{R}^n) \to \mathcal{S}(\mathbb{R}^n)$ satisfying

$$T(f \cdot g) = T(f) * T(g)$$

for all $f, g \in \mathcal{S}(\mathbb{R}^n)$ is just a slight modification of the Fourier transform: there exists a diffeomorphism $\omega : \mathbb{R}^n \to \mathbb{R}^n$ such that either $T(f) = \mathcal{F}(f \circ \omega)$ for all $f \in \mathcal{S}(\mathbb{R}^n)$ or $T(f) = \overline{\mathcal{F}(f \circ \omega)}$ for all $f \in \mathcal{S}(\mathbb{R}^n)$.

Assuming in addition that T also maps convolutions into products,

$$T(f * g) = T(f) \cdot T(g)$$

for all $f, g \in \mathcal{S}(\mathbb{R}^n)$, the diffeomorphism ω is given by a linear map $A \in GL(n, \mathbb{R})$ with $|\det(A)| = 1$, i.e., $T(f) = \mathcal{F}(f \circ A)$ for all $f \in \mathcal{S}(\mathbb{R}^n)$ or $T(f) = \overline{\mathcal{F}(f \circ A)}$ for all $f \in \mathcal{S}(\mathbb{R}^n)$. Details are found in the papers by the above authors and Alesker [AAM], [AAFM] and [AFM].

© Springer Nature Switzerland AG 2018
H. König, V. Milman, *Operator Relations Characterizing Derivatives*,
https://doi.org/10.1007/978-3-030-00241-1_1

Note that T is not assumed to be linear or continuous. Nevertheless the real linearity and the continuity of T are a consequence of the result.

A priori the Fourier transform is an analytic construction. However, it is essentially uniquely recovered by the basic properties which we mentioned. Starting with the simple formula for a bijective map exchanging products and convolutions, we get an operation which has a rich structure and extremely useful properties.

There are some other properties which also characterize the Fourier transform, e.g., the Poisson summation formula, cf. Faifman [F1], [F2].

The exponential function $e = \exp : \mathbb{R} \to \mathbb{R}$ on the real line is, up to multiples, characterized by its *functional equation*

$$e(x + y) = e(x) \cdot e(y)$$

for all $x, y \in \mathbb{R}^n$, if measurability of e is assumed, see Aczél [A]. In comparison, the Fourier transform $T = \mathcal{F} : \mathcal{S}(\mathbb{R}^n) \to \mathcal{S}(\mathbb{R}^n)$ is characterized, up to diffeomorphism and complex conjugation, by being bijective and satisfying the *operator functional equation*

$$T(f \cdot g) = T(f) * T(g)$$

for all $f, g \in \mathcal{S}(\mathbb{R}^n)$. Therefore we recover a classical transform in analysis by an elementary relation, namely the above operator functional equation. Note that the operator T is not assumed to satisfy *any* regularity condition like continuity or measurability as in the case of the above map e.

Following a similar approach, we will study in this book the question to which extent the derivative is characterized by properties like the *Leibniz rule operator equation*

$$T(f \cdot g) = T(f) \cdot g + f \cdot T(g)$$

or the *chain rule operator equation*

$$T(f \circ g) = T(f) \circ g \cdot T(g)$$

on classical function spaces like the spaces C^k of k-times continuously differentiable functions, $T : C^k \to C$, $f, g \in C^k$. We will determine all solutions of either one equation and also of various extensions of them. In most cases, we will a priori assume neither continuity nor linearity or another algebraic property of the operator T. However, a posteriori, a natural type of continuity of T will be a consequence of the result.

Simple additional *initial* conditions like $T(-2\,\mathrm{Id}) = -2$ in the case of the chain rule will guarantee that T is actually the derivative, $Tf = f'$, and, in particular, linear, see Chapter 4.

Returning to the Fourier transform, suppose that $T : \mathcal{S}(\mathbb{R}^n) \to \mathcal{S}(\mathbb{R}^n)$ is bijective and satisfies $T(f \cdot g) = T(f) * T(g)$. By the properties of the Fourier transform, $J := \mathcal{F} \circ T$ then satisfies $J(f \cdot g) = J(f) \cdot J(g)$ for all $f, g \in \mathcal{S}(\mathbb{R}^n)$. To prove the result for T, it therefore suffices to determine all bijective multiplicative maps $J : \mathcal{S}(\mathbb{R}^n) \to \mathcal{S}(\mathbb{R}^n)$, $J(f \cdot g) = J(f) \cdot J(g)$, i.e., solve another simple operator functional equation on a classical function space of analysis. In this case there is a diffeomorphism $\omega : \mathbb{R}^n \to \mathbb{R}^n$ such that either $Jf = f \circ \omega$ for all $f \in \mathcal{S}(\mathbb{R}^n)$ or $Jf = \overline{f \circ \omega}$ for all $f \in \mathcal{S}(\mathbb{R}^n)$, cf. [AFM]. Bijective multiplicative maps on relevant functions spaces of analysis were studied before in the papers of Milgram [M] and of Mrčun and Šemrl [Mr], [MS].

Another transformation which is important in analysis and geometry is the Legendre transform $\mathcal{L}$. Let $\mathcal{C}_n$ denote the class of all lower-semi-continuous functions $\phi : \mathbb{R}^n \to \mathbb{R} \cup \{\pm\infty\}$ and fix some scalar product $\langle \cdot, \cdot \rangle$ on $\mathbb{R}^n$. The *Legendre transform* of ϕ, also called *Legendre–Fenchel transform* in higher dimensions, is given by

$$\mathcal{L}(\phi)(x) = \sup[\langle x, y \rangle - \phi(y)], \quad \phi \in \mathcal{C}_n, \ x \in \mathbb{R}^n.$$

Then $\mathcal{L}(\phi) \in \mathcal{C}_n$, $\mathcal{L}$ is an involution , i.e., $\mathcal{L}^2(\phi) = \phi$, and $\mathcal{L}$ is order-reversing, i.e., $\phi \leq \psi$ implies $\mathcal{L}(\phi) \geq \mathcal{L}(\psi)$ for all $\phi, \psi \in \mathcal{C}_n$. Being an involution and order-reversing are the most basic properties of a "duality" relation, which is a natural operation having many other interesting and very useful consequences. In fact, they nearly characterize the Legendre transform. By a result of Artstein-Avidan and Milman [AM], for any order-reversing involution $T : \mathcal{C}_n \to \mathcal{C}_n$ there is a symmetric linear map $B \in GL(n, \mathbb{R})$ and there are $v_0 \in \mathbb{R}^n$ and $c_0 \in \mathbb{R}$ such that T has the form $T(\phi) = \mathcal{L}(\phi \circ B + v_0) + \langle \cdot, v_0 \rangle + c_0$, $\phi \in \mathcal{C}_n$. So up to affine transformations, T is the Legendre transform.

The general problem considered in this book, whether basic constructions or operations in analysis or geometry are essentially characterized by very simple properties like order-reversion or some functional operator equations, was actually motivated by the question what "duality" or "polarity" means in convex geometry and convex analysis. Let $\mathcal{K}_n$ denote the class of closed convex bodies with 0 in its interior. For $K \in \mathcal{K}_n$, the *polar body* $K^\circ \in \mathcal{K}_n$ is given by

$$K^\circ = \{x \in \mathbb{R}^n \mid \text{ for all } y \in K : \langle x, y \rangle \leq 1\}.$$

Then the map $K \mapsto K^\circ$ from $\mathcal{K}_n$ to itself is an involution which is order-reversing, $K \subset L$ implying $K^\circ \supset L^\circ$ for all $K, L \in \mathcal{K}_n$. Gruber [Gr] (in a different language), Böröczky, Schneider [BS] and Artstein-Avidan, Milman [AM] (in different setups) showed that conversely any involution $T : \mathcal{K}_n \to \mathcal{K}_n$ which is order-reversing, $K \subset L$ implying $T(K) \supset T(L)$, is actually the polar map, up to linear transformations: There exists $B \in GL(n, \mathbb{R})$ such that T is given by $T(K) = (B(K))^\circ$ for all $K \in \mathcal{K}_n$. The result for the Legendre transform is a corresponding duality result for convex functions instead of convex bodies.

Let us turn back to analysis. The main attention in this book will be given to the study of properties of the derivative, like the Leibniz or the chain rule, and to the question to what extent any of these operator functional equations will nearly characterize the derivative, or what other solutions they admit. We also consider characterizations of the Laplacian and other second-order derivative operations.

It is interesting to compare these classical operations in analysis with functional equations which the exponential function or the logarithm satisfy. The logarithm $\log : \mathbb{R}^+ \to \mathbb{R}$ sends products to sums, $\log(xy) = \log(x) + \log(y)$ for $x, y \in \mathbb{R}^+$. However, on a linear class of functions an analogous non-trivial operation T satisfying $T(f \cdot g) = T(f) + T(g)$ does not exist: considering $g = 0$, one finds that $Tf = 0$ for all f, i.e., $T = 0$. Let us change this operator equation slightly by allowing some "tuning" operators A_1, A_2 which will act on a larger linear space of functions (of "lower" order). For example, consider $T : C^1 \to C$ and $A_1, A_2 : C \to C$ such that

$$T(f \cdot g) = T(f) \cdot A_1(g) + A_2(f) \cdot T(g) \tag{1.1}$$

for all $f, g \in C^1$. Then T still maps in some sense products to sums, with some correction by the tuning operators A_1 and A_2. Clearly, if $A_1 = A_2 = \mathrm{Id}$, we just get the *Leibniz rule equation*, or simply *Leibniz equation*,

$$T(f \cdot g) = T(f) \cdot g + f \cdot T(g), \ \ f, g \in C^1.$$

In addition to the derivative, $T(f) = f'$, also the entropy operation $T(f) = f \ln|f|$ satisfies this equation in C^1 or C, reflecting a logarithmic behavior. The general solution of the Leibniz rule equation turns out to be a linear combination of the derivative and the entropy operation. We prove this in Chapter 3. In this extended interpretation, the derivative operation is an *analogue of the logarithm on linear spaces of functions*. We also determine the solutions of the more general equation (1.1) in Chapter 3.

Another algebraically inspired, interesting aspect of the derivative is illustrated by the chain rule equation

$$T(f \circ g) = T(f) \circ g \cdot T(g) \tag{1.2}$$

for all $f, g \in C^k(\mathbb{R})$, $k \in \mathbb{N}$. In this case T maps the composition $f \circ g = f(g)$ to a "compound" product $T(f) \circ g \cdot T(g)$. Since the information on the left-hand side of the equation involves the composition $f \circ g$ and not individually f and g, on the right-hand side also the composition with g is needed, when f appears, to yield meaningful solutions. A simple product equation $T(f \circ g) = T(f) \cdot T(g)$ only admits the trivial solutions $T = 0$ and $T = 1$. The solutions of the chain rule operator equation are classified in Chapter 4. There are other solutions besides the derivative (and its powers), but the derivative can be characterized by the chain rule and an additional initial condition, e.g., $T(-2\,\mathrm{Id}) = -2$.

Suppose the terms in the chain rule equation (1.2) for T are positive. Then $P := \log T$ satisfies

$$P(f \circ g) = P(f) \circ g + P(g), \ \ f, g \in C^1(\mathbb{R}), \tag{1.3}$$

mapping compositions to a "compound sum". Note that equation (1.3) makes sense for all, and not only for positive functions. Again, compositions with g are needed on both sides when terms with f appear.

On linear classes of functions like $C^1(\mathbb{R})$ or $C(\mathbb{R})$, the solutions of equation (1.3) are easily described: there is a continuous function $H \in C(\mathbb{R})$ such that

$$Pf = H \circ f - H.$$

This solution by itself is not very interesting. So let us add on the right-hand side of (1.3) some "tuning" operators, as we did in (1.1). This yields the operator equation

$$T(f \circ g) = T(f) \circ g \cdot A_1(g) + A_2(f) \circ g \cdot T(g) \tag{1.4}$$

with three operators T, A_1, A_2.

One solution of this equation is well-known, namely the second derivative $T(f) = D^2(f) = f''$, with $A_1(f) = (f')^2$ and $A_2(f) = f'$. Natural domains are $C^2(\mathbb{R})$ for T and $C^1(\mathbb{R})$ for A_1, A_2, so A_1, A_2 may be considered of "lower order". In our interpretation, this second-order chain rule equation appears after a logarithmic operation is applied to the first-order chain rule, which then is appropriately "tuned".

We study the solutions of equation (1.4) in Chapter 9 under a mild condition of non-degeneration, and determine all triples of operators (T, A_1, A_2) on suitable $C^k(\mathbb{R})$-spaces which lead to nontrivial solutions.

The operators (T, A_1, A_2) are intertwined by (1.4), and there are fewer types of solutions than one might imagine at first. There are non-trivial solutions for T on $C^k(\mathbb{R})$-spaces for $k \in \{0, 1, 2, 3\}$, with appropriately chosen tuning operators A_1, A_2. On $C^k(\mathbb{R})$-spaces for $k \geq 4$ there are no further solutions, i.e., solutions which might depend on the fourth or higher derivatives. The only solution for $k = 0$ was already described above, $Tf = H \circ f - H$, with $A_1 = A_2 = \mathbf{1}$.

For $k = 1$ there are three different families of solutions, where all operators act on C^1. For $k = 2$, in addition to the solutions mentioned for $k = 0, 1$, there is very little diversity for the operators A_1, A_2. They are again defined on C^1 with $A_1(f) = f' \cdot A_2(f)$ and $A_2(f) = |f'|^p \{\operatorname{sgn} f'\}$ for a suitable $p \geq 1$. The term $\{\operatorname{sgn} f'\}$ may appear here or not, yielding two solutions. The main operator T is described by the above value of p and two continuous parameter functions c, H, $c \neq 0$, namely

$$Tf = (cf'' + [H \circ f - H] \cdot f') \cdot |f'|^{p-1} \{\operatorname{sgn} f'\}.$$

So for $H = 0$, we essentially get the second derivative. Suitable additional initial conditions determine the form of T. Requiring, e.g., $T(\operatorname{Id}^2) = 2$ and $T(\operatorname{Id}^3) = 6 \operatorname{Id}$, with $\operatorname{Id}^l(x) = x^l$, $l \in \mathbb{N}$, yields $T(f) = f''$, $A_1(f) = f'^2$ and $A_2(f) = f'$.

The case $k = 3$, in addition to the solutions for $k = 0, 1, 2$, leads to solutions in terms of the Schwarzian derivative S. In this case $A_1(f) = f'^2 A_2(f)$, where A_2

has the same form as for $k = 2$ but with $p \geq 2$. The most interesting solution is $T(f) = f'^2 S(f)$, $A_1(f) = f'^4$ and $A_2(f) = f'^2$, cf. Chapter 9.

One main step of our method is the localization technique, which allows to reduce operator equations to functional equations. We show, e.g., in the case of the chain rule equation (1.2) that any non-degenerate solution $Tf(x)$ is determined by some function F of the variables x, $f(x)$ and the derivatives of f at x up to order k, if $f \in C^k(\mathbb{R})$ and $x \in \mathbb{R}$. No regularity of this function F is known at the outset but has to be proved later. We show that the operator equation for T then turns into a functional equation for F, the solutions of which have to be determined. Usually then we have to prove the continuity or regularity of the coefficient functions appearing in the structure of the solutions of F. Functional equations and regularity results for them are studied in Chapter 2, in preparation for later application in subsequent chapters.

We already mentioned that various of our results are proved under some condition of non-degeneration. There are two different forms and reasons for this type of assumption.

One of them is a very weak form of surjectivity of the operator. This together with the operator equation will often yield in the final result that T is actually surjective. For example, the assumption in Theorem 4.1 for the chain rule equation only requires as non-degeneration condition that T is not the zero operator on the half-bounded functions, allowing a complete description of all solutions.

A very different type of non-degeneration is required when two or three different operators appear in the equation, such as in (1.1). We then need, e.g., that a tuning operator A will not be proportional on some open interval to the operator T, cf. Theorem 3.7, or not be proportional to the identity, cf. Theorem 7.2. By these conditions of non-degeneration we avoid a "resonance" behavior of two different operators, which often has the consequence that they are not localized. In the case of equation (1.1) there is, e.g., the following non-localized solution

$$T(f)(x) = f(x) - f(x+1), \quad A_1(f)(x) = A_2(f)(x) = \frac{1}{2}(f(x) + f(x+1)),$$

where the operators T, A_1 and A_2 act from $C(\mathbb{R})$ to itself. For functions with small support around x, T here acts as identity and A_1 and A_2 are homothetic to the identity. These effects typically appear with Leibniz rule type equations which are studied in Chapters 3 and 7. The exact form of the non-degeneration condition differs from one chapter to the other, but stays the same in each chapter.

In some cases we may avoid the assumption of non-degeneration and prove theorems about the general structure of the solutions of equations like (1.1) without localization. These results are found in Chapter 8.

Interestingly enough, the equations we consider in this book show some unexpected stability or even rigidity. Perturbing the Leibniz rule equation by a "small" additive term yields solutions which are perturbations of the original equation. In the case of the chain rule, we even have rigidity: the perturbed solutions have the same solutions as the original equation. The chain rule equation allows no reasonable additive perturbation. This is shown in Theorems 5.6 and 5.8.

In the case of the chain rule, the rigidity even allows us to study the solutions of the *inequality*

$$T(f \circ g) \leq T(f) \circ g \cdot T(g), \quad f, g \in C^1(\mathbb{R}).$$

To completely describe the solutions of this operator inequality, under some non-degeneration condition and a weak continuity assumption, we have to prove localization and classify certain submultiplicative functions on the real line, which by itself is a curious result, cf. Theorems 6.1 and 6.2.

Let us consider not necessarily small additive "perturbations" of the Leibniz rule. Suppose, e.g., that we add to the Leibniz equation a product of two copies of a "lower-order" operator A,

$$T(f \cdot g) = Tf \cdot g + f \cdot Tg + Af \cdot Ag,$$

$f, g \in C^k(\mathbb{R})$. This equation is not only motivated by a perturbation of the (first order) Leibniz rule, but, in fact, reflects the behavior of the second derivative $T = D^2$. Indeed, choosing $A = \sqrt{2}\, D$, the equation is satisfied for these operators (T, A). The natural domain for T is $C^2(\mathbb{R})$, for A it is $C^1(\mathbb{R})$. Thus A is of "lower order" than T.

This point of view leads to higher-order Leibniz rule type equations determining derivatives of any order, cf. Section 3 of Chapter 5. Moreover, it may be considered for functions on $\mathbb{R}^n$, too. The equations then yield characterizations of the Laplacian under natural assumptions, e.g., orthogonal invariance and annihilation of affine functions. We investigate this in Chapter 7.

In most of the results on operator equations for one operator T in this book we do not make any continuity or regularity assumption on the operator T. A posteriori, the theorems imply that the operator T is actually continuous in a natural way. In the proofs we use the fact that the image of the operators is contained in spaces of continuous functions. We feel, however, that the main reason for the automatic continuity of the solution operators is a consequence of the *nonlinearity* of the equations, like the chain rule equation. Using the axiom of choice and Hamel bases, it is of course easy to construct non-continuous and even non-measurable solutions of *linear* equations on infinite-dimensional spaces. However, this is not the case for non-linear equations.

Much of the material of the book is based on papers of the authors and their coauthors. However, various theorems shown in this book extend published results or relax the assumptions made there. The proofs of most results are provided in detail. The book is addressed to a general mathematical audience.

We are very grateful to Mrs. M. Hercberg for her invaluable help in typing and editing the text. The book could not have appeared without her efforts. We also thank several colleagues for helpful discussions on various aspects of the book, in particular S. Alesker, S. Artstein-Avidan, J. Bernstein, P. Domański, D. Faifman, R. Farnsteiner and L. Polterovich.

Chapter 2

Regular Solutions of Some Functional Equations

The derivative D is an operator which acts as a map from the continuously differentiable functions $C^1(\mathbb{R})$ on $\mathbb{R}$ to the continuous functions $C(\mathbb{R})$. It satisfies the Leibniz and the chain rule

$$D(f \cdot g) = Df \cdot g + f \cdot Dg,$$
$$D(f \circ g) = (Df) \circ g \cdot Dg, \qquad f, g \in C^1(\mathbb{R}).$$

In this book, we show that operators $T : C^1(\mathbb{R}) \to C(\mathbb{R})$ obeying either the Leibniz or the chain rule operator equation

$$T(f \cdot g) = Tf \cdot g + f \cdot Tg, \tag{2.1}$$
$$T(f \circ g) = (Tf) \circ g \cdot Tg, \qquad f, g \in C^1(\mathbb{R}) \tag{2.2}$$

are close to the standard derivative. Actually, we completely establish the form of the solutions of either equation. We also consider more general operator equations modeling second-order derivatives or the Laplacian. Only very mild conditions are imposed on the map T.

The basic question mentioned already in the introduction is: Are classical operators in analysis like differential operators characterized by very simple properties such as (2.1) or (2.2), and additional initial conditions, e.g., $T(-2\,\mathrm{Id}) = -2$?

Chapters 3 and 4 will be devoted to determine and describe all solutions of either equation (2.1) or (2.2). The first step in solving equations like (2.1) and (2.2) is to show that the operator T is *localized*, i.e., that there is a function $F : \mathbb{R}^3 \to \mathbb{R}$, such that

$$Tf(x) = F(x, f(x), f'(x)), \quad f \in C^1(\mathbb{R}), \ x \in \mathbb{R}.$$

At this point, the function F and its possible regularity is unknown, but the operator equation for T translates into a functional equation for F, in the above

© Springer Nature Switzerland AG 2018
H. König, V. Milman, *Operator Relations Characterizing Derivatives*,
https://doi.org/10.1007/978-3-030-00241-1_2

cases into either

$$F(x, \alpha_0\beta_0, \alpha_1\beta_0 + \alpha_0\beta_1) = F(x, \alpha_0, \alpha_1)\beta_0 + F(x, \beta_0, \beta_1)\alpha_0,$$

or

$$F(x, z, \alpha_1\beta_1) = F(y, z, \alpha_1)F(x, y, \beta_1),$$

for all $x, y, z, \alpha_0, \beta_0, \alpha_1, \beta_1 \in \mathbb{R}$.

Functional equations, of course, are a classic subject, and there is a vast literature on the topic, cf., e.g., the books of Aczél [A], Aczél, Dhombres [AD], Járai [J], Székelyhidi [Sz] or the recent book by Rassias, Thandapani, Ravi, Senthil Kumar [RTRS]. Much less is known about the operator equations which we will discuss in this book, and the specific functional equations which they generate.

In this chapter, we determine the solutions of a few functional equations which originate by localization and various reduction steps from the operator equations we will study, identifying the representing function F up to some parametric functions. To be self-contained, we provide the proofs of these results, even though most of them are found in, e.g., [A] or [AD] or in more generality in [J] or [Sz]. Some of the proofs are new, and we present them in more detail. In this chapter we do not outline the general theory of functional equations as done, e.g., in [J] or [Sz], but rather only solve those functional equations which will be relevant in later chapters.

To show the regularity of the parameter functions occurring in the representing function F, we prove some new general continuity results under assumptions which are easily verified in the case of the operator equations which we investigate. A general reference when solutions of functional equations are smooth is Járai [J].

2.1   Regularity results for additive and multiplicative equations

We start with the classical question when additive functions are linear.

Proposition 2.1. *Let $f : \mathbb{R} \to \mathbb{R}$ be measurable and additive, i.e., satisfy the* Cauchy *equation*

$$f(x + y) = f(x) + f(y), \quad x, y \in \mathbb{R}.$$

Then f is linear: there is $c \in \mathbb{R}$ such that $f(x) = cx$ for all $x \in \mathbb{R}$.

Clearly, additive functions satisfy $f(rx) = rf(x)$ for all $r \in \mathbb{Q}$. Thus, continuous additive functions are linear, $f(x) = cx$ with $c = f(1)$, as already noted by Cauchy.

Proof. Fix $x \neq 0$ and define functions $\varphi, \psi : \mathbb{R} \to \mathbb{R}$ by

$$\varphi(t) := f(t) - \frac{f(x)}{x}t, \quad \psi(t) := \frac{1}{1 + |\varphi(t)|}, \quad t \in \mathbb{R}.$$

By assumption φ and ψ are measurable with $0 \leq \psi \leq 1$. Hence, ψ is integrable on finite intervals. Note that $\varphi(x) = 0$, $\varphi(t + x) = \varphi(t) + \varphi(x) = \varphi(t)$, and $\psi(t + x) = \psi(t)$. Thus φ and ψ are periodic with period x. Therefore,

$$\int_0^x \psi(t)dt = \frac{1}{2} \int_0^{2x} \psi(t)dt = \int_0^x \psi(2t)dt,$$

$$0 = \int_0^x \big(\psi(t) - \psi(2t)\big)dt = \int_0^x \frac{|\varphi(t)|}{(1 + |\varphi(t)|)(1 + 2|\varphi(t)|)}dt,$$

using $|\varphi(2t)| = 2|\varphi(t)|$. We conclude that $\varphi = 0$ almost everywhere, i.e., $f(t) = \frac{f(x)}{x}t$ for almost all $t \in \mathbb{R}$. In particular, for $x = 1$, $f(t) = f(1)t$ for almost all $t \in \mathbb{R}$. Hence, for any $x \neq 0$, there is $0 \neq t_0 \in \mathbb{R}$ such that $f(t_0) = \frac{f(x)}{x}t_0$ and $f(t_0) = f(1)t_0$. Hence, $\frac{f(x)}{x} = \frac{f(t_0)}{t_0} = f(1)$, $f(x) = f(1)x$ for all $x \neq 0$. Obviously, this also holds for $x = 0$. $\qquad\square$

In general, additive functions are not linear: Let $X \subset \mathbb{R}$ be a Hamel basis of $\mathbb{R}$ over $\mathbb{Q}$ (assuming the axiom of choice) and $g : X \to \mathbb{R}$ be an arbitrary function. Any $x \in \mathbb{R}$ can be written uniquely as $x = \sum_{i \in J} \lambda_i x_i$, $x_i \in X$, $\lambda_i \in \mathbb{Q}$, J a finite index set. Define $f : \mathbb{R} \to \mathbb{R}$ by

$$f(x) = \sum_{i \in J} g(x_i)\lambda_i x_i, \quad x = \sum_{i \in J} \lambda_i x_i.$$

Then f is additive but not linear, unless g is constant. These pathological functions need to be unbounded on any small interval.

Proposition 2.2. *Let $I \in \mathbb{R}$ be a non-empty open interval and $f : \mathbb{R} \to \mathbb{R}$ be additive and bounded on I. Then f is linear, $f(x) = cx$ with $c \in \mathbb{R}$.*

Proof. Let $|I| \geq \delta > 0$ and $M := \sup_{x \in I} |f(x)|$. Then for any $t \in \mathbb{R}$ with $|t| < \delta$ there are $x, y \in I$ with $t = x - y$,

$$|f(t)| = |f(x - y)| = |f(x) - f(y)| \leq 2M.$$

Using the additivity again, we find for any $s \in \mathbb{R}$ with $|s| < \delta/n$ that $|f(s)| \leq 2M/n$. Let $u \in \mathbb{R}$ be arbitrary. Then, for any $n \in \mathbb{N}$, there is $r \in \mathbb{Q}$ with $|u - r| < \delta/n$. We find

$$\big|f(u) - uf(1)\big| = \big|f(u) - f(r) + rf(1) - uf(1)\big|$$
$$\leq \big|f(u - r)\big| + |r - u|f(1) \leq \big(2M + \delta f(1)\big)/n,$$

which yields $f(u) = f(1)u$ for all $u \in \mathbb{R}$. $\qquad\square$

The multiplicative analogue of Proposition 2.1 is

Proposition 2.3. *Let* $K : \mathbb{R} \setminus \{0\} \to \mathbb{R}$ *be measurable, not identically zero and multiplicative, i.e.,*

$$K(uv) = K(u)K(v), \quad u, v \in \mathbb{R}.$$

Then there is $p \in \mathbb{R}$ *such that, for all* $u \in \mathbb{R}$, *either* $K(u) = |u|^p$ *or* $K(u) = |u|^p \operatorname{sgn}(u)$.

Proof. Since K is not identically zero, $K(u) \neq 0$ if $u \neq 0$. Therefore, we may define $f : \mathbb{R} \to \mathbb{R}$ by $f(x) = \ln|K(e^x)|$. Then, for any $x, y \in \mathbb{R}$, $f(x + y) = f(x) + f(y)$. Since f is measurable, too, by Proposition 2.1 there is $p \in \mathbb{R}$ such that $f(x) = px$ for all $x \in \mathbb{R}$. Hence, $|K(u)| = u^p$ for any $u > 0$. Since $K(u) = K(\sqrt{u})^2 > 0$, we get $K(u) = u^p$ for $u > 0$. Further, $K(-1)^2 = K(1)^2 = K(1) = 1$ implies that $K(-1) \in \{+1, -1\}$. Then $K(-u) = K(-1)K(u)$ implies that $K(u) = |u|^p$ or $K(u) = |u|^p \operatorname{sgn}(u)$, depending on whether $K(-1) = 1$ or $K(-1) = -1$. $\qquad\square$

For the complex version of this result, we assume continuity. For $z \in \mathbb{C} \setminus \{0\}$, let $\operatorname{sgn} z := \frac{z}{|z|}$. Also put $\operatorname{sgn} 0 := 0$.

Proposition 2.4. *Let* $f : \mathbb{C} \to \mathbb{C}$ *be continuous, not identically zero and multiplicative,*

$$f(zw) = f(z)f(w), \quad z, w \in \mathbb{C}.$$

Then there are $p \in \mathbb{C}$ *with* $\operatorname{Re}(p) \geq 0$ *and* $m \in \mathbb{Z}$ *such that*

$$f(z) = |z|^p (\operatorname{sgn} z)^m, \quad z \in \mathbb{C}.$$

We prove Proposition 2.4 by applying the following proposition which we need later not only for functions defined on $\mathbb{C}$ but on $\mathbb{C}^n$. For $z = (z_j)_{j=1}^n$, $d = (d_j)_{j=1}^n \in \mathbb{C}^n$, we denote by $\langle \cdot, \cdot \rangle$ the linear form – not the scalar product – on $\mathbb{C}^n$, $\langle d, z \rangle = \sum_{j=1}^n d_j z_j$. Moreover we put $\bar{z} = (\bar{z}_j)_{j=1}^n$.

Proposition 2.5. *Let* $n \in \mathbb{N}$ *and suppose that* $F : \mathbb{C}^n \to \mathbb{C} \setminus \{0\}$ *is continuous and satisfies*

$$F(z + w) = F(z) \cdot F(w), \quad z, w \in \mathbb{C}^n.$$

Then there are $c, d \in \mathbb{C}^n$ *such that*

$$F(z) = \exp(\langle c, z \rangle + \langle d, \bar{z} \rangle), \quad z \in \mathbb{C}^n.$$

Proof of Proposition 2.5. Write $z \in \mathbb{C}^n$ as $z = x + iy$, $x, y \in \mathbb{R}^n$ and F in polar decomposition form,

$$F(z) = G(x + iy) \exp(iH(x + iy)).$$

where $G : \mathbb{C}^n \to \mathbb{R}_{>0}$ is continuous and $H : \mathbb{C}^n \to \mathbb{R}$ may be chosen to be continuous, too, since it may be constructed from continuous branches. Note that H is defined on $\mathbb{C}^n$ and not on n-fold products of strips, so that it does not yield an injective representation of F. (E.g., for $n = 1$ and $F(z) = \exp(2z)$, we would have $H(x + iy) = 2y$ and we would not identify $2y = +\pi$ and $-\pi$ for $y = +\frac{\pi}{2}$ and $y = -\frac{\pi}{2}$.) Then, for all $x, y, u, v \in \mathbb{R}^n$,

$$G\big((x + u) + i(y + v)\big) = G(x + iy)G(u + iv),$$
$$H\big((x + u) + i(y + v)\big) = H(x + iy) + H(u + iv) + 2\pi k,$$

for some $k \in \mathbb{Z}$ which is independent of x, y, u, v since H is continuous. Define $\Phi : \mathbb{R}^{2n} \to \mathbb{R}$ by either $\Phi(x, y) := \ln G(x + iy)$ or $\Phi(x, y) := H(x + iy) + 2\pi k$. Then Φ is continuous and additive,

$$\Phi(x + u, y + v) = \Phi(x, y) + \Phi(u, v).$$

Selecting $u = y = 0$ and renaming v as y, we get $\Phi(x, y) = \Phi(x, 0) + \Phi(0, y)$ and similarly $\Phi(x + u, 0) = \Phi(x, 0) + \Phi(u, 0)$. If $x = (x_j)_{j=1}^n = \sum_{j=1}^n x_j e_j$, where (e_j) denotes the canonical unit vector basis in $\mathbb{R}^n$, we have by additivity $\Phi(x, 0) = \sum_{j=1}^n \Phi(x_j e_j, 0)$. Proposition 2.1 yields that there are $\alpha_j, \beta_j \in \mathbb{R}$ such that $\Phi(x_j e_j, 0) = \alpha_j x_j$ and $\Phi(0, y_j e_j) = \beta_j y_j$. Hence with $\alpha = (\alpha_j)_{j=1}^n$, $\beta = (\beta_j)_{j=1}^n$, $a := \frac{1}{2}(\alpha - i\beta)$ and $b := \frac{1}{2}(\alpha + i\beta) \in \mathbb{C}^n$,

$$\Phi(x, y) = \langle \alpha, x \rangle + \langle \beta, y \rangle = \langle a, z \rangle + \langle b, \bar{z} \rangle.$$

This means that $G(z) = \exp(\Phi(x, y)) = \exp(\langle a, z \rangle + \langle b, \bar{z} \rangle)$, and with different vectors $\tilde{a}, \tilde{b} \in \mathbb{C}^n$, $H(z) = \langle \tilde{a}, z \rangle + \langle \tilde{b}, \bar{z} \rangle - 2\pi k$, so that

$$F(z) = \exp(\langle c, z \rangle + \langle d, \bar{z} \rangle), \quad c := a + i\tilde{a}, \ d := b + i\tilde{b} \in \mathbb{C}^n. \qquad \square$$

Proof of Proposition 2.4. We have $f(w) \neq 0$ for $w \neq 0$ since $f \not\equiv 0$. Define $F : \mathbb{C} \to \mathbb{C} \setminus \{0\}$ by $F(z) := f(\exp z)$. Then F is continuous and

$$F(z + w) = F(z)F(w), \quad z, w \in \mathbb{C}.$$

By Proposition 2.5 with $n = 1$, $F(z) = \exp(cz + d\bar{z})$, hence $f(w) = w^c \bar{w}^d$, $w \in \mathbb{C}$. For $w \neq 0$, let $\text{sgn}(w) := \frac{w}{|w|}$. Then $f(w) = |w|^p \text{sgn}(w)^q$ with $p = c + d \in \mathbb{C}$ and $q = c - d \in \mathbb{C}$. Since f is continuous, q has to be an integer, $q = m \in \mathbb{Z}$. Since f is bounded near zero, $\text{Re}(p) \geq 0$ is required. $\qquad \square$

In later applications of Proposition 2.1, the measurable additive function f will actually depend on parameters or independent variables, so the linearity factor c will depend on these parameters. To prove the continuous dependence of c on the variables, we use the following result. Before formulating it, we introduce some notations. Let $\mathbb{N}_0 := \mathbb{N} \cup \{0\}$. For $n \in \mathbb{N}, k \in \mathbb{N}_0, I \subset \mathbb{R}^n$ open, let

$$C^k(I, \mathbb{R}) := \{f : I \to \mathbb{R} \mid f \text{ is } k\text{-times continuously differentiable}\}$$

and $C^\infty(I, \mathbb{R}) := \bigcap_{k \in \mathbb{N}} C^k(I, \mathbb{R})$, $C(I, \mathbb{R}) := C^0(I, \mathbb{R})$. Let $l \in \mathbb{N}, f \in C^l(I, \mathbb{R})$. By Schwarz' theorem, the l-th derivative $f^{(l)}(x)$ of f at $x \in I$ can be represented by the $M(n, l) := \binom{n+l-1}{n-1}$ independent l-th order partial derivatives

$(\frac{\partial^l f(x)}{\partial x_{i_1} \cdots \partial x_{i_l}})_{1 \le i_1 \le \cdots \le i_l \le n}$. For $k \in \mathbb{N}$, let $N(n,k) := \sum_{l=0}^{k-1} M(n,l) = \binom{n+k-1}{n}$. Then, using this representation of derivatives, we put

$$J_k(x,f) := \big(f(x), \ldots, f^{(k-1)}(x)\big) \in \mathbb{R}^{N(n,k)}, \quad f \in C^{k-1}(I,\mathbb{R}), \ x \in I.$$

Theorem 2.6. *Let $n \in \mathbb{N}$, $k \in \mathbb{N}_0$ and $I \subset \mathbb{R}^n$ be an open set, possibly unbounded. Let $B : I \times \mathbb{R}^{N(n,k)} \to \mathbb{R}$ be a function satisfying*

(a) $B(x, v_1 + v_2) = B(x, v_1) + B(x, v_2), \quad x \in I, \ v_i \in \mathbb{R}^{N(n,k)}.$

(b) $B(\cdot, J_k(\cdot\, ; f))$ *is a continuous function from I to $\mathbb{R}$ for all $f \in C^\infty(I,\mathbb{R})$.*

Then there is a continuous function $c : I \to \mathbb{R}^{N(n,k)}$ so that

$$B(x, v) = \langle c(x), v \rangle, \quad x \in I, \ v \in \mathbb{R}^{N(n,k)}.$$

By $\langle\, \cdot\,, \cdot\,\rangle$ we denote the standard scalar product on the appropriate $\mathbb{R}^N$-space, here $N = N(n,k)$. Then

$$B\big(x, J_k(x,f)\big) = \sum_{l=0}^{k-1} \langle c_l(x), f^{(l)}(x) \rangle, \quad x \in I, \ f \in C^{k-1}(I,\mathbb{R}),$$

with continuous functions $c_l : I \to \mathbb{R}^{M(n,l)}$.

For $k = 0$, the variable v and $J_k(\cdot\,; f)$ are not present in (a) and (b).

Proof. To keep the notation simple, we give the proof only in dimension $n = 1$, although the arguments in higher dimensions follow the same basic idea. For $n = 1$, we may assume that I is an open interval. We proceed by induction on $k = N(1,k)$. For $k = 0$ there is nothing to prove. Assume $k \in \mathbb{N}$ and that the result holds for $k - 1$.

(i) Define $A := \{x \in I \mid B(x, \cdot, 0, \ldots, 0) : \mathbb{R} \to \mathbb{R} \text{ is discontinuous}\}$. We claim that A has no accumulation points in I. Assume to the contrary that $x_m \in A \to x_\infty \in I$. We may assume that (x_m) is strictly monotone, say decreasing, so that $x_m > x_{m+1} > x_\infty$. Fix a smooth, non-negative cut-off function $\psi \in C^\infty(\mathbb{R})$ with $\psi|_{\mathbb{R} \setminus [-1,1]} = 0$, $\max \psi = \psi(0) = 1$ and $\psi^{(l)}(0) = 0$ for all $l \in \mathbb{N}$. Denote $c_l := \max |D^l \psi|$. For $m \in \mathbb{N}$, let

$$\delta_m := \min \big(\tfrac{1}{2} \min \{ |x_m - x_j| : 1 \le j \le \infty, \ m \ne j \}, \ \tfrac{1}{2^m} \big).$$

By assumption (a), $B(x_m, \cdot, 0 \ldots, 0) : \mathbb{R} \to \mathbb{R}$ is an additive function which is discontinuous for each $m \in \mathbb{N}$. By Proposition 2.2 it must be unbounded on $(0, \epsilon)$ for any $\epsilon > 0$. Therefore, we may choose $0 < y_m < \exp\big(-\tfrac{1}{\delta_m}\big)$ with $|B(x_m, y_m, 0, \ldots, 0)| > 1$. Define

$$g_m(x) := y_m \psi\left(\frac{x - x_m}{\delta_m}\right), \quad x \in I.$$

Then $g_m \in C^\infty(I)$ with $g_m(x_m) = y_m$, $g_m^{(l)}(x_m) = 0$ for all $l \in \mathbb{N}$ and $g_m(x) = 0$ for all $x \in I$ with $|x - x_m| > \delta_m$. Moreover, $|D^l g_m| \leq c_l y_m \delta_m^{-l}$. Define $g := \sum_{m \in \mathbb{N}} g_m$. We find, for any $l \in \mathbb{N}_0$,

$$\sum_{m \in \mathbb{N}} |D^l g_m| \leq c_l \sum_{m \in \mathbb{N}} y_m \delta_m^{-l} \leq c_l \sum_{m \in \mathbb{N}} \delta_m^{-l} \exp\left(-\frac{1}{\delta_m}\right) < \infty,$$

so that $g \in C^\infty(I)$. Note that $g(x_m) = y_m$ since we have by definition of δ_j for any $m \neq j$ that $|x_m - x_j| \geq 2\delta_j$ so that $g_j(x_m) = 0$. Since $x_m \to x_\infty$ and $y_m \to 0$, we have by continuity that $g(x_\infty) = 0$. Also $g^{(l)}(x_m) = 0$ for all $l \in \mathbb{N}$, and again by continuity $g^{(l)}(x_\infty) = 0$ for all $l \in \mathbb{N}$. Since $B(\cdot, J_k(\cdot, g))$ is a continuous function by assumption (b),

$$B(x_m, J_k(x_m, g)) \longrightarrow B(x_\infty, J_k(x_\infty, g)) = B(x_\infty, 0, \ldots, 0) = 0.$$

However, $|B(x_m, J_k(x_m, g))| = |B(x_m, y_m, 0, \ldots, 0)| > 1$, which is a contradiction. Therefore, A has no accumulation points in I and its complement in I is dense in I.

(ii) We next claim that A is empty. Take any $x_0 \in I$. By (i) there is a sequence (x_m) with $x_m \notin A$, $x_m \to x_0$. For all $y_0 \in \mathbb{R}$, $B(\cdot, y_0, 0, \ldots, 0)$ is continuous on $\mathbb{R}$, applying (b) to the constant function $f(x) = y_0$, and therefore, $B(x_m, y_0, 0, \ldots, 0) \to B(x_0, y_0, 0, \ldots, 0)$. Hence, $B(x_m, \cdot, 0, \ldots, 0) \to B(x_0, \cdot, 0, \ldots, 0)$ pointwise. This implies that $B(x_0, \cdot, 0, \ldots, 0)$ is a measurable function, being the pointwise limit of continuous functions. By (a), $B(x_0, \cdot, 0, \ldots, 0)$ is additive so that Proposition 2.1 yields that $B(x_0, \cdot, 0, \ldots, 0)$ is linear and hence continuous so that $x_0 \notin A$. Hence, $A = \emptyset$.

We conclude that $B(x, y, 0, \ldots, 0) = c_0(x)y$ for some function $c_0 : I \to \mathbb{R}$. Since $c_0(x) = B(x, 1, 0, \ldots, 0)$, c_0 is continuous by assumption (b). Finally write

$$B(x, y_0, \ldots, y_{k-1}) = B(x, y_0, 0, \ldots, 0) + B(x, 0, y_1, \ldots, y_{k-1})$$
$$= c_0(x)y_0 + B(x, 0, y_1, \ldots, y_{k-1}).$$

Note that conditions (a), (b) also hold for $B(x, 0, y_1, \ldots, y_{k-1})$ as a function from $I \times \mathbb{R}^{k-1}$ to $\mathbb{R}$. Thus, by induction assumption, $B(x, 0, y_1, \ldots, y_{k-1}) = \sum_{j=1}^{k-1} c_j(x)y_j$, $c_j \in C(I)$. Hence,

$$B(x, y_0, \ldots, y_{k-1}) = \sum_{j=0}^{k-1} c_j(x)y_j = \langle c(x), y \rangle$$

with $c(x) = (c_j(x))_{j=0}^{k-1}$, $y = (y_j)_{j=0}^{k-1}$. $\qquad\square$

Theorem 2.6 will be used in the next chapter to analyze the solutions of the Leibniz rule operator equation. We will also study perturbations of the Leibniz rule equation. To show that the solutions of the perturbed equations are perturbations of the solutions of the unperturbed Leibniz rule equation, we need a more technical variant of Theorem 2.6 in dimension $n = 1$ which we will apply in Chapter 5.

Proposition 2.7. *Let $k \in \mathbb{N}$, $I \subset \mathbb{R}$ be an open set and $B, \widetilde{B}, \Psi : I \times \mathbb{R}^k \to \mathbb{R}$ be functions, Ψ measurable, and $M : I \to \mathbb{R}_+$ be a locally bounded function such that*

(i) $\widetilde{B}(x,v) = B(x,v) + \Psi(x,v)$, $x \in I$, $v \in \mathbb{R}^k$.

(ii) $\widetilde{B}(x, v_1 + v_2) = \widetilde{B}(x, v_1) + \widetilde{B}(x, v_2)$, $x \in I$, $v_1, v_2 \in \mathbb{R}^k$.

(c) $B(\cdot, J_k(\cdot, f))$ *is a continuous function from I to $\mathbb{R}$ for all $f \in C^\infty(\mathbb{R})$.*

(d) $\sup\{|\Psi(x,v)| \mid v \in \mathbb{R}^k\} \le M(x) < \infty$, $x \in I$.

Then $\widetilde{B}(x, \cdot)$ is linear for all $x \in I$, i.e., there is $c(x) \in \mathbb{R}^k$ such that $\widetilde{B}(x,v) = \langle c(x), v \rangle$ for all $v \in \mathbb{R}^k$.

Proof. (i) We adapt the previous proof and first claim that

$$A := \{ x \in I \mid \widetilde{B}(x, \cdot, 0, \ldots, 0) : \mathbb{R} \to \mathbb{R} \text{ is discontinuous} \}$$

has no accumulation point in I. If this would be false, there would be a sequence of pairwise disjoint, say strictly decreasing points $x_m \in A$ with $x_m \to x_\infty \in I$. Since M is locally bounded,

$$K := \max \big(M(x_\infty), \sup\{ M(x_m) \mid m \in \mathbb{N} \} \big) < \infty.$$

Since $\widetilde{B}(x_m, \cdot, 0, \ldots, 0)$ is discontinuous and additive, by Proposition 2.2, it attains arbitrarily large values in any neighborhood of zero. Again, choosing δ_m and $0 < y_m < \exp(-1/\delta_m)$ as in the previous proof, such that $|\widetilde{B}(x_m, y_m, 0, \ldots, 0)| > 3K + 1$, we define $g \in C^\infty(I)$ as before with

$$g(x_m) = y_m, \quad g(x_\infty) = 0, \quad g^{(l)}(x_m) = g^{(l)}(x_\infty) = 0,$$

for all $m, l \in \mathbb{N}$. By assumption (c)

$$B(x_m, y_m, 0, \ldots, 0) = B\big(x_m, J_k(x_m, g) \big)$$
$$\longrightarrow B\big(x_\infty, J_k(x_\infty, g) \big) = B\big(x_\infty, 0, \ldots, 0 \big).$$

But $\widetilde{B}(x_\infty, \cdot)$ is additive, hence $\widetilde{B}(x_\infty, 0) = 0$. Since $B = \widetilde{B} - \Psi$ and $|\Psi(x_m, \cdot)| \le K$, we arrive at the contradiction

$$2K < \lim_{m \to \infty} \big| B(x_m, y_m, 0, \ldots, 0) \big| = \big| B(x_\infty, 0, \ldots, 0) \big| \le K.$$

(ii) Fix an arbitrary point $x_0 \in I$. By (i) there are $x_m \notin A$ with $x_m \to x_0$. Therefore, $\widetilde{B}(x_m, \cdot, 0, \ldots, 0)$ is continuous for all $m \in \mathbb{N}$ and, by assumption (c), $B(\cdot, y_0, 0, \ldots, 0)$ is continuous for any $y_0 \in \mathbb{R}$. Thus

$$B(x_m, y_0, 0, \ldots, 0) \longrightarrow B(x_0, y_0, 0, \ldots, 0).$$

Hence $B(x_0, \cdot, 0, \ldots, 0)$ is the pointwise limit of the functions

$$B(x_m, \cdot, 0, \ldots, 0) = \widetilde{B}(x_m, \cdot, 0, \ldots, 0) - \Psi(x_m, \cdot, 0, \ldots, 0),$$

therefore measurable, so that

$$\left|\widetilde{B}(x_0, \cdot, 0, \dots, 0)\right| \leq K + \left|B(x_0, \cdot, 0, \dots, 0)\right|,$$

i.e., $\widetilde{B}(x_0, \cdot, 0, \dots, 0)$ is additive and bounded by a measurable function. By a result of Kestelman [Ke] – similar to Proposition 2.2 but slightly more general – $\widetilde{B}(x_0, \cdot, 0, \dots, 0)$ is linear, i.e.,

$$\widetilde{B}(x_0, y_0, 0, \dots, 0) = c_0(x_0) y_0.$$

Induction on k using

$$\widetilde{B}(x_0, y_0, \dots, y_{k-1}) = \widetilde{B}(x_0, y_0, 0, \dots, 0) + \widetilde{B}(x_0, 0, y_1, \dots, y_{k-1})$$

ends the proof. □

In the case of the chain rule operator equation studied in chapter 4, we will need different regularity results, yielding the regularity of a function from the property that certain differences of the function are regular.

Proposition 2.8. (a) *Let* $L : \mathbb{R} \to \mathbb{R}$ *be a function such that for any* $b \in \mathbb{R}$

$$\varphi(x) := L(x) - L(bx), \quad x \in \mathbb{R}$$

defines a continuous function $\varphi \in C(\mathbb{R})$. *Then* L *is the pointwise limit of continuous functions and hence measurable.*

(b) *Let* $0 < a \leq 1$ *and* $L \in C(\mathbb{R})$ *be a continuous function such that*

$$\psi(x) := L(x) - a L\left(\tfrac{x}{2}\right), \quad x \in \mathbb{R}$$

defines a C^1-*function* $\psi \in C^1(\mathbb{R})$. *Then* L *is a* C^1-*function,* $L \in C^1(\mathbb{R})$.

Proof. (i) For $b = 1/2$, $\varphi(x) = L(x) - L(x/2)$ is continuous and for $n \in \mathbb{N}$

$$\sum_{j=0}^{n-1} \left(\varphi\left(\frac{x}{2^j}\right) - \varphi\left(\frac{1}{2^j}\right) \right) = (L(x) - L(1)) + \left(L\left(\frac{1}{2^n}\right) - L\left(\frac{x}{2^n}\right) \right).$$

For $b = x$, $\widetilde{\varphi}(y) = L(y) - L(xy)$ is continuous in $y = 0$, hence,

$$\lim_{n \to \infty} \left(L\left(\frac{1}{2^n}\right) - L\left(\frac{x}{2^n}\right) \right) = \widetilde{\varphi}(0) = 0.$$

Therefore, the limit exists for $n \to \infty$ in the above equation and

$$L(x) = L(1) + \sum_{j=0}^{\infty} \left(\varphi\left(\frac{x}{2^j}\right) - \varphi\left(\frac{1}{2^j}\right) \right).$$

Hence L is the pointwise limit of continuous functions.

(ii) Fix $M > 0$ and let $x, x_1 \in [-M, M]$. For any $n \in \mathbb{N}$

$$\sum_{j=0}^{n-1} a^j \left(\psi \left(\frac{x}{2^j} \right) - \psi \left(\frac{x_1}{2^j} \right) \right) = (L(x) - L(x_1)) - a^n \left(L \left(\frac{x}{2^n} \right) - L \left(\frac{x_1}{2^n} \right) \right).$$

Since L is continuous, the last term on the right-hand side tends to 0 for $n \to \infty$. Since $\psi \in C^1(\mathbb{R})$, ψ' is uniformly continuous in $[-M, M]$ and bounded in modulus, say by N. Let $\epsilon > 0$. Then there is $\delta > 0$ such that for all $y, z \in [-M, M]$ with $|y - z| < \delta$, we have $|\psi'(y) - \psi'(z)| < \epsilon/2$. Assume $|x - x_1| < \delta$. Then, by the mean-value theorem,

$$\psi \left(\frac{x}{2^j} \right) - \psi \left(\frac{x_1}{2^j} \right) = \psi' \left(\frac{x(j)}{2^j} \right) \frac{x - x_1}{2^j},$$

for some $x(j)$ between x and x_1. Since $\left| \frac{x(j)}{2^j} - \frac{x_1}{2^j} \right| \leq |x - x_1| < \delta$, we find

$$\left| \sum_{j=1}^{n-1} a^j \cdot \frac{\psi \left(\frac{x}{2^j} \right) - \psi \left(\frac{x_1}{2^j} \right)}{x - x_1} - \sum_{j=0}^{n-1} \left(\frac{a}{2} \right)^j \psi' \left(\frac{x_1}{2^j} \right) \right|$$

$$= \left| \sum_{j=0}^{n-1} \left(\frac{a}{2} \right)^j \left(\psi' \left(\frac{x(j)}{2^j} \right) - \psi' \left(\frac{x_1}{2^j} \right) \right) \right| \leq \frac{\epsilon}{2} \sum_{j=0}^{n-1} \left(\frac{a}{2} \right)^j \leq \epsilon.$$

Moreover,

$$\left| a^n \frac{L \left(\frac{x}{2^n} \right) - L \left(\frac{x_1}{2^n} \right)}{x - x_1} \right| = \left| \sum_{j=n}^{\infty} a^j \cdot \frac{\psi \left(\frac{x}{2^j} \right) - \psi \left(\frac{x_1}{2^j} \right)}{x - x_1} \right|$$

$$= \left| \sum_{j=n}^{\infty} \left(\frac{a}{2} \right)^j \psi' \left(\frac{x(j)}{2^j} \right) \right| \leq N \sum_{j=n}^{\infty} \left(\frac{a}{2} \right)^j \longrightarrow 0,$$

uniformly in $x, x_1 \in [-M, M]$ for $n \to \infty$. Therefore,

$$L'(x_1) = \lim_{x \to x_1} \frac{L(x) - L(x_1)}{x - x_1} = \sum_{j=0}^{\infty} \left(\frac{a}{2} \right)^j \psi' \left(\frac{x_1}{2^j} \right)$$

exists and $\psi' \in C(\mathbb{R})$ implies $L' \in C(\mathbb{R})$, $L \in C^1(\mathbb{R})$. $\square$

2.2   Functional equations with two unknown functions

In this section we discuss the solutions of some functional equations which involve two unknown functions. It is an interesting subject by itself which was studied intensively, cf., e.g., the books by Aczél [A], Aczél, Dhombres [AD] and Székelyhidi [Sz]. We will use these results in Chapters 7, 8 and 9 to study operator equations which are inspired by the Leibniz rule or by the chain rule of the second order. Several theorems in this section are special cases of results in [Sz]. Our intention here is to give direct proofs.

The second derivative D^2 satisfies the Leibniz and the chain rule type formulas

$$D^2(f \cdot g) = D^2 f \cdot g + f \cdot D^2 g + 2Df \cdot Dg,$$
$$D^2(f \circ g) = (D^2 f \circ g) \cdot (Dg)^2 + (Df) \circ g \cdot D^2 g, \qquad f, g \in C^2(\mathbb{R}).$$

To understand the structure of these equations, we will later consider a more general setting: We will study operators $T : C^2(\mathbb{R}) \to C(\mathbb{R})$ and $A, A_1, A_2 : C^1(\mathbb{R}) \to C(\mathbb{R})$ satisfying one of the following equations

$$T(f \cdot g) = Tf \cdot g + f \cdot Tg + Af \cdot Ag,$$
$$T(f \circ g) = Tf \circ g \cdot A_1 g + (A_2 f) \circ g \cdot Tg, \qquad f, g \in C^2(\mathbb{R}).$$

Under mild assumptions, it will turn out that there are not too many choices of operators (T, A) or (T, A_1, A_2) satisfying any one of these operator equations. To solve them, after localization, we have to find the solutions of some specific functional equations which involve two unknown functions.

We now discuss the solutions of these functional equations. The results of this section will only be used later in Chapters 7, 8 and 9.

Proposition 2.9. *Let $m \in \mathbb{N}$ and assume that $F, B : \mathbb{R}^m \to \mathbb{R}$ are functions such that for any $\alpha, \beta \in \mathbb{R}^m$,*

$$F(\alpha + \beta) = F(\alpha) + F(\beta) + B(\alpha)B(\beta). \qquad (2.3)$$

Then there are additive functions $c, d : \mathbb{R}^m \to \mathbb{R}$ and $\gamma \in \mathbb{R}$ such that F and B have one of the following three forms:
Either

(a) $F(\alpha) = -\gamma^2 + d(\alpha)$, $B(x) = \gamma$,

or

(b) $F(\alpha) = \frac{1}{2}c(\alpha)^2 + d(\alpha)$, $B(\alpha) = c(\alpha)$,

or

(c) $F(\alpha) = \gamma^2(\exp(c(\alpha)) - 1) + d(\alpha)$, $B(\alpha) = \gamma(\exp(c(\alpha)) - 1)$,

for any $\alpha \in \mathbb{R}^m$.
Conversely, these functions satisfy equation (2.3).

Proof. (i) If $B = 0$, then F is additive and we are in case (a) with $\gamma = 0$. Therefore, we may assume that $B \neq 0$. Choose $a \in \mathbb{R}^m$ with $B(a) \neq 0$. For $\alpha \in \mathbb{R}^m$, define functions $f, b : \mathbb{R}^m \to \mathbb{R}$ by

$$f(\alpha) := F(\alpha + a) - F(\alpha) - F(a), \quad b(\alpha) := B(\alpha + a) - B(\alpha).$$

Then by (2.3)

$$f(\alpha + \beta) = f(\alpha) + b(\alpha)B(\beta), \quad \alpha, \beta \in \mathbb{R}^m. \tag{2.4}$$

For $\alpha = 0$, $f(\beta) = f(0) + b(0)B(\beta)$. Inserting this back into (2.3), we find

$$b(0)\big(B(\alpha + \beta) - B(\alpha)\big) = b(\alpha)B(\beta). \tag{2.5}$$

Suppose first $b(0) = 0$. Since $B(a) \neq 0$, (2.5) implies that $b \equiv 0$ identically and that $f = f(0)$ is a constant function. Since $f(\alpha) = B(a)B(\alpha)$ by (2.3), also B is constant, $B = f(0)/B(a) =: \gamma$. Let $d(\alpha) := F(\alpha) + \gamma^2$. Then by (2.3)

$$d(\alpha + \beta) = F(\alpha + \beta) + \gamma^2 = \big(F(\alpha) + F(\beta) + \gamma^2\big) + \gamma^2 = d(\alpha) + d(\beta),$$

i.e., d is additive on $\mathbb{R}^m$ and F and B have the form given in (a).

(ii) Assume now $b(0) \neq 0$. Putting $\alpha = 0$ in (2.5), we find that $B(0) = 0$. Moreover,

$$B(\alpha + \beta) = B(\alpha) + \frac{b(\alpha)}{b(0)}B(\beta). \tag{2.6}$$

Suppose first that b is a constant function. Then $c(\alpha) := B(\alpha)$ is additive and $d(a) := F(\alpha) - \frac{1}{2}c(\alpha)^2$ satisfies

$$\begin{aligned}
d(\alpha + \beta) &= F(\alpha + \beta) - \tfrac{1}{2}\big(c(\alpha) + c(\beta)\big)^2 \\
&= \big(F(\alpha) + F(\beta) + B(\alpha)B(\beta)\big) - \tfrac{1}{2}c(\alpha)^2 - \tfrac{1}{2}c(\beta)^2 - c(\alpha)c(\beta) \\
&= d(\alpha) + d(\beta).
\end{aligned}$$

Hence, d is additive and F and B have the form given in (b).

(iii) Now assume $b(0) \neq 0$ and that b is not constant. Choose $\alpha_0 \in \mathbb{R}^m$ with $b(\alpha_0) \neq b(0)$. Since the left-hand side of (2.6) is symmetric in α and β, we have

$$B(\alpha) + \frac{b(\alpha)}{b(0)}B(\beta) = B(\beta) + \frac{b(\beta)}{b(0)}B(\alpha).$$

For $\beta = \alpha_0$, $B(\alpha) = \frac{B(\alpha_0)}{b(\alpha_0) - b(0)}(b(\alpha) - b(0))$, and by (2.4),

$$f(\alpha) - f(0) = b(0)B(\alpha) = \gamma\big(b(\alpha) - b(0)\big), \tag{2.7}$$

with $\gamma := b(0)B(\alpha_0)/(b(\alpha_0) - b(0))$. For $\gamma = 0$, $B = 0$, and we are again in case (a).

So assume $\gamma \neq 0$. Then, by (2.4) and (2.7),

$$
\begin{aligned}
\gamma\big(b(\alpha+\beta) - b(0)\big) &= f(\alpha+\beta) - f(0) \\
&= f(\alpha) - f(0) + b(\alpha)B(\beta) \\
&= \gamma\big(b(\alpha) - b(0)\big) + \frac{b(\alpha)}{b(0)}\gamma\big(b(\beta) - b(0)\big) \\
&= \gamma\left(\frac{b(\alpha)b(\beta)}{b(0)} - b(0)\right).
\end{aligned}
$$

Hence, $\widetilde{b}(\alpha) := b(\alpha)/b(0)$ satisfies $\widetilde{b}(\alpha+\beta) = \widetilde{b}(\alpha)\widetilde{b}(\beta)$, $\widetilde{b}(\alpha) = \widetilde{b}\left(\frac{\alpha}{2}\right)^2 > 0$. Note that $\widetilde{b}(\alpha) \neq 0$, since otherwise $\widetilde{b}(0) = \widetilde{b}(\alpha)\widetilde{b}(-\alpha) = 0$, but $\widetilde{b}(0) = 1$. Therefore, $c(\alpha) := \ln \widetilde{b}(\alpha)$ is additive and $b(\alpha) = b(0)\exp(c(\alpha))$. This yields $B(\alpha) = \gamma(\exp(c(\alpha)) - 1)$. Put similarly as above $d(\alpha) := F(\alpha) - \gamma^2(\exp(c(\alpha)) - 1)$. Then (2.3) and the additivity of c yield

$$
\begin{aligned}
d(\alpha+\beta) &= \big(F(\alpha) + F(\beta) + B(\alpha)B(\beta)\big) - \gamma^2\big(\exp(c(\alpha))\exp(c(\beta)) - 1\big) \\
&= d(\alpha) + d(\beta),
\end{aligned}
$$

i.e., d is additive. Therefore, we have the solution given in (c),

$$
F(\alpha) = \gamma^2\big(\exp(c(\alpha)) - 1\big) + d(\alpha). \qquad \square
$$

In the case $m = 1$, we need a multiplicative analogue of Proposition 2.9.

Proposition 2.10. *Assume that $F, B : \mathbb{R} \to \mathbb{R}$ are functions such that, for any $\alpha, \beta \in \mathbb{R}$,*

$$
F(\alpha\beta) = F(\alpha)\beta + F(\beta)\alpha + B(\alpha)\,B(\beta). \tag{2.8}
$$

Then there are additive functions $c, d : \mathbb{R} \to \mathbb{R}$, and there is $\gamma \in \mathbb{R}$ such that F and B have one of the following three forms:

(a) $F(\alpha) = \alpha\,(c(\ln|\alpha|) - \gamma^2)$, $B(\alpha) = \gamma\alpha$;

(b) $F(\alpha) = \alpha\big(\frac{1}{2}c(\ln|\alpha|)^2 + d(\ln|\alpha|)\big)$, $B(\alpha) = \alpha c(\ln|\alpha|)$;

(c) $F(\alpha) = \alpha\big(\gamma^2[\{\operatorname{sgn}\alpha\}\exp(c(\ln|\alpha|)) - 1] + d(\ln|\alpha|)\big)$,
$\quad$ $B(\alpha) = \alpha\gamma[\{\operatorname{sgn}\alpha\}\exp(c(\ln|\alpha|)) - 1]$.

In (c), there are two possibilities, with $\operatorname{sgn}\alpha$ present in both F and B and the other one with $\operatorname{sgn}\alpha$ replaced by 1.

Conversely, these functions satisfy equation (2.8).

Proof. (i) For $a \in \mathbb{R}$, define $f(a) := F(\exp(a))/\exp(a)$, $g(a) := B(\exp(a))/\exp(a)$. Then (2.8) implies

$$
f(a+b) = f(a) + f(b) + g(a)g(b), \quad a, b \in \mathbb{R}.
$$

The solutions of this equation $(m = 1)$ were given in Proposition 2.9, e.g., in case (b) with additive functions $c, d : \mathbb{R} \to \mathbb{R}$,

$$f(a) = \tfrac{1}{2}c(a)^2 + d(a), \quad g(a) = c(a).$$

Then for $\alpha > 0$, $a := \ln \alpha$, so that $\alpha = \exp(a)$,

$$F(\alpha) = \alpha \left(\tfrac{1}{2}c(\ln \alpha)^2 + d(\ln \alpha) \right), \quad B(\alpha) = \alpha c(\ln \alpha).$$

The cases (a) and (c) are similar, which yields Proposition 2.10 if $\alpha > 0$.

(ii) We will now determine $F(\alpha)$ and $B(\alpha)$ for negative α. In all cases except one, F and B turn out to be odd functions. The exceptional one is the case of the third solution when the $\operatorname{sgn}\alpha$-term appears. Unfortunately, this requires distinguishing several cases in the basic equation (2.9) below. Choosing $\beta = -1$ in (2.8) and exchanging α and $-\alpha$, we find

$$F(\alpha) + F(-\alpha) = F(-1)\alpha + B(-1)B(\alpha) = -F(-1)\alpha + B(-1)B(-\alpha),$$
$$\text{i.e., } B(-1)B(-\alpha) = B(-1)B(\alpha) + 2F(-1)\alpha.$$

For $\alpha = 1$, $B(-1)^2 = B(-1)B(1) + 2F(-1)$. Hence,

$$\begin{aligned} B(-1)B(-\alpha) &= B(-1)\big[B(\alpha) + (B(-1) - B(1))\alpha\big], \\ F(\alpha) + F(-\alpha) &= B(-1)\left[B(\alpha) + \tfrac{1}{2}(B(-1) - B(1))\alpha\right]. \end{aligned} \tag{2.9}$$

If $B(-1) = 0$, also $F(-1) = 0$ and (2.9) implies that $F(-\alpha) = -F(\alpha)$ and, using (2.8),

$$\begin{aligned} B(-\alpha)B(\beta) &= F(-\alpha\beta) - F(-\alpha)\beta + F(\beta)\alpha \\ &= -F(\alpha\beta) + F(\alpha)\beta + F(\beta)\alpha = -B(\alpha)B(\beta), \end{aligned}$$

i.e., F and B are odd functions, which means that in cases (a), (b) and (c) $\ln \alpha$ has to be replaced by $\ln |\alpha|$ for $\alpha < 0$.

(iii) Now assume $B(-1) \neq 0$. In cases (b), (c), we know $B(1) = F(1) = 0$. Then by (2.9)

$$B(-\alpha) = B(\alpha) + B(-1)\alpha, \quad F(\alpha) + F(-\alpha) = B(-1)\left[B(\alpha) + \tfrac{1}{2}B(-1)\alpha\right]. \tag{2.10}$$

Using (2.10) for $\alpha\beta$ instead of α and (2.8), we find

$$\begin{aligned} B(-1)\left[B(\alpha\beta) + \tfrac{1}{2}B(-1)\alpha\beta\right] &= F(\alpha\beta) + F(-\alpha\beta) \\ &= \big(F(\alpha) + F(-\alpha)\big)\beta + \big(B(\alpha) + B(-\alpha)\big)B(\beta) \\ &= B(-1)\left[B(\alpha)\beta + \tfrac{1}{2}B(-1)\alpha\beta\right] + 2B(\alpha)B(\beta) + B(-1)B(\beta)\alpha \\ &= 2\big(B(\alpha) + \tfrac{1}{2}B(-1)\alpha\big)\big(B(\beta) + \tfrac{1}{2}B(-1)\beta\big). \end{aligned}$$

Therefore, $\varphi(\alpha) := \frac{2}{B(-1)}B(\alpha) + \alpha$ is multiplicative, $\varphi(\alpha\beta) = \varphi(\alpha)\varphi(\beta)$. For positive $\alpha > 0$, this occurs only in case (c) when

$$B(\alpha) = \alpha\gamma\big[\exp(c(\ln\alpha)) - 1\big].$$

This identifies $\gamma = \frac{1}{2}B(-1)$, and for $\alpha < 0$ we have

$$B(\alpha) = \alpha\gamma\big[\operatorname{sgn}\alpha \cdot \exp(c(\ln|\alpha|)) - 1\big],$$

from the multiplicity of φ, where the term $\operatorname{sgn}\alpha$ has to be present since otherwise $B(-1) = 0$. For $\alpha < 0$, we get from (2.10) and the known form of $F(-\alpha)$ for $-\alpha = |\alpha| > 0$

$$
\begin{aligned}
F(\alpha) &= -F(-\alpha) + 2\gamma\big(B(\alpha) + \gamma\alpha\big) \\
&= \alpha\big[\gamma^2(\exp(c(\ln|\alpha|)) - 1) + d(\ln|\alpha|)\big] \\
&\quad + 2\gamma\big[-\gamma\alpha(\exp(c(\ln|\alpha|)) + 1) + \gamma\alpha\big] \\
&= \alpha\big[\gamma^2(\operatorname{sgn}\alpha\exp(c(\ln|\alpha|)) - 1) + d(\ln|\alpha|)\big].
\end{aligned}
$$

In this case B and F are not odd, in the other cases of (b) and (c) they are odd.

(iv) It remains to consider case (a) for $\alpha < 0$, when $B(-1) \neq 0$. Then $B(1) = \gamma$ and (2.9) yields for $\alpha > 0$ that $B(-\alpha) = B(-1)\alpha$ and

$$F(\alpha) + F(-\alpha) = \tfrac{1}{2}B(-1)\big(\gamma + B(-1)\big)\alpha.$$

Using this for $\alpha\beta$ instead of α and (2.8) we find

$$
\begin{aligned}
\tfrac{1}{2}B(-1)\big(\gamma + B(-1)\big)\alpha\beta &= F(\alpha\beta) + F(-\alpha\beta) \\
&= \big(F(\alpha) + F(-\alpha)\big)\beta + \big(B(\alpha) + B(-\alpha)\big)B(\beta) \\
&= \tfrac{1}{2}B(-1)\big(\gamma + B(-1)\big)\alpha\beta + \big(\gamma + B(-1)\big)\alpha B(\beta),
\end{aligned}
$$

hence, $B(-1) = -\gamma$, $B(-\alpha) = -\gamma\alpha = -B(\alpha)$, $F(-\alpha) = -F(\alpha)$, so that B and F are odd functions, which means, in the formula of (a), that $\ln\alpha$ has to replaced by $\ln|\alpha|$ for $\alpha < 0$. $\qquad\square$

In Chapter 3 we will need the solution of a functional equation which resembles the addition formula for the sin function. We first consider the complex case.

Proposition 2.11. *Let $n \in \mathbb{N}$ and $F, B : \mathbb{C}^n \to \mathbb{C}$ be continuous functions satisfying*

$$F(z + w) = F(z) \cdot B(w) + F(w) \cdot B(z), \quad z, w \in \mathbb{C}^n. \tag{2.11}$$

Suppose F is not identically zero. Then there are vectors $c_1, c_2, d_1, d_2 \in \mathbb{C}^n$ and there are $k \in \mathbb{C} \setminus \{0\}$ and $\epsilon_1, \epsilon_2 \in \{0, 1\}$, with ϵ_1, ϵ_2 not both zero, such that F and B have one of the following two forms:

(a) $F(z) = (\langle c_1, z \rangle + \langle c_2, \bar{z} \rangle) \exp(\langle d_1, z \rangle + \langle d_2, \bar{z} \rangle)$,
$B(z) = \exp(\langle d_1, z \rangle + \langle d_2, \bar{z} \rangle)$;

(b) $F(z) = \frac{1}{2k}(\epsilon_1 \exp(\langle c_1, z \rangle + \langle c_2, \bar{z} \rangle) - \epsilon_2 \exp(\langle d_1, z \rangle + \langle d_2, \bar{z} \rangle))$,
$B(z) = \frac{1}{2}(\epsilon_1 \exp(\langle c_1, z \rangle + \langle c_2, \bar{z} \rangle) + \epsilon_2 \exp(\langle d_1, z \rangle + \langle d_2, \bar{z} \rangle))$, $z \in \mathbb{C}^n$.

Conversely, these functions satisfy equation (2.11).

In the real case we get

Corollary 2.12. *Let* $F, B : \mathbb{R}^n \to \mathbb{R}$ *be continuous functions satisfying*

$$F(\alpha + \beta) = F(\alpha)B(\beta) + F(\beta)B(\alpha), \quad \alpha, \beta \in \mathbb{R}^n.$$

Suppose F *is not identically zero. Then there are vectors* $b, c, d \in \mathbb{R}^n$ *and there is* $a \in \mathbb{R}$ *such that* F *and* B *have one of the following four forms:*

(a) $F(\alpha) = \langle b, \alpha \rangle \exp(\langle d, \alpha \rangle)$, $B(\alpha) = \exp(\langle d, \alpha \rangle)$;

(b) $F(\alpha) = a \exp(\langle c, \alpha \rangle) \sin(\langle d, \alpha \rangle)$, $B(\alpha) = \exp(\langle c, \alpha \rangle) \cos(\langle d, \alpha \rangle)$;

(c) $F(\alpha) = a \exp(\langle c, \alpha \rangle) \sinh(\langle d, \alpha \rangle)$, $B(\alpha) = \exp(\langle c, \alpha \rangle) \cosh(\langle d, \alpha \rangle)$;

(d) $F(\alpha) = a \exp(\langle d, \alpha \rangle)$, $B(\alpha) = \frac{1}{2} \exp(\langle d, \alpha \rangle)$, $\alpha \in \mathbb{R}^n$.

Conversely, these functions satisfy the above functional equation.

Proof of Proposition 2.11. (i) Fix $t \in \mathbb{C}^n \setminus \{0\}$. We claim that F, B and $B(\cdot + t)$ are linearly dependent functions. For all $x, y \in \mathbb{C}^n$

$$F(x+t)B(y)+B(x+t)F(y) = F(x+y+t) = F(x)B(y+t)+B(x)F(y+t). \quad (2.12)$$

Since F is not identically zero, by (2.11) also B is not identically zero. Hence there is $y_1 \in \mathbb{C}^n$ such that $B(y_1) \neq 0$. Choosing $y = y_1$, equation (2.12) shows that $F(\cdot + t)$ is a linear combination of F, B and $B(\cdot + t)$ with coefficients depending on the values $B(y_1)$, $F(y_1)$, $B(y_1 + t)$ and $F(y_1 + t)$. Inserting this back into (2.12) yields for all $x, y \in \mathbb{C}^n$

$$F(x)\big(B(y)B(y_1 + t) - B(y_1)B(y + t)\big) + B(x)\big(B(y)F(y_1 + t) - B(y_1)F(y + t)\big)$$
$$+ B(x + t)\big(B(y_1)F(y) - B(y)F(y_1)\big) = 0. \quad (2.13)$$

Suppose $B(y_1)F(y) - B(y)F(y_1) = 0$ holds for all $y \in \mathbb{C}^n$. Then $F = \frac{F(y_1)}{B(y_1)}B$, and already F and B are linearly dependent. Else there is $y_2 \in \mathbb{C}^n$ such that $B(y_1)F(y_2) - B(y_2)F(y_1) \neq 0$, and equation (2.13) shows that F, B and $B(\cdot + t)$ are linearly dependent.

(ii) Assume that $B = kF$ for some $k \in \mathbb{C}$. Then $F(x + y) = 2kF(x)F(y)$, and $k \neq 0$ since F is not identically zero. Proposition 2.5 implies that there are $c_1, c_2 \in \mathbb{C}^n$ such that $F(z) = \frac{1}{2k} \exp(\langle c_1, z \rangle + \langle c_2, \bar{z} \rangle)$, $B(z) = \frac{1}{2} \exp(\langle c_1, z \rangle + \langle c_2, \bar{z} \rangle)$. This is a solution of type (b) with $\epsilon_2 = 0$.

(iii) We may now assume that B and F are linearly independent. Then by (i) there are functions $c_1, c_2 : \mathbb{C}^n \to \mathbb{C}$ such that

$$B(x + t) = c_1(t)F(x) + c_2(t)B(x), \quad x, t \in \mathbb{C}^n. \tag{2.14}$$

The left-hand side is symmetric in x and t. Applying it to $x + y + t$, we get an equation similar to (2.12). The arguments in (i) then show that c_2, B and F are linearly dependent: there are $b_1, b_2 \in \mathbb{C}$ such that

$$c_2(x) = b_1 B(x) + b_2 F(x).$$

Inserting this back into (2.14) and using the symmetry in (x, t), we find

$$c_1(t)F(x) + \big(b_1 B(t) + b_2 F(t)\big)B(x) = B(x + t)$$
$$= c_1(x)F(t) + \big(b_1 B(x) + b_2 F(x)\big)B(t),$$
$$c_1(x) - b_2 B(x) = \frac{c_1(t) - b_2 B(t)}{F(t)} F(x) =: b_3 F(x),$$

for any fixed t with $F(t) \neq 0$. Hence $c_1(x) = b_2 B(x) + b_3 F(x)$, and again by (2.14)

$$B(x + t) = \big(b_2 B(t) + b_3 F(t)\big)F(x) + \big(b_1 B(t) + b_2 F(t)\big)B(x).$$

Insert this and formula (2.11) for $F(x+t)$ into (2.12) to find, after some calculation,

$$\big((1 - b_1)B(t) - b_2 F(t)\big)\big(F(x)B(y) - F(y)B(x)\big) = 0,$$

for all $x, y, t \in \mathbb{C}^n$. Since B and F are linearly independent, we first conclude that $(1 - b_1)B(t) = b_2 F(t)$ for all t, and then that $b_1 = 1$, $b_2 = 0$. Therefore, $c_1 = b_3 F$, $c_2 = B$, and (2.14) yields

$$B(x + t) = b_3 F(t)F(x) + B(t)B(x), \quad x, t \in \mathbb{C}^n.$$

Take $k \in \mathbb{C}$ with $k^2 = b_3$. Using (2.11) again, we find

$$\big(B(x + y) \pm kF(x + y)\big) = \big(B(x) \pm kF(x)\big)\big(B(y) \pm kF(y)\big),$$

so that $f := B \pm kF$ solves the equation $f(x + y) = f(x)f(y)$. Since $f \not\equiv 0$, by Proposition 2.5, there are $c_1, c_2, d_1, d_2 \in \mathbb{C}^n$ such that

$$B(z) + kF(z) = \exp(\langle c_1, z \rangle + \langle c_2, \bar{z} \rangle),$$
$$B(z) - kF(z) = \exp(\langle d_1, z \rangle + \langle d_2, \bar{z} \rangle),$$

which gives solution (b) with $\epsilon_1 = \epsilon_2 = 1$, if $k \neq 0$.

(iv) If $k = 0$, again by Proposition 2.5, $B(z) = \exp(\langle d_1, z \rangle + \langle d_2, \bar{z} \rangle)$ for suitable $d_1, d_2 \in \mathbb{C}$. Define $G(z) := \frac{F(z)}{B(z)}$. Since $B(z + w) = B(z)B(w)$, equation (2.11) yields

$$G(z + w) = G(z) + G(w), \quad z, w \in \mathbb{C}^n.$$

Hence G is additive and continuous. As in the proof of Proposition 2.5 there are $c_1, c_2 \in \mathbb{C}^n$ such that $G(z) = \langle c_1, z \rangle + \langle c_2, \bar{z} \rangle$, which yields with $F(z) = G(z)B(z)$ the form of F and B given in part (a). $\qquad \square$

Proof of Corollary 2.12. Extend $F, B : \mathbb{R}^n \to \mathbb{R}$ to $\widetilde{F}, \widetilde{B} : \mathbb{C}^n \to \mathbb{C}$ by $\widetilde{F}(z) :=$ $F(\Re z)$, $\widetilde{B}(z) := B(\Re z)$ with $\Re z = (\Re z_j)_{j=1}^n$ if $z = (z_j)_{j=1}^n$. Here $\Re$ denotes the real part, and below $\Im$ will stand for the imaginary part. Then $\widetilde{F}, \widetilde{B}$ satisfy (2.11) and are real valued. The functions B and F in part (a) of Proposition 2.11 are real valued if and only if $c_1 = \bar{c}_2$ and $d_1 = \bar{d}_2$ yielding the solution in (a), when restricted to $\mathbb{R}^n$, with $b = 2\Re c_1$ and $d = 2\Re d_1$.

The formula for B in part (b) of Proposition 2.11 with $\epsilon_1 = \epsilon_2 = 1$ is real valued if and only if either $c_1 = \bar{c}_2$ and $d_1 = \bar{d}_2$ or $c_1 = \bar{d}_2$ and $c_2 = \bar{d}_1$. In the first case one gets a solution of type (c) with vectors $c = \Re(c_1 + d_1)$, $d = \Re(c_1 - d_1)$, in the second case a solution of type (b) with $c = \Re(c_1 + c_2)$ and $d = \Im(c_1 + c_2)$. In the first case k needs to be real, in the second case purely imaginary. The last solution (d) originates from (b) in Proposition 2.11 for $\epsilon_1 = 1$, $\epsilon_2 = 0$ (or $\epsilon_1 = 0$, $\epsilon_2 = 1$). $\qquad\square$

In Chapter 9 we need a multiplicative one-dimensional analogue of Corollary 2.12 which is the following result.

Proposition 2.13. *Let* $F, B : \mathbb{R} \to \mathbb{R}$ *be continuous functions satisfying*

$$F(xy) = F(x)B(y) + F(y)B(x), \quad x, y \in \mathbb{R}. \tag{2.15}$$

Suppose F is not identically zero. Then there are constants $a, b, c, d \in \mathbb{R}$, $c, d > 0$, so that F and B have one of the following four forms:

(a) $F(x) = b(\ln|x|)|x|^d\{\operatorname{sgn} x\}$, $B(x) = |x|^d\{\operatorname{sgn} x\}$;

(b) $F(x) = b|x|^d \sin(a\ln|x|)\{\operatorname{sgn} x\}$, $B(x) = |x|^d \cos(d\ln|x|)\{\operatorname{sgn} x\}$;

(c) $F(x) = \frac{b}{2}(|x|^c[\operatorname{sgn} x] - |x|^d\{\operatorname{sgn} x\})$, $B(x) = \frac{1}{2}(|x|^c[\operatorname{sgn} x] + |x|^d\{\operatorname{sgn} x\})$;

(d) $F(x) = b|x|^d\{\operatorname{sgn} x\}$, $B(x) = \frac{1}{2}|x|^d\{\operatorname{sgn} x\}$, $x \in \mathbb{R}$.

Here the terms $\{\operatorname{sgn} x\}$ and $[\operatorname{sgn} x]$ may be present or not, simultaneously in F and B. If a sgn-factor is not present, the corresponding value of c or d could be 0, too, Conversely, these functions satisfy the above functional equation.

Proof. (i) Let $\widetilde{F}(\alpha) := F(\exp \alpha)$, $\widetilde{B}(\alpha) := B(\exp \alpha)$. Then $\widetilde{F}(\alpha+\beta) = \widetilde{F}(\alpha)\widetilde{B}(\beta) + \widetilde{B}(\alpha)\widetilde{F}(\beta)$. Hence $(\widetilde{F}, \widetilde{B})$ have one of the four forms given in Corollary 2.12. Then for $x > 0$, substituting $\alpha = \ln x = \ln|x|$, (F, B) have the form given in Proposition 2.13 with $\operatorname{sgn} x = 1$.

(ii) It remains to determine $F(x)$ and $B(x)$ for $x \le 0$. In the first three cases $F(1) = 0$. Then $0 = F(1) = F((-1)^2) = 2F(-1)B(-1)$. Assume first that $F(-1) = 0$. Then $F(x) = F(-x)B(-1) = F(x)B(-1)^2$, hence $B(-1)^2 = 1$, $B(-1) \in \{1, -1\}$. Thus F is even or odd, depending on whether $B(-1) = 1$ or $B(-1) = -1$. Using $F(x) = F(-x)B(-1)$, the functional equation implies for any $x, y \in \mathbb{R}$

$$F(x)B(-y) + B(-1)F(y)B(x) = F(x)B(-y) + F(-y)B(x) = F(-xy)$$
$$= B(-1)F(xy) = B(-1)[F(x)B(y) + F(y)B(x)].$$

Therefore $F(x)B(-y) = F(x)B(-1)B(y)$ which yields $B(-y) = B(-1)B(y)$. Hence F and B are both even or both odd. This implies the formulas for F and B for negative x in the first three cases. Since F and B and the right-hand sides are continuous, the values at zero are obtained by taking the limit for $x \to 0$ on both sides.

In the last case $F(1) =: b \neq 0$. Equation (2.15) yields for $y = 1$ that $F(x) = F(x)B(1) + bB(x)$. Since $B \not\equiv 0$, we conclude that $B(1) \neq 1$ and $F(x) = \lambda B(x)$ with $\lambda := \frac{b}{1-B(1)} \neq 0$. Inserting this into (2.15), we get $B(xy) = 2B(x)B(y)$, so that $2B$ is multiplicative on $\mathbb{R}$. By Proposition 2.3, $B(x) = \frac{1}{2}|x|^d\{\operatorname{sgn} x\}$, $F(x) = \frac{\lambda}{2}|x|^d\{\operatorname{sgn} x\}$, so that $b = \frac{\lambda}{2}$. $\qquad\square$

2.3   Notes and References

The classical result for measurable additive functions, Proposition 2.1, is due to Fréchet [Fr]. The paper [Fr] is written in Esperanto. Alternative proofs were given by Banach [B] and Sierpinski [S]. The proofs in [Fr] and [B] use the axiom of choice, the one in [S] does not require it. The simple proof presented here is due to Alexiewicz and Orlicz [AO].

The proof of Proposition 2.2 follows Kestelman [Ke], where the linearity of additive functions is shown under the even weaker assumption that f is bounded from above by a measurable function on a set of positive Lebesgue measure. This stronger result is used in the proof of Proposition 2.7.

Proposition 2.3 on measurable multiplicative functions is found, e.g., in Aczél [A], Section 2.1.2.

Proposition 2.5 is shown by Aczél [A] in Section 5.1.1, Theorem 3, in the case of $n = 1$. The generalization to $n > 1$ is straightforward. The result also holds if F is assumed to be only measurable instead of being continuous, cf. Aczél, Dhombres [AD], Theorem 5 of Section 5.1 ($n = 1$). The proof is slightly more elaborate than in the continuous case.

Since Proposition 2.4 follows directly from Proposition 2.5, it is also true if the non-zero function f is only assumed to be multiplicative and measurable.

Theorem 2.6 is due to Faifman, see the Appendix of [KM1].

Proposition 2.8 is a slight extension of Lemma 19 in [AKM].

Proposition 2.9 is a special case of Theorem 10.4. in Székelyhidi [Sz], which is illustrated by the functional equation (10.6b) in this book. Theorem 10.4. also covers solutions of functional equations with more than two unknown functions. In the case $m = 1$, Proposition 2.9 is related to some functional equations in Section 3.1.3 of Aczél [A] and in Chapter 15, Theorem 1 of Aczél, Dhombres [AD] to which this result could be reduced. Our direct proof uses ideas of Section 3.1.3 of Aczél [A].

Proposition 2.11 can be found in Székelyhidi [Sz], Theorem 12.2., as an application of his general theory of functional equations on topological abelian groups, cf. also Theorem 10.4. in [Sz]. We gave a direct proof which was inspired by the book of Aczél [A], where the case $n = 1$ is considered in Section 4.2.5, Theorem 2 and its Corollary. For Corollary 2.12 in the case $n = 1$ cf. Aczél [A], p. 180.

Chapter 3

The Leibniz Rule

We will show that the derivative as a map on classical function spaces of analysis is characterized by the Leibniz rule as well as the chain rule. This is a consequence of results in this and the next chapter. We first study the solutions of the Leibniz rule equation as a map on the k-times continuously differentiable functions C^k. There are many examples of derivations in algebra and differential geometry generalizing the Leibniz rule for the derivative of products of functions. However, on C^k there are only few examples of derivations. A priori, we assume neither linearity nor continuity of the derivations which we characterize. However, the continuity of the operator is a consequence of the results. Various solutions are actually non-linear.

3.1 The Leibniz rule in C^k

To formulate the basic result, we use the following notation:

Let $I \subset \mathbb{R}$ be an open set. In particular, $I = (-\infty, a), (a, b), (b, \infty)$ with $a, b \in \mathbb{R}$ or $I = \mathbb{R}$ are natural choices. For $k \in \mathbb{N}_0 := \mathbb{N} \cup \{0\}$ let

$$C^k(I) := \{ f : I \to \mathbb{R} \mid f \text{ is } k\text{-times continuously differentiable on } I \}.$$

We denote the continuous functions also by $C(I) := C^0(I)$ and put $C^\infty(I) = \bigcap_{k \in \mathbb{N}} C^k(I)$. The basic result for the Leibniz rule operator equation is

Theorem 3.1 (Leibniz rule). *Let $k \in \mathbb{N}_0$ and $I \subset \mathbb{R}$ be an open set. Suppose that $T : C^k(I) \to C(I)$ is an operator satisfying the* Leibniz rule equation

$$T(f \cdot g) = Tf \cdot g + f \cdot Tg, \quad f, g \in C^k(I). \tag{3.1}$$

Then there are continuous functions $c, d \in C(I)$ such that, if $k \in \mathbb{N}$,

$$Tf = c\, f \ln |f| + d\, f', \quad f \in C^k(I). \tag{3.2}$$

Conversely, any map T given by (3.2) satisfies (3.1). For $k = 0$, if $T : C(I) \to C(I)$ satisfies (3.1), there is $c \in C(I)$ such that $Tf = c\, f \ln |f|$.

© Springer Nature Switzerland AG 2018
H. König, V. Milman, *Operator Relations Characterizing Derivatives*,
https://doi.org/10.1007/978-3-030-00241-1_3

Since $\lim_{x\to 0}\ x\ln|x| = 0$, $0\ln|0|$ should be read as 0.

Remarks. (a) The formulas (3.1) and (3.2) are meant pointwise, e.g., (3.2):

$$(Tf)(x) = c(x)f(x)\ln|f(x)| + d(x)f'(x), \quad f \in C^k(I), \ x \in I.$$

Thus the solutions of the Leibniz rule are linear combinations of the derivative and the "entropy solution" $f\ln|f|$ which acts as a "derivative" on spaces of continuous functions. Note that neither continuity nor linearity is imposed on the operator T; in fact, $Tf = f\ln|f|$ is a non-linear solution.

(b) For $k \geq 2$, there are not more solutions than for $k = 1$. Hence, $T : C^k(I) \to C(I)$ naturally extends by the same formula to $T : C^1(\mathbb{R}) \to C(\mathbb{R})$. Therefore $C^1(I)$ is the "natural domain" for the Leibniz formula among the $C^k(I)$-spaces.

(c) If T also maps $C^2(I)$ into $C^1(I)$, it has the form $Tf = d\,f'$ with $d \in C^1(I)$, since in general $f\ln|f| \notin C^1(I)$ for $f \in C^2(I)$. "Initial" conditions like $T(\text{Id}) = 1$ and $T(2\,\text{Id}) = 2$ together with (3.1) also imply that $Tf = f'$ is the derivative.

(d) If the image of T does not consist of continuous or at least measurable functions, there are different solutions of the Leibniz rule equation. Let $F(\mathbb{R})$ denote the space of all functions $f : \mathbb{R} \to \mathbb{R}$, and $H : \mathbb{R} \to \mathbb{R}$ be an additive but not linear function, as constructed after Proposition 2.1. Let $c \in F(\mathbb{R})$ and define $T : C(\mathbb{R}) \to F(\mathbb{R})$ by

$$Tf(x) = c(x)f(x)H\big(\ln|f(x)|\big), \quad f \in C(\mathbb{R}), \ x \in \mathbb{R},$$

with $Tf(x) := 0$ if $f(x) = 0$. Then T satisfies the Leibniz rule

$$T(f \cdot g) = Tf \cdot g + f \cdot Tg.$$

(e) For $k \geq 2$, there are more solutions of (3.1) on the *positive* C^k-tfunctions than those given in (3.2), cf. Corollary 3.4.

The proof of Theorem 3.1 consists of two steps. The first is to show *localization*, i.e., that T is defined pointwise in the sense that there is a function $F : I \times \mathbb{R}^{k+1} \to \mathbb{R}$ such that for all $f \in C^k(I)$ and $x \in I$

$$Tf(x) = F(x, f(x), \dots, f^{(k)}(x)).$$

At that point no regularity of F is known. The operator equation (3.1) then is equivalent to a functional equation for the *representing* function F. The second step of the proof is to analyze the structure of F and to prove the continuity of the coefficient functions occurring there, by using the fact that the image of T consists of continuous functions. In the case of Theorem 3.1, we have to show that F does not depend on the variables $\alpha_j = f^{(j)}(x)$ for $j \geq 2$ and that the functions c, d in (3.2) are continuous. To find the solutions of other operator equations in later chapters, we will use the same basic strategy in the proofs, although with very different representing functions.

To prove Theorem 3.1, we first show that T is "localized on intervals".

Lemma 3.2. *Suppose $T : C^k(I) \to C(I)$ satisfies (3.1). Then $T(1) = T(-1) = 0$. If $J \subset I$ is open and $f_1, f_2 \in C^k(I)$ satisfy $f_1|_J = f_2|_J$, then $Tf_1|_J = Tf_2|_J$.*

Proof. For any $f \in C^k(I)$, $T(f) = T(f \cdot 1) = T(f) \cdot 1 + T(1) \cdot f$, which implies $T(1) = 0$. Moreover $0 = T(1) = T((-1)^2) = -2T(-1)$, $T(-1) = 0$. If $J \subset I$ is open and $f_1|_J = f_2|_J$, let $x \in J$ be arbitrary and choose $g \in C^k(I)$ with $g(x) = 1$ and $\operatorname{supp} g \subset J$. Then $f_1 \cdot g = f_2 \cdot g$ and hence by (3.1)

$$f_1 \cdot Tg + Tf_1 \cdot g = T(f_1 \cdot g) = T(f_2 \cdot g) = f_2 \cdot Tg + Tf_2 \cdot g,$$

which implies $Tf_1(x) = Tf_2(x)$ for any $x \in J$, yielding $Tf_1|_J = Tf_2|_J$. $\qquad\square$

Localization on intervals always implies pointwise localization.

Proposition 3.3. *Let $k \in \mathbb{N}_0$ and $I \subset \mathbb{R}$ be an open set. Suppose $T : C^k(I) \to C(I)$ satisfies, for all open intervals $J \subset I$, that*

$$\left[f_1|_J = f_2|_J \implies Tf_1|_J = Tf_2|_J, \quad f_1, f_2 \in C^k(I) \right]. \tag{3.3}$$

Then there is a function $F : I \times \mathbb{R}^{k+1} \to \mathbb{R}$ such that

$$Tf(x) = F\big(x, f(x), f'(x), \ldots, f^{(k)}(x)\big) \tag{3.4}$$

holds for all $x \in I$ and $f \in C^k(I)$. It suffices to have (3.3) only for all intervals J of the form $J = (-\infty, x) \cap I$ and $J = (x, \infty) \cap I$ with $x \in I$.

Proof. Let $x_0 \in I$ be arbitrary but fixed. For any $f \in C^k(I)$, let g be the Taylor polynomial of order k at x_0. Let $J_1 := (-\infty, x_0) \cap I$ and $J_2 := (x_0, \infty) \cap I$ and define

$$h(x) := \begin{cases} f(x), & x \in \overline{J_1}, \\ g(x), & x \in J_2. \end{cases}$$

Then $h \in C^k(I)$ and $f|_{J_1} = h|_{J_1}$, $h|_{J_2} = g|_{J_2}$. By assumption $Tf|_{J_1} = Th|_{J_1}$ and $Th|_{J_2} = Tg|_{J_2}$. Since Tf, Th and Tg are continuous functions and $\{x_0\} = \overline{J_1} \cap \overline{J_2}$, we find $Tf(x_0) = Th(x_0) = Tg(x_0)$. Since g only depends on $(x_0, f(x_0), \ldots, f^{(k)}(x_0))$, so does $Tg(x_0)$. Therefore, $Tf(x_0) = Tg(x_0)$ only depends on these values, i.e., there is a function $F : I \times \mathbb{R}^{k+1} \to \mathbb{R}$ such that

$$Tf(x_0) = F\big(x_0, f(x_0), \ldots, f^{(k)}(x_0)\big),$$

for any $f \in C^k(I)$, $x_0 \in I$. $\qquad\square$

Proof of Theorem 3.1. (i) We will first show that for any $f > 0$, $\frac{Tf}{f}$ depends linearly on $\ln f$ and its derivatives, and then that no derivatives of order ≥ 2 show up in the formula for T. By Lemma 3.2 and Proposition 3.3 there is a function $F : I \times \mathbb{R}^{k+1} \to \mathbb{R}$ such that, for any $f \in C^k(I)$ and $x \in I$,

$$Tf(x) = F\big(x, f(x), f'(x), \ldots, f^{(k)}(x)\big).$$

Define a map $S : C^k(I) \to C(I)$ by

$$Sg(x) := T(\exp(g))(x)/\exp(g)(x), \quad g \in C^k(I), \ x \in I.$$

Then $Sg(x) = F(x, \exp(g)(x), \ldots, \exp(g)^{(k)}(x))/\exp(g)(x)$ depends only on $x, g(x)$ and all derivatives of g up to $g^{(k)}(x)$. Hence, there is a function $G : I \times \mathbb{R}^{k+1} \to \mathbb{R}$ such that

$$Sg(x) = G\big(x, g(x), \ldots, g^{(k)}(x)\big), \quad g \in C^k(I), \ x \in I.$$

For any $g_1, g_2 \in C^k(I)$, by the Leibniz rule equation on $C^k(I)$,

$$S(g_1 + g_2) = T(e^{g_1} \cdot e^{g_2})/(e^{g_1} \cdot e^{g_2}) = T(e^{g_1})/e^{g_1} + T(e^{g_2})/e^{g_2} = Sg_1 + Sg_2.$$

Since for any $\alpha = (\alpha_j)_{j=0}^k$, $\beta = (\beta_j)_{j=0}^k \in \mathbb{R}^{k+1}$ and $x \in I$, there are $g_1, g_2 \in C^k(I)$ with $g_1^{(j)}(x) = \alpha_j$, $g_2^{(j)}(x) = \beta_j$ for all $j \in \{0, \ldots, k\}$, we have

$$G(x, \alpha + \beta) = G(x, \alpha) + G(x, \beta), \quad x \in I, \ \alpha, \beta \in \mathbb{R}^{k+1}.$$

Since $Sg = T(e^g)/e^g$ is a continuous function on I, we also know that

$$G(x, g(x), \ldots, g^{(k)}(x))$$

is a continuous function of $x \in I$ for all $g \in C^k(I)$. By Theorem 2.6, there is a continuous function $c : I \to \mathbb{R}^{k+1}$ so that $G(x, \alpha) = \langle c(x), \alpha \rangle = \sum_{j=0}^k c_j(x)\alpha_j$, writing $c = (c_j)_{j=0}^k$, with continuous coefficient functions $c_j \in C(I)$.

For $f \in C^k(I)$, $f > 0$, let $g := \ln f$. Then $f = \exp g$ and

$$Tf(x) = f(x)S(\ln f)(x) = f(x)\sum_{j=0}^k c_j(x)(\ln f)^{(j)}(x). \tag{3.5}$$

Conversely, this formula defines a map on the *strictly positive* functions into the continuous functions satisfying the Leibniz rule since

$$\big(\ln(fg)\big)^{(j)} = (\ln f)^{(j)} + (\ln g)^{(j)}, \quad f, g \in C^k(I).$$

(ii) Let us now consider the Leibniz rule for $T : C^k(I) \to C(I)$ when the functions are negative. Suppose $f \in C^k(I)$ and $x \in I$ are given with $f(x) < 0$. Then there is an open interval $J \in I$, $x \in J$ with $f|_J < 0$. Choose $g \in C^k(I)$ with

$g < 0$ on I and $f|_J = g|_J$. Then $Tf(x) = Tg(x)$. To determine $Tf(x)$, we may therefore assume that $f < 0$ on I. Then $f = -|f|$ and by the Leibniz rule and Lemma 3.2

$$T(f) = T(-|f|) = -T(|f|) + |f|T(-\mathbf{1}) = -T(|f|).$$

Using (3.5), we find

$$Tf = -T(|f|) = -|f| \sum_{j=0}^{k} c_j (\ln |f|)^{(j)}$$

$$= f \sum_{j=0}^{k} c_j (\ln |f|)^{(j)}, \qquad f \in C^k(I).$$

To be defined on $C^k(I)$, Tf needs to be continuous also for f and x with $f(x) = 0$. However, for $j \geq 2$, $f(\ln |f|)^{(j)}$ is of order $O(|f|^{-(j-1)})$ as $|f| \searrow 0$, if $f' \neq 0$. Therefore, using localization, in the above formula $c_2 = \cdots = c_k = 0$ is required for $T : C^k(I) \to C(I)$ to be well defined.

   To be more specific, let $k \geq 2$, $x_0 \in I$ and choose $\epsilon_0 > 0$ with $(x_0 - 2\epsilon_0, x_0 + 2\epsilon_0) \subset I$ and consider $f(x) := x - x_0$. Let $0 < \epsilon < \epsilon_0$ and h be a strictly positive function with $h|_{(x_0+\epsilon,\infty)\cap I} = f|_{(x_0+\epsilon,\infty)\cap I}$, i.e., h has to bend upwards for $x < x_0 + \epsilon$ in a smooth way. Applying the above formula for h, we get for $Tf(x_0 + \epsilon) = Th(x_0 + \epsilon)$

$$Tf(x_0 + \epsilon) = Th(x_0 + \epsilon) = c_0(x_0 + \epsilon)\epsilon \ln \epsilon + \sum_{j=1}^{k} c_j(x_0 + \epsilon)(-1)^{j-1}(j-1)!\, \epsilon^{1-j}.$$

Since Tf and $c_0, \ldots, c_k$ are continuous functions, this implies for $\epsilon \to 0$ that $c_k(x_0) = \cdots = c_2(x_0) = 0$. This means that

$$Tf = c_0 f \ln |f| + c_1 f'.$$

This also holds when f has isolated zeros x, $f(x) = 0$, since $\lim_{y \to 0} y \ln |y| = 0$. Note that $Tf(x) = 0$ in this case since we have continuous functions on both sides. This is true by continuity of Tf, too, if x is a limit of isolated zeros of f. If $f|_J$ is zero on a non-trivial interval $J \subset I$, $Tf|_J = 0$. □

Corollary 3.4. *Let $k \in \mathbb{N}$ and $I \subset \mathbb{R}$ be an open set. Suppose that $T : C^k(I) \to C(I)$ satisfies the Leibniz rule equation (3.1). Then there are continuous functions $c_0, \ldots, c_k \in C(I)$ such that for every strictly positive function $f \in C^k(I)$, $f > 0$ and all $x \in I$*

$$Tf(x) = f(x) \sum_{j=0}^{k} c_j(x)\, (\ln f)^{(j)}(x).$$

Conversely, T defined this way satisfies equation (3.1) for all positive functions $f \in C^k(I)$.

This is a corollary to the proof of Theorem 3.1, which yielded (3.5) for positive functions $f > 0$. Note, however, that we need T to be defined and to satisfy (3.1) for *all functions* $f \in C^k(I)$, and not only for the strictly positive ones, since in the proof of Lemma 3.2 the operator T is applied to functions $f_1 g = f_2 g$ which are zero on a large part of the set I. For $k \geq 2$, there are more solution operators T on the positive functions than on all functions. For $k = 1$, we just recover (3.2).

3.2 The Leibniz rule on $\mathbb{R}^n$

Theorem 3.1 gives the solutions of the Leibniz rule on $I \subset \mathbb{R}$. It has an analogue for functions on n-dimensional domains $I \subset \mathbb{R}^n$. For $n \in \mathbb{N}$, $k \in \mathbb{N}_0$, open sets $I \subset \mathbb{R}^n$ and finite-dimensional real Banach spaces X let

$$C^k(I, X) := \{f : I \to X \mid f \text{ is } k\text{-times continuously differentiable on I}\},$$

with $C(I, X) := C^0(I, X)$ denoting the continuous functions. In this section, we include the image space X of functions in the notation $C^k(I, X)$ to indicate whether X is, e.g., $\mathbb{R}$ or $\mathbb{R}^n$. Let $L(\mathbb{R}^n, \mathbb{R}^n)$ denote the continuous linear maps for $\mathbb{R}^n$ into itself. The derivative $T = D$ maps $C^1(I, \mathbb{R})$ into $C(I, \mathbb{R}^n)$. The following theorem extends Theorem 3.1 to this n-dimensional setting. We did not directly state the result in the more general form, since its proof is a bit more elaborate and requires further notations.

Theorem 3.5. *Let $n \in \mathbb{N}$, $k \in \mathbb{N}_0$ and $I \subset \mathbb{R}^n$ be an open set. Suppose that $T : C^k(I, \mathbb{R}) \to C(I, \mathbb{R}^n)$ satisfies the Leibniz rule*

$$T(f \cdot g) = Tf \cdot g + f \cdot Tg, \quad f, g \in C^k(I, \mathbb{R}).$$

Then there are continuous functions $c \in C(I, \mathbb{R}^n)$ and $d \in C(I, L(\mathbb{R}^n, \mathbb{R}^n))$ such that for all $f \in C^k(I, \mathbb{R})$ and all $x \in I$

$$Tf(x) = c(x)f(x) \ln|f(x)| + d(x)(f'(x)).$$

For $k = 0$, d should be zero. Conversely, any such map T satisfies the Leibniz rule.

Note that on the right-hand side of the Leibniz formula we have pointwise multiplications of scalar and $\mathbb{R}^n$-valued functions. In the result, $d(x)$ is a matrix operating on the vector $f'(x)$, and $c(x)$ is a vector multiplying the scalar entropy expression $f(x) \ln|f(x)|$ for any $x \in I$.

For $k \geq 2$ there are no more solutions than for $k = 1$. Therefore T extends by the same formula to $C^1(I, \mathbb{R})$, so that $C^1(I, \mathbb{R})$ is the "natural" domain of T. If $d = 0$, T even extends to $C(I, \mathbb{R})$.

The Leibniz rule immediately implies $T\mathbf{1} = 0$ for the function $\mathbf{1}$ on $I \subset \mathbb{R}^n$. If $J \subset I$ is open and $f_1, f_2 \in C^k(I, \mathbb{R})$ satisfy $f_1|_J = f_2|_J$, we claim that $Tf_1|_J = Tf_2|_J$: Let $x \in J$ be arbitrary and choose $g \in C^k(I, \mathbb{R})$ with $g(x) = 1$ and support of g in J. Then $f_1 \cdot g = f_2 \cdot g$ and hence by the Leibniz rule $(f_1 - f_2) \cdot Tg = (Tf_1 - Tf_2) \cdot g$, so $Tf_1(x) = Tf_2(x)$, $Tf_1|_J = Tf_2|_J$. Therefore we have localization on (small) open sets. We now show that this implies pointwise localization, as in the 1-dimensional case.

For $0 \le l \le k$, the l-th derivative $f^{(l)}(x)$ of $f \in C^k(I, \mathbb{R})$, $I \subset \mathbb{R}^n$ open, at $x \in I$ is an l-multilinear form $f^{(l)}(x) : \underbrace{\mathbb{R}^n \times \cdots \times \mathbb{R}^n}_{l} \to \mathbb{R}$ which we may identify with the vector of all l-th order partial derivatives of f at x, a vector in $\mathbb{R}^{n^l}$. By Schwarz' theorem, the iterated partial derivatives do not depend on the order of taking them, so that we have only $M(n, l) := \binom{n+l-1}{n-1}$ different l-th order partial derivatives, indexed by $(\frac{\partial^l f(x)}{\partial x_{i_1} \cdots \partial x_{i_l}})_{1 \le i_1 \le \cdots \le i_l \le n}$. As in Theorem 2.6, we will identify $f^{(l)}(x)$ with this vector in $\mathbb{R}^{M(n,l)}$ to allow for independent choices of the values of these derivatives. Together the function and all derivatives of order $\le k$ constitute

$$N(n, k) := \sum_{l=0}^{k} M(n, l) = \binom{n+k}{n}$$

independent variables. In this setup, we have:

Proposition 3.6. *Let $m, n \in \mathbb{N}$, $k \in \mathbb{N}_0$, $I \subset \mathbb{R}^n$ be open and $T : C^k(I, \mathbb{R}) \to C(I, \mathbb{R}^m)$ be an operator. Suppose that for all open subsets $J \subset I$ and all $f_1, f_2 \in C^k(I, \mathbb{R})$ with $f_1|_J = f_2|_J$ we have that $Tf_1|_J = Tf_2|_J$. Then there is a function $F : I \times \mathbb{R}^{N(n,k)} \to \mathbb{R}^m$ such that*

$$Tf(x) = F\big(x, f(x), f'(x), \ldots, f^{(k)}(x)\big)$$

for all $f \in C^k(I, \mathbb{R})$ and $x \in I$.

Proof. Fix $x_0 = (x_{0i})_{i=1}^n \in I$. By assumption, $Tf_1(x_0) = Tf_2(x_0)$ for every two functions $f_1, f_2 \in C^k(I, \mathbb{R})$ which coincide on a small open neighborhood of x_0 in I. To prove that $Tf(x_0)$ depends only on $(x_0, f(x_0), \ldots, f^{(k)}(x_0))$, we may therefore assume that I is a (possibly small) open cube or ball centered at x_0. Let $f \in C^k(I, \mathbb{R})$. Define, for $x = (x_i)_{i=1}^n \in I$ and $i \in \{1, \ldots, n\}$ the i-th partial k-th order Taylor approximation to f at x_0 by

$$h_i(x) := \sum_{l=0}^{k} \frac{1}{l!} f^{(l)}(x_{01}, \ldots, x_{0i}, x_{i+1}, \ldots, x_n)((x - x_0)_{[i]}, \ldots, (x - x_0)_{[i]}),$$

where $(x - x_0)_{[i]} := (x_1 - x_{01}, \ldots, x_i - x_{0i}, 0, \ldots, 0) \in \mathbb{R}^n$. Here we consider $f^{(l)}$ as an l-multilinear form from $\mathbb{R}^n \times \cdots \mathbb{R}^n$ to $\mathbb{R}$. Note that $h := h_n$ is the k-th order Taylor approximation to f at x_0. Let $h_0 := f$. Then the functions h_0 and h_1 join C^k-smoothly at the intersection of the hyperplane $x_1 = x_{01}$ with I, since by definition of $(x - x_0)_{[1]}$ only the iterated derivatives with respect to x_1 occur non-trivially in h_1. Similarly h_{i-1} and h_i join C^k-smoothly at the intersection of the hyperplane $x_i = x_{0i}$ with I, for all $i \in \{2, \ldots, n\}$. Therefore, putting

$$g_i(x) := \begin{cases} h_{i-1}(x), & x \in I, \ x_i < x_{0i}, \\ h_i(x), & x \in I, \ x_i \geq x_{0i} \end{cases}$$

for $i \in \{1, \ldots, n\}$, we have that $g_i \in C^k(I, \mathbb{R})$. On $J_i^- := \{x \in I \mid x_i < x_{0i}\}$ and $J_i^+ := \{x \in I \mid x_i > x_{0i}\}$, we have $h_{i-1}|_{J_i^-} = g_i|_{J_i^-}$, $g_i|_{J_i^+} = h_i|_{J_i^+}$. Hence, also using that the image of T consists of continuous functions,

$$(Th_{i-1})(x_0) = (Tg_i)(x_0) = (Th_i)(x_0),$$

since $x_0 \in \overline{J_i^-} \cap \overline{J_i^+}$. We conclude

$$(Tf)(x_0) = (Th_1)(x_0) = \cdots = (Th_n)(x_0) = (Th)(x_0).$$

However, h only depends on $(x_0, f(x_0), f'(x_0), \cdots, f^{(k)}(x_0))$. Therefore, there exists a function of these parameters which determines $Tf(x_0)$. Identifying $f^{(l)}(x_0)$ with vectors of iterated partial derivatives in $\mathbb{R}^{M(n,l)}$ as described before, this means that there is a function $F : I \times \mathbb{R}^{N(n,k)} \to \mathbb{R}^m$ such that

$$Tf(x_0) = F\big(x_0, f(x_0), f'(x_0), \cdots, f^{(k)}(x_0)\big)$$

for all $x_0 \in I$, $f \in C^k(I, \mathbb{R})$, with $N(n,k) := \sum_{l=0}^{k} M(n,l)$. $\square$

Proof of Theorem 3.5. We adapt the proof of Theorem 3.1 to the multidimensional setting. By Proposition 3.6 for $m = n$ and the localization on (small) open sets which we proved before formulating Proposition 3.6, there is a function $F : \mathbb{R}^{N(n,k)} \to \mathbb{R}^n$ such that for all $f \in C^k(I, \mathbb{R})$, $x \in I$

$$Tf(x) = F(x, f(x), f'(x), \ldots, f^{(k)}(x)).$$

Define $S : C^k(I, \mathbb{R}) \to C(I, \mathbb{R}^n)$ by

$$Sg(x) := T(\exp(g))(x)/\exp(g)(x), \quad g \in C^k(I, \mathbb{R}), \ x \in I.$$

Then $Sg(x) = F(x, \exp(g)(x), \ldots, \exp(g)^{(k)}(x))/\exp(g)(x)$ depends only on $x, g(x)$ and all derivatives of g up to $g^{(k)}(x)$. Therefore there is a function $G : I \times \mathbb{R}^{N(n,k)} \to \mathbb{R}^n$ such that

$$Sg(x) = G(x, g(x), \ldots, g^{(k)}(x)), \quad g \in C^k(I, \mathbb{R}), \ x \in I.$$

For any $g_1, g_2 \in C^k(I, \mathbb{R})$ by the Leibniz rule

$$\begin{aligned} S(g_1 + g_2) &= T(\exp(g_1) \cdot \exp(g_2))/(\exp(g_1) \cdot \exp(g_2)) \\ &= T(\exp(g_1))/\exp(g_1) + T(\exp(g_2))/\exp(g_2) = Sg_1 + Sg_2, \end{aligned}$$

i.e., S is additive in the function and derivative variables. We split any $\alpha \in \mathbb{R}^{N(n,k)}$ as $\alpha = (\alpha_l)_{l=0}^k$ where $\alpha_l \in \mathbb{R}^{M(n,l)}$. Then for any $x \in I$ and any $\alpha = (\alpha_l)_{l=0}^k$ and $\beta = (\beta_l)_{l=0}^k \in \mathbb{R}^{N(n,k)}$ there are functions $g_1, g_2 \in C^k(I, \mathbb{R})$ such that $g_1^{(l)}(x) = \alpha_l$ and $g_2^{(l)}(x) = \beta_l$ for all $l \in \{0, \dots, k\}$. Recall that all iterated partial derivatives with indices $1 \le i_1 \le \cdots \le i_l \le n$ can be chosen independently. Therefore the additivity of S is equivalent to the additivity of G in the sense that

$$G(x, \alpha + \beta) = G(x, \alpha) + G(x, \beta), \quad x \in I, \ \alpha, \beta \in \mathbb{R}^{N(n,k)}.$$

Since $Sg = T(\exp(g))/\exp(g)$ is a continuous function, we have that $G(x, g(x), \cdots, g^{(k)}(x))$ is a continuous function of x for all $g \in C^k(I, \mathbb{R})$. By Theorem 2.6, applied with k instead of $k - 1$ to any coordinate function $G_i : I \to \mathbb{R}$ of $G = (G_i)_{i=1}^n$ (with respect to the canonical unit vector basis of $\mathbb{R}^n$) separately, there is a continuous function $c : I \to L(\mathbb{R}^{N(n,k)}, \mathbb{R}^n)$ so that

$$G(x, \alpha) = c(x)(\alpha) = \sum_{l=0}^k c_l(x)(\alpha_l), \quad x \in I, \ \alpha = (\alpha_l)_{l=0}^k \in \mathbb{R}^{N(n,k)},$$

with direct sum splitting $c(x) = \sum_{l=0}^k c_l(x)$, $c_l \in L(\mathbb{R}^{M(n,l)}, \mathbb{R}^n)$. The direct sum splitting of c is a result of the coordinatewise application of Theorem 2.6.

For $f \in C^k(I, \mathbb{R})$ with $f > 0$, let $g := \ln f$. Then $f = \exp(g)$ and

$$Tf(x) = f(x)\, S(\ln f)(x) = f(x) \sum_{l=0}^k c_l(x)((\ln f)^{(l)}(x)). \tag{3.6}$$

Here the l-th derivative of $\ln f \in C^k(I, \mathbb{R})$ at x is identified with a vector in $\mathbb{R}^{M(n,l)}$. For $l \ge 2$, in the regular derivative sense

$$(\ln f)^{(l)}(x) = (\frac{f'}{f})^{(l-1)}(x) = (-1)^{l-1}(l-1)!(\frac{f'(x)}{f(x)})^l + P_l(f(x), \dots, f^{(l)}(x)),$$

where $f'(x)^l$ is the (tensor product) l-multilinear form

$$f'(x)^l(y_1, \dots, y_l) = \prod_{j=1}^l \langle f'(x), y_j \rangle, \quad y_1, \dots, y_l \in \mathbb{R}^n,$$

and P_l is a sum of quotients of terms containing powers of $f(x)$ of order $\le l - 1$ in the denominator and tensor product terms of derivatives in the numerator. Therefore for $f(x) \searrow 0$, the order of singularity of $f(x)\,(\ln f)^{(l)}(x)$ is $\frac{f'(x)^l}{f(x)^{l-1}}$, if $f'(x) \neq 0$, up to terms of smaller growth. Since Tf is continuous and hence bounded on compact sets of I also for functions having zeros in I, in (3.6) we need $c_k(x) = \cdots = c_2(x) = 0$, $x \in I$. To be more precise, suppose that $k \ge 2$, that $x = 0 \in I$ for simplicity of notation and that the cube of side-length $\epsilon_0 > 0$ centered at 0 is contained in I. Choose any $b = (b_i)_{i=1}^n \in (\mathbb{R}_{>0})^n$ and consider $f(x) := \langle b, x \rangle$ and $I_\epsilon := \{x = (x_i)_{i=1}^n \in I \mid x_i > \frac{\epsilon}{2}, i \in \{1, \ldots, n\}\}$ for any $0 < \epsilon < \epsilon_0$. Let $\mathbf{1} := (1)_{i=1}^n \in \mathbb{R}^n$. Then $f|_{I_\epsilon} \geq \frac{\epsilon}{2} \langle b, \mathbf{1} \rangle > 0$ and

$$\frac{\partial^l}{\partial x_{i_1} \cdots \partial x_{i_l}} (\ln f)(x) = (-1)^{l-1}(l-1)! \frac{\prod_{j=1}^l b_{i_j}}{\langle b, x \rangle^l}$$

for $x \in I_\epsilon$, $l \in \mathbb{N}$. Put $\psi_l(b) := (-1)^{l-1}(l-1)! \, (\prod_{j=1}^l b_{i_j})_{1 \leq i_1 \leq \cdots \leq i_l \leq n}$. Let $h \in C^k(I, \mathbb{R})$ be a smooth strictly positive extension of $f|_{I_\epsilon}$ to I. By localization, $Tf(\epsilon\mathbf{1}) = Th(\epsilon\mathbf{1})$ since $\epsilon\mathbf{1} \in I_\epsilon$. Applying (3.6) to h yields at the point $\epsilon\mathbf{1}$ with $h|_{I_\epsilon} = f|_{I_\epsilon}$

$$Tf(\epsilon\mathbf{1}) = Th(\epsilon\mathbf{1}) = c_0(\epsilon\mathbf{1}) \, \langle b, \epsilon\mathbf{1} \rangle \, \ln(\langle b, \epsilon\mathbf{1} \rangle)$$

$$+ \sum_{l=1}^k c_l(\epsilon\mathbf{1})(\psi_l(b)) \, \langle b, \epsilon\mathbf{1} \rangle^{-(l-1)}.$$

Since Tf, $c_0, \ldots, c_k$ are continuous at 0, we get for $\epsilon \to 0$ that $c_k(0)(\psi_k(b)) = 0$ for any $b \in (\mathbb{R}_{>0})^n$. This implies $c_k(0) = 0$. Recall that $c_k \in L(\mathbb{R}^{M(n,k)}, \mathbb{R}^n)$. If $k \geq 3$, we find successively in the same way $c_{k-1}(0) = 0, \ldots, c_2(0) = 0$. Therefore $c_2 = 0, \ldots, c_k = 0$ on I and hence

$$Tf(x) = c_0(x)f(x)\ln f(x) + c_1(x)(f'(x))$$

for positive C^k-functions f. Note here that $c_0(x)$ can be identified with a vector in $\mathbb{R}^n$ and $c_1(x) \in L(\mathbb{R}^n, \mathbb{R}^n)$. For general $f \in C^k(I, \mathbb{R})$, which may be also negative or zero, it has to be modified to

$$Tf(x) = c_0(x)f(x)\ln|f(x)| + c_1(x)(f'(x)).$$

This is shown similarly as in part (ii) of the proof of Theorem 3.1 by proving that T is odd, $T(-f) = -T(f)$. $\square$

3.3 An extended Leibniz rule

We study in this section some families of operator equations to which the Leibniz rule belongs. These families turn out to be very rigid, in the sense that they admit only very few "isolated" solutions, in our view a manifestation of the exceptional role which the derivative plays in analysis.

We return to functions of one variable. Looking at derivations from a more general point of view, we keep the operator $T : C^k(I) \to C(I)$, $k \in \mathbb{N}$, but replace the identity operation on the right-hand side of the Leibniz rule by some more general operators $A_1, A_2 : C^k(I) \to C(I)$ and study the solutions of the extended Leibniz rule operator equation

$$T(f \cdot g) = Tf \cdot A_1 g + A_2 f \cdot Tg, \quad f, g \in C^k(I).$$

Thus $A_1 = A_2 = \mathrm{Id}$ is the classical case of the Leibniz rule. Choosing $A_1 f = A_2 f = 1$ for all $f \in C^k(I)$ would result in the equation $T(f \cdot g) = Tf + Tg$ mapping products to sums, as the logarithm does on the positive reals. However, choosing $g = 0$, we conclude immediately that this equation only admits the trivial solution $T = 0$. Therefore, adding operators A_1, A_2 to the formula plays a "tuning" role, helping to create reasonable operators T which in some sense map products to sums on classical function spaces.

The maps A_1, A_2 should be rather different from T since, for $A_1 = A_2 = \frac{1}{2}T$, we would have the multiplicative equation $T(f \cdot g) = Tf \cdot Tg$, where bijective solutions $T : C^k(I) \to C^k(I)$ have a very different form, e.g., for $k = 0$, $Tf(x) = |f(u^{-1}(x))|^{p(x)} \{\mathrm{sgn}\, f(u^{-1}(x))\}$ where $u : I \to I$ is a homeomorphism, cf. Milgram [M], or for $k \in \mathbb{N}$, $Tf(x) = f(u^{-1}(x))$, where $u : I \to I$ is a diffeomorphism, cf. Mrčun, Šemrl [MS] or Artstein-Avidan, Faifman, Milman [AFM].

Though, for $A_1 = A_2 =: A$, the operators T and A are closely intertwined by the equation $T(f \cdot g) = Tf \cdot Ag + Af \cdot Tg$, there is more variability when solving an operator equation for two unknown operators. Typically we have to impose a weak assumption of "non-degeneration", to guarantee that the operators are localized and avoid examples like the above proportional one or the following:

Example. Define $T : C^k(\mathbb{R}) \to C(\mathbb{R})$ and $A : C^k(\mathbb{R}) \to C(\mathbb{R})$ by

$$Tf(x) := f(x) - f(x+1), \quad Af(x) := \frac{1}{2}(f(x) + f(x+1)).$$

Then for all $f, g \in C^k(\mathbb{R})$, $T(f \cdot g) = Tf \cdot Ag + Af \cdot Tg$ since the mixed terms cancel. This means that both operators are not localized. Here for functions with small support $\mathrm{supp}\, f \subset (-\frac{1}{2}, \frac{1}{2})$, we have $Tf(x) = 2Af(x) = f(x)$ for all $x \in (-\frac{1}{2}, \frac{1}{2})$. To be able to prove localization, we have to avoid that T and A are "locally homothetic", i.e., homothetic on functions with small support. To exclude this type of "resonance" situation between T and A, we introduce the following condition for the pair (T, A).

Definition. Let $k \in \mathbb{N}$, $I \subset \mathbb{R}$ be an open set and $T, A : C^k(I) \to C(I)$ be operators. The pair (T, A) is C^k-*non-degenerate* if, for every open interval $J \subset I$ and $x \in J$, there are functions $g_1, g_2 \in C^k(I)$ with support in J such that $z_i := (Tg_i(x), Ag_i(x)) \in \mathbb{R}^2$ are linearly independent in $\mathbb{R}^2$ for $i = 1, 2$. We also assume that, for every $x \in \mathbb{R}$, there is $g \in C^k(\mathbb{R})$ with $Tg(x) = 0$ and $Ag(x) \neq 1$.

The first condition here is weaker than asking that T and A are not proportional.

We will assume a weak continuity assumption to simplify the proof of the main theorem.

Definition. For $k \in \mathbb{N}$, a map $A : C^k(I) \to C(I)$ is *pointwise continuous* provided that, for any sequence $(f_n)_{n \in \mathbb{N}}$ of $C^k(I)$-functions and $f \in C^k(I)$ such that $f_n^{(j)} \to f^{(j)}$ converge uniformly on all compact subsets of I for all $j \in \{0, \dots, k\}$, we have pointwise convergence $\lim_{n \to \infty} A f_n(x) = A f(x)$ for every $x \in I$.

We now state the main result for the extended Leibniz rule equation.

Theorem 3.7 (Extended Leibniz rule). *Let $k \in \mathbb{N}_0$. Assume that $I \subset \mathbb{R}$ is an open interval and that $T, A_1, A_2 : C^k(I) \to C(I)$ are operators satisfying*

$$T(f \cdot g) = Tf \cdot A_1 g + A_2 f \cdot Tg, \quad f, g \in C^k(I). \tag{3.7}$$

Suppose that (T, A_1) are C^k-non-degenerate and that T, A_1 and A_2 are pointwise continuous. Then T, A_1 and A_2 are localized.

There are three possible families of solutions for T and A_1, A_2, given by the formulas below. They might be defined on disjoint subsets I_1, I_2 and I_3 of the interval I, being combined to yield a globally non-degenerate solution so that T and A_1, A_2 have ranges in the continuous functions on I.

More precisely, there are three pairwise disjoint subsets I_1, I_2, I_3 of I, one or two of them possibly empty, with I_2, I_3 open, such that $I = I_1 \cup I_2 \cup I_3$, and there are functions $a, d_0, \dots, d_k, p : I \to \mathbb{R}$ with $p > 0$ which are continuous on $I \setminus N$ where $N := \partial I_2 \cup \partial I_3$, and functions $\gamma \in C(I)$ and $q \in C(I_3)$ with $q > 0$ such that $A_1 - A_2 = 2\gamma T$ on $C^k(I)$, and putting $A := \frac{1}{2}(A_1 + A_2)$, we have for all $f \in C^k(I)$ and $x \in I_1$,

$$Tf(x) = a(x)\left(\sum_{l=0}^{k} d_l(x) \, (\ln|f|)^{(l)}(x)\right)|f(x)|^{p(x)}\{\operatorname{sgn} f(x)\}, \tag{3.8}$$

$$Af(x) = |f(x)|^{p(x)}\{\operatorname{sgn} f(x)\},$$

and for $x \in I_2$,

$$Tf(x) = a(x)\sin\left(\sum_{l=0}^{k} d_l(x) \, (\ln|f|)^{(l)}(x)\right)|f(x)|^{p(x)}\{\operatorname{sgn} f(x)\},$$

$$Af(x) = \cos\left(\sum_{l=0}^{k} d_l(x) \, (\ln|f|)^{(l)}(x)\right)|f(x)|^{p(x)}\{\operatorname{sgn} f(x)\}, \tag{3.9}$$

and for $x \in I_3$,

$$Tf(x) = \frac{1}{2}a(x)\left(|f(x)|^{p(x)}\{\operatorname{sgn} f(x)\} - |f(x)|^{q(x)}[\operatorname{sgn} f(x)]\right),$$

$$Af(x) = \frac{1}{2}\left(|f(x)|^{p(x)}\{\operatorname{sgn} f(x)\} + |f(x)|^{q(x)}[\operatorname{sgn}(x)]\right). \tag{3.10}$$

The terms $\{\operatorname{sgn} f(x)\}$ and $[\operatorname{sgn} f(x)]$ may be present in both formulas for T and A or not at all, yielding different solutions.

The solution (3.8) requires that $p(x) \geq \max\{l \leq k \mid d_l(x) \neq 0\}$ to guarantee that the range of T consists of continuous functions.

In (3.10), $p(x) = 0$ or $q(x) = 0$ are allowed, too, if the corresponding sign-terms do not occur.

Conversely, let $A_1 := A + \gamma T$, $A_2 := A - \gamma T$ where T and A are given by the above formulas. Then (T, A_1, A_2) satisfy (3.7).

Remarks. (i) Theorem 3.7 shows that basically only three different types of combinations of operators (T, A_1, A_2) satisfying the extended Leibniz rule (3.7) are possible. For $k > 1$, the first one is similar to the one for positive functions in Corollary 3.4. Note that $(\ln|f|)^{(k)}|f|^p = a_k|f|^{p-k}(f')^k + Q_{k,p}$ where, for $p \geq k$, $Q_{k,p}$ is a polynomial in the function f and its derivatives, so that $Tf(x)$ is well defined by (3.8) for $p \geq p(x)$ (in the limit) also for functions f having zeros in x, and equation (3.8) provides the solution in this situation, too. In (3.8), Tf depends linearly on the highest derivative $f^{(k)}$, although with a factor which is a power of f, e.g., for $k = 2$, $Tf = ff'' - (f')^2$, $Af = f^2$.

(ii) For $k = 1$, the first solution is similar to the one of the Leibniz rule in Theorem 3.1, namely $Tf = c_0 f \ln|f| + c_1 f'$. Since (3.7) reminds of the addition formula for the sin-function when logarithmic arguments occur, the second solution is not surprising, cf. Proposition 2.13.

(iii) Note that only very few tuning operators A yield possible solutions of (3.7), and that they then determine the main operator T to a large extent. E. g. choosing A to be given by $Af = |f|^p\{\operatorname{sgn} f\}$, we get that Tf is a linear combination of terms $(\ln|f|)^{(l)}|f|^p\{\operatorname{sgn} f\}$.

(iv) The following example shows that the three solutions in Theorem 3.7 may be combined on different subintervals of I to form a non-degenerate solution.

Example. Let $I := (-1, 1)$ and $f \in C(I)$. Define maps T, A on $C(I)$ by

$$Tf(x) := \begin{cases} \frac{1}{x}\sin(x\ln|f(x)|)\, f(x), & x \in (-1, 0), \\ \ln|f(x)|\, f(x), & x = 0, \\ \frac{1}{x}(|f(x)|^x - 1)\, f(x), & x \in (0, 1), \end{cases}$$

$$Af(x) := \begin{cases} \cos(x\ln|f(x)|)\, f(x), & x \in (-1, 0), \\ f(x), & x = 0, \\ \frac{1}{2}(|f(x)|^x + 1)\, f(x), & x \in (0, 1). \end{cases}$$

On $I_1 := \{0\}$, the pair (T, A) has the form of the first solution (3.8), on $I_2 := (-1, 0)$ the form of the second solution (3.9) and on $I_3 := (0, 1)$ the form of the third solution (3.10). Note, however, that for $x \to 0$, $d(x) = x \to 0$, $p_3(x) - q(x) =$

$x \to 0$ and that $c_2(x) = c_3(x) = \frac{1}{x}$ have a singularity at 0. Nevertheless, Tf and Af define continuous functions on I since $\lim_{y\to 0} \frac{\sin(y)}{y} = 1$ and

$$\lim_{x\to 0} \frac{1}{x}(|f(x)|^x - 1) = \ln|f(x)| \text{ for } f(x) \neq 0.$$

For $f(x) = 0$, there is nothing to prove. Therefore T and A map $C(I)$ into $C(I)$ and satisfy (3.7). The solution is non-degenerate at zero: Just choose functions g_1, g_2 with small support and $g_1(0) = 3, g_2(0) = 2$. Then $(g_i(0) \ln g_i(0), g_i(0)) \in \mathbb{R}^2$ are linearly independent for $i = 1, 2$.

(v) It is also possible to combine the two solutions involving derivative terms, as the following example shows.

Example. Let $I := (-1, 1)$, $p > 1$ and $f \in C^1(I)$. Define maps T, A on $C^1(I)$ by

$$Tf(x) := \begin{cases} \frac{1}{x}\sin(x\frac{f'(x)}{f(x)})\,|f(x)|^p, & x \in (-1,0), \\ \frac{f'(x)}{f(x)}\,|f(x)|^p, & x \in [0,1), \end{cases}$$

$$Af(x) := \begin{cases} \cos(x\frac{f'(x)}{f(x)})\,|f(x)|^p, & x \in (-1,0), \\ |f(x)|^p, & x \in [0,1). \end{cases}$$

On $[0, 1)$, the solution is of the first type (3.8), with $(\ln|f|)' = \frac{f'}{f}$; it could be defined on $\mathbb{R}$ as well. But $p \geq 1$ is required here. On $(-1, 0)$, the solution is of the second type (3.9) and requires only $p > 0$ to yield continuous functions. For $x \to 0$, $d_1(x) = x$ tends to zero and $a(x) = 1/x$ has a singularity. This behavior is needed to join the other solution in a continuous way. We note that there is a delicate point about the continuity at zero. Both solutions are well defined for $p = 1$. However, choosing $p = 1$ does not yield a solution T with range in the continuous functions. Simply take $f(x) = x$. Then for $p = 1$, $Tf(x) = 1$ for $x \geq 0$ while $Tf(x) = \sin(1)$ for $x < 0$; Tf is not continuous at 0. However, for any $p > 1$, the range of T consists of continuous functions, since

$$\left|\frac{f'(x)}{f(x)}|f(x)|^p - \frac{1}{x}\sin(x\frac{f'(x)}{f(x)})|f(x)|^p\right| \leq 2|f(x)|^{p-1}|f'(x)|$$

as easily seen using $|\sin(t)| \leq |t|$, and this tends to zero as f tends to zero.

(vi) Let $S : C^k(I) \to C(I)$ satisfy the Leibniz rule and $M : C^k(I) \to C(I)$ be multiplicative. Then the pointwise product $T := S \cdot M : C^k(I) \to C(I)$ satisfies equation (3.7) with A being given by $A(f) := f \cdot M(f)$, $f \in C^k(I)$. The solution (3.8) is of this form.

Additional conditions will guarantee in the case $k = 1$ that the solutions have a simple form:

Corollary 3.8. *Assume that $T, A_1, A_2 : C^1(I) \to C(I)$ satisfy (3.7), with $k = 1$, $T \not\equiv 0$, and that (T, A_1) are C^1-non-degenerate and pointwise continuous. Let $A := \frac{1}{2}(A_1 + A_2)$. Suppose further that T maps $C^\infty(I)$ into $C^\infty(I)$.*
Then there are $n, m \in \mathbb{N}_0$ and a function $c \in C^\infty(I)$ such that the solution of (3.7) has one of the following two forms: either

$$Tf = c \, f' \, f^n, \quad Af = f^{n+1},$$

or

$$Tf = c \, (f^n - f^m), \quad Af = \frac{1}{2}(f^n + f^m),$$

for any $f \in C^1(I)$. If additionally $0 \in I$, $T2 = 0$ and $T(2\,\mathrm{Id}) = 2$, we have

$$Tf = f', \quad Af = f.$$

Corollary 3.9. *Assume that $T, A_1, A_2 : C^1(I) \to C(I)$ satisfy (3.7), with $k = 1$, $T \not\equiv 0$, and that (T, A_1) are C^1-non-degenerate and pointwise continuous. Let $A := \frac{1}{2}(A_1 + A_2)$. Suppose further that T maps linear functions into polynomials. Then there are $n, m \in \mathbb{N}_0$ and a polynomial function c such that the solution of (3.7) has one of the following two forms:*
either

$$Tf = c \, f' \, f^n, \quad Af = f^{n+1},$$

or

$$Tf = c \, (f^n - f^m), \quad Af = \frac{1}{2}(f^n + f^m),$$

for any $f \in C^1(I)$. If additionally $T2 = 0$ and $T(2\,\mathrm{Id}) = 2$, we have

$$Tf = f', \quad Af = f.$$

In both corollaries, there is $\gamma \in C(I)$ such that $A_1 = A + \gamma T$ and $A_2 = A - \gamma T$.

Note that the second solution in both corollaries may be extended to any $f \in C(I)$.

We now turn to the proof of Theorem 3.7. We again start by showing that T is localized.

Lemma 3.10. *Under the assumptions of Theorem 3.7, we have*

(i) $T(0) = T(1) = 0$ and $A_1(1) = A_2(1) = 1$.

(ii) *If $J \subset I$ is open and $f_1, f_2 \in C^k(I)$ are such that $f_1|_J = f_2|_J$, then $(Tf_1)|_J = (Tf_2)|_J$ and $(A_i f_1)|_J = (A_i f_2)|_J$ for $i = 1, 2$.*

Proof. (i) Choosing $f = 0$ in (3.7), we find for any $x \in I$ and $g \in C^k(I)$

$$T(0)(x)\big(1 - A_1 g(x)\big) = A_2(0)(x) T g(x).$$

By the C^k-non-degeneracy assumption, there is $g \in C^k(I)$ with $A_1 g(x) \neq 1$ and $T g(x) = 0$. Hence, $T(0)(x) = 0$, $T(0) = 0$. Therefore, $0 = A_2(0)(x) T g(x)$ for all $g \in C^k(I)$ which also yields $A_2(0) = 0$. Taking $g = 0$ in (3.7), we get

$$T f(x) A_1(0)(x) = T(0)(x)\big(1 - A_2 f(x)\big) = 0,$$

for all $x \in I$, $f \in C^k(I)$. Hence also $A_1(0) = 0$.

Next, choose $f = 1$ in (3.7) to find

$$T g(x)(1 - A_2(\mathbf{1})(x)) = T(\mathbf{1})(x) A_1 g(x), \quad x \in I, \ g \in C^k(I).$$

By C^k-non-degeneracy, there are functions $g_1, g_2 \in C^k(I)$ such that $(T g_i(x), A_1 g_i(x)) \in \mathbb{R}^2$ are linearly independent for $i = 1, 2$. Therefore the previous equation with $g = g_1$ and $g = g_2$ implies $A_2(\mathbf{1}) = \mathbf{1}$, $T(\mathbf{1}) = 0$. Taking $g = \mathbf{1}$ in (3.7), we find similarly

$$T f(x)(1 - A_1(\mathbf{1})(x)) = T(\mathbf{1})(x) A_2 f(x) = 0,$$

for all $f \in C^k(I)$. This yields $A_1(\mathbf{1}) = \mathbf{1}$.

(ii) Let $J \subset I$ be given and $f_1, f_2 \in C^k(I)$ with $f_1|_J = f_2|_J$. Let $g \in C^k(I)$ with $\operatorname{supp} g \subset J$. Then $f_1 \cdot g = f_2 \cdot g$. By (3.7)

$$T f_1 \cdot A_1 g + A_2 f_1 \cdot T g = T(f_1 \cdot g) = T(f_2 \cdot g)$$
$$= T f_2 \cdot A_1 g + A_2 f_2 \cdot T g,$$
$$\big(T f_1(x) - T f_2(x)\big) \cdot A_1 g(x) = \big(A_2 f_2(x) - A_2 f_1(x)\big) \cdot T g(x), \quad x \in I.$$

For a given $x \in J$, choose $g_1, g_2 \in C^k(I)$ with support in J such that $(T g_i(x), A_1 g_i(x)) \in \mathbb{R}^2$ are linearly independent for $i \in 1, 2$. The previous equation then yields for $g = g_1$ and $g = g_2$ that $T f_1(x) = T f_2(x)$, $A_2 f_1(x) = A_2 f_2(x)$, i.e., $T f_1|_J = T f_2|_J$, $A_2 f_1|_J = A_2 f_2|_J$. The argument for $A_1 f_1|_J = A_1 f_2|_J$ is similar. □

Proof of Theorem 3.7. (i) Assume that (T, A_1, A_2) satisfy the extended Leibniz rule (3.7). Then for all $f, g \in C^k(I)$ and $x \in I$, using the symmetry in f and g,

$$T(f \cdot g)(x) = T f(x) A_1 g(x) + A_2 f(x) T g(x) = T g(x) A_1 f(x) + A_2 g(x) T f(x),$$

hence $T f(x)(A_1 g(x) - A_2 g(x)) = T g(x)(A_1 f(x) - A_2 f(x))$. If $A_1 \not\equiv A_2$, there is $g \in C^k(I)$ and $x \in I$ such that $A_1 g(x) \neq A_2 g(x)$. Then $T g(x) \neq 0$ since otherwise $T f(x) = 0$ for all $f \in C^k(I)$ which would contradict the assumption of non-degeneration of (T, A_1), and therefore $A_1 f(x) - A_2 f(x) = 2\gamma(x) T f(x)$ holds for all $f \in C^k(I)$, where $\gamma(x) := \frac{A_1 g(x) - A_2 g(x)}{2 T g(x)}$. Since $T f, A_1 f, A_2 f$ are continuous functions, so is γ. Thus $A_1 - A_2 = 2\gamma T$. Clearly, $A_1 f(x) = A_2 f(x)$ is possible for some x or all $x \in I$, with $\gamma(x) = 0$. Put $A := \frac{1}{2}(A_1 + A_2)$. Then $A_1 = A + \gamma T$, $A_2 = T - \gamma T$. Equation (3.7) holds for (T, A) if A_1 and A_2 are replaced by the one operator A.

In the following, we write equation (3.7) with T and A and analyze the structure of these operators.

(ii) By Lemma 3.10 and Proposition 3.3 there are functions $\widetilde{F}, \widetilde{B} : I \times \mathbb{R}^{k+1} \to \mathbb{R}$ such that for all $f \in C^k(I)$ and $x \in I$

$$Tf(x) = \widetilde{F}(x, f(x), \dots, f^{(k)}(x)), \quad Af(x) = \widetilde{B}(x, f(x), \dots, f^{(k)}(x)).$$

We introduce operators $S, R : C^k(I) \to C(I)$ by $Sh := T(\exp h)$, $Rh := A(\exp h)$ for all $h \in C^k(I)$. Since the derivatives of $\exp h$ of order l can be written as a function of h and its derivatives of order $\leq l$, the operators S and R are localized as well, i.e., there exist functions $F, B : I \times \mathbb{R}^{k+1} \to \mathbb{R}$ such that for all $h \in C^k(I)$ and $x \in I$

$$Sh(x) = F(x, h(x), \dots, h^{(k)}(x)), \quad Rh(x) = B(x, h(x), \dots, h^{(k)}(x)).$$

Equation (3.7) yields for $h_1, h_2 \in C^k(I)$

$$
\begin{aligned}
S(h_1 + h_2) = T(\exp h_1 \exp h_2) &= T(\exp h_1)A(\exp h_2) + A(\exp h_1)T(\exp h_2) \\
&= S(h_1)R(h_2) + R(h_1)S(h_2). \quad (3.11)
\end{aligned}
$$

Let $\alpha = (\alpha_j)_{j=0}^k$, $\beta = (\beta_j)_{j=0}^k \in \mathbb{R}^{k+1}$ and $x \in I$ be arbitrary. Choose $h_1, h_2 \in C^k(I)$ with $h_1^{(j)}(x) = \alpha_j$ and $h_2^{(j)}(x) = \beta_j$ for all $j \in \{0, \dots, k\}$. Then the operator equation (3.11) is equivalent to the functional equation for F and B

$$F(x, \alpha + \beta) = F(x, \alpha)B(x, \beta) + F(x, \beta)B(x, \alpha) \quad (3.12)$$

for all $\alpha, \beta \in \mathbb{R}^{k+1}$, $x \in I$.

We claim that for any fixed $x \in I$, $B(x, \cdot)$ and $F(x, \cdot)$ are continuous functions on $\mathbb{R}^{k+1}$. To verify this, take a sequence $\alpha_n = (\alpha_{n,j})_{j=0}^k \in \mathbb{R}^{k+1}$ and $\alpha \in \mathbb{R}^{k+1}$ such that $\alpha_n \to \alpha$ in $\mathbb{R}^{k+1}$. Consider the functions $h_n(t) := \sum_{j=0}^k \frac{\alpha_{n,j}}{j!}(t-x)^j$, $h(t) := \sum_{j=0}^k \frac{\alpha_j}{j!}(t-x)^j$. Then $h_n^{(l)} \to h^{(l)}$ and $f_n := \exp(h_n)^{(l)} \to f := \exp(h)^{(l)}$ converge uniformly on all compact subsets of I for any $l \in \{0, \dots, k\}$. By the assumption of pointwise continuity , we have $Af_n(x) \to Af(x)$ and $Tf_n(x) \to Tf(x)$ for all $x \in I$. This means

$$
\begin{aligned}
B(x, \alpha_{n,0}, \dots, \alpha_{n,k}) &= Af_n(x) \to Af(x) = B(x, \alpha_0, \dots, \alpha_k), \\
F(x, \alpha_{n,0}, \dots, \alpha_{n,k}) &= Tf_n(x) \to Tf(x) = F(x, \alpha_0, \dots, \alpha_k).
\end{aligned}
$$

Therefore for all $x \in I$, $B(x, \cdot)$ and $F(x, \cdot)$ are continuous functions which satisfy (3.12). The solutions of (3.12) were studied in Chapter 2, Corollary 2.12.

(iii) We now determine the form of Tf and Af for strictly positive functions $f > 0$. By Corollary 2.12 for $n = k + 1$ there are vectors $b(x), c(x), d(x) \in \mathbb{R}^{k+1}$ and $a(x) \in \mathbb{R}$ such that $F(x, \cdot)$ and $B(x, \cdot)$ have one of the following forms

(a) $F(x, \alpha) = \langle b(x), \alpha \rangle \exp(\langle c(x), \alpha \rangle)$, $B(x, \alpha) = \exp(\langle c(x), \alpha \rangle)$;

(b) $F(x, \alpha) = a(x) \exp(\langle c(x), \alpha \rangle) \sin(\langle d(x), \alpha \rangle)$,
    $B(x, \alpha) = \exp(\langle c(x), \alpha \rangle) \cos(\langle d(x), \alpha \rangle)$;

(c) $F(x, \alpha) = a(x) \exp(\langle c(x), \alpha \rangle) \sinh(\langle d(x), \alpha \rangle)$,
    $B(x, \alpha) = \exp(\langle c(x), \alpha \rangle) \cosh(\langle d(x), \alpha \rangle)$;

(d) $F(x, \alpha) = a(x) \exp(\langle c(x), \alpha \rangle)$, $B(x, \alpha) = \frac{1}{2} \exp(\langle c(x), \alpha \rangle)$, $\alpha \in \mathbb{R}^n$.

Since $A(\mathbf{1}) = \mathbf{1}$ by Lemma 3.10, $1 = A(\mathbf{1})(x) = R(0)(x) = B(x, 0)$. Therefore the last case (d) is impossible here since in that case $B(x, 0) = \frac{1}{2}$.

For positive functions $f \in C^k(I)$, $f > 0$, let $h := \ln f$, $f = \exp h$, so that in case (a) with $b = (b_l)_{l=0}^k$, $c = (c_l)_{l=0}^k$

$$Af(x) = R(\ln f)(x) = B(x, (\ln f)(x), \ldots, (\ln f)^{(k)}(x))$$

$$= \exp\left(\sum_{l=0}^k c_l(x)(\ln f)^{(l)}(x) \right),$$

$$Tf(x) = S(\ln f)(x) = F(x, (\ln f)(x), \ldots, (\ln f)^{(k)}(x))$$

$$= \left(\sum_{l=0}^k b_l(x)(\ln f)^{(l)}(x) \right) \exp\left(\sum_{l=0}^k c_l(x)(\ln f)^{(l)}(x) \right). \tag{3.13}$$

Depending on $x \in I$, one of the formulas (a), (b) or (c) might apply. Let I_1, I_2 and I_3, respectively, denote the subsets of I where $Tf(x)$, $Af(x)$ is determined by (a), (b) and (c), respectively. For (a) and $f > 0$, we just wrote down the formulas in (3.13). However, the sets are restricted by the requirement that Tf and Af have to be continuous functions for all $f \in C^k(I)$. Suppose that the interior of the domain I_1 where (3.13) gives the solution – for $f > 0$ – is not empty. Let us show that the functions $c_0, \ldots, c_k$ and $b_0, \ldots, b_k$ have to be continuous in the interior of I_1. Indeed, starting with constant functions f, the continuity of Af and Tf yields that c_0 and b_0 are continuous. Then choosing linear functions, it follows that c_1 and b_1 are continuous. Repeat the argument with polynomials of successively higher degree.

Since T and A are localized and have to be well defined also for functions having zeros in the interior of I_1, the formula for Af should never become singular, i.e., unbounded when $f \searrow 0$. The argument for this is exactly the same as in the proof of Theorem 3.1. However, $(\ln f)^{(l)}$ is of order $(\frac{f'}{f})^l$, if $f' \neq 0$ and $l \in \mathbb{N}$, up to terms of smaller order. Therefore we must have $c_1 = \cdots = c_k = 0$ in (3.13) on I_1. Put $p(x) := c_0(x)$. Then for $f > 0$, $x \in I_1$,

$$Af(x) = f(x)^{p(x)}, \quad Tf(x) = \left(\sum_{l=0}^{k} b_l(x)(\ln f)^{(l)}(x) \right) f(x)^{p(x)}. \qquad (3.14)$$

The continuity of Tf for all f requires that $p(x) \geq \max\{l \leq k \mid b_l(x) \neq 0\} =: P(x)$. If $P(x) = 0$, we need $p(x) > 0$. In this case, (3.14) provides a solution of (3.7) for positive f.

The case (b) for T and A on I_2 yields the formula

$$Af(x) = \exp\left(\sum_{l=0}^{k} c_l(x)(\ln f)^{(l)}(x) \right) \cos\left(\sum_{l=0}^{k} d_l(x)(\ln f)^{(l)}(x) \right),$$

with continuous coefficient functions c_l, d_l on I_2. Continuity for functions with zeros requires that $c_1 = \cdots = c_k = 0$. Then with $p(x) := c_0(x)$, for $f > 0$, $x \in I_2$,

$$Af(x) = \cos\left(\sum_{l=0}^{k} d_l(x)(\ln f)^{(l)}(x) \right) f(x)^{p(x)},$$

$$Tf(x) = a(x) \sin\left(\sum_{l=0}^{k} d_l(x)(\ln f)^{(l)}(x) \right) f(x)^{p(x)}, \qquad (3.15)$$

where $p(x) > 0$ is required and a is continuous in I_2. In the last case (c)

$$Af(x) = \exp\left(\sum_{l=0}^{k} c_l(x)(\ln f)^{(l)}(x) \right) \cosh\left(\sum_{l=0}^{k} d_l(x)(\ln f)^{(l)}(x) \right),$$

and here necessarily $c_1 = \cdots = c_k = 0$ and $d_1 = \cdots = d_k = 0$. Then with $p(x) := c_0(x) + d_0(x)$ and $q(x) := c_0(x) - d_0(x)$, $Af(x) = \frac{1}{2}(f(x)^{p(x)} + f(x)^{q(x)})$, $p(x) \geq 0$, $q(x) \geq 0$, yielding for $f > 0$, $x \in I_3$

$$Af(x) = \frac{1}{2}\left(f(x)^{p(x)} + f(x)^{q(x)} \right), \quad Tf(x) = a(x)\left(f(x)^{p(x)} - f(x)^{q(x)} \right). \qquad (3.16)$$

To be non-degenerate, the solution on I_2 given by (3.15) requires that some of the continuous functions d_l are non-zero at any $x \in I_2$, and the one on I_3 given (3.16) requires that $p(x) \neq q(x)$ for any $x \in I_3$. They can be joined to another one of the three solutions only when the d_l or $p - q$ tend to zero and at the same time $|a|$ becomes unbounded. Hence, by continuity of the parameter functions, the subsets I_2 and I_3 are open. Of course, any of the sets I_1, I_2 or I_3 could be empty; the solution may be given on all of I by just one of the formulas, this being the most natural case. However, I_1 is not necessarily open. In the first example in Remark (ii) after Theorem 3.7 we had $I_1 = \{0\}$.

(iv) It remains to determine the formulas for $Tf(x)$ and $Af(x)$ when $f \in C^k(I)$ is negative or zero. Since Af and Tf are continuous and the coefficient functions are continuous on their domains, the localized formulas (3.14), (3.15),(3.16) extend by continuity to $Tf(x)$ and $Af(x)$ when $f(x) = 0$ and x is an isolated zero of f or a limit of isolated zeros. If $x \in J \subset I$, J open and $f|_J = 0$, we know by Lemma 3.10 that $Tf(x) = 0$.

Suppose now that $f \in C^k(I)$ and $x \in I$ are such that $f(x) < 0$. We may assume that $f < 0$ on the full set I, since $Tf(x)$ and $Af(x)$ are determined locally near x with $f(x) < 0$. For constant functions $f(x) = \alpha_0$, $g(x) = \beta_0$, we have

$$Tf(x) = \widetilde{F}(x, \alpha_0, 0, \ldots, 0), \quad Af(x) = \widetilde{B}(x, \alpha_0, 0, \ldots, 0).$$

Therefore the extended Leibniz rule (3.7) yields

$$\widetilde{F}(x, \alpha_0\beta_0, 0, \ldots, 0) = \widetilde{F}(x, \alpha_0, 0, \ldots, 0)\widetilde{B}(x, \beta_0, 0, \ldots, 0)$$
$$+ \widetilde{B}(x, \alpha_0, 0, \ldots, 0)\widetilde{F}(x, \beta_0, 0, \ldots, 0).$$

Proposition 2.13 gives the possible solutions of this functional equation. They imply for constant functions f having negative values, too, that one of the following three cases can occur:

$$Tf = b(\ln|f|)|f|^p\{\operatorname{sgn} f\}, \quad Af = |f|^p\{\operatorname{sgn} f\},$$
$$Tf = b\sin(d\ln|f|)|f|^p\{\operatorname{sgn} f\}, \quad Af = \cos(d\ln|f|)|f|^p\{\operatorname{sgn} f\},$$
$$Tf = b(|f|^p\{\operatorname{sgn} f\} - |f|^q[\operatorname{sgn} f]), \quad Af = \frac{1}{2}(|f|^p\{\operatorname{sgn} f\} + |f|^q[\operatorname{sgn} f]),$$

leaving out the variable x. The fourth solution in Proposition 2.13 is not applicable since there $B(\mathbf{1}) = \frac{1}{2} \neq 1$.

In the first two cases and in the last case when both sgn f-terms are present or both are absent, we have $T(-\mathbf{1}) = 0$ and $A(-\mathbf{1}) \in \{\mathbf{1}, -\mathbf{1}\}$. Then by (3.7), $T(-f) = Tf \, A(-\mathbf{1}) + Af \, T(-\mathbf{1}) = Tf \, A(-\mathbf{1})$. Hence T is even or odd, depending on whether $A(-\mathbf{1}) = 1$ or $A(-\mathbf{1}) = -1$. For A, we have similarly $A(-f) = Af \, A(-\mathbf{1})$, by the same arguments as in the proof of Proposition 2.13. In the last case, when the sgn f-terms are different, T and A are neither even nor odd. The determination of $T(-f)$ and $A(-f)$ in this case is similar to the last case in the proof of Proposition 2.13. Using this, formulas (3.14), (3.15) and (3.16) yield formulas (3.8), (3.9) and (3.10) in Theorem 3.7 for general functions $f \in C^k(I)$.

Conversely, the operators T and A defined by these formulas satisfy (3.7). To check this, e.g., in the case of (3.9), use the addition formula for the sin-function and $(\ln|fg|)^{(l)} = (\ln|f|)^{(l)} + (\ln|g|)^{(l)}$. This ends the proof of Theorem 3.7. $\quad\square$

Proof of Corollary 3.8. The operator T defined by (3.9) does not map C^∞-functions to C^∞-functions, since – possibly large order – derivatives of Tf will become singular in points where f has zeros. The operator given by (3.8) for $k = 1$ has the form

$$Tf = (bf\ln|f| + af')\,|f|^q\{\operatorname{sgn} f\},$$

$q = p-1$. Choosing for f constant or linear functions, we conclude that $a, b, q \in C^\infty$ is required. Since $|f|^q\{\text{sgn } f\}$ has to be a C^∞-function for any C^∞-function f, we moreover need that $|f|^q\{\text{sgn } f\} = f^n$ for a suitable $n \in \mathbb{N}_0$. If b would not be zero, a suitable derivative of Tf would have a singularity of order $\ln |f|$ when $|f| \searrow 0$. Hence $Tf = af'f^n$ in the case of (3.8). Similarly, the solution (3.10) maps C^∞-functions into C^∞-functions if and only if $Tf = a(f^n - f^m)$ for suitable $n, m \in \mathbb{N}_0$ and $a \in C^\infty$. Both solutions cannot be combined on disjoint subsets partitioning I since f' cannot be continuously approximated by differences $f^N - f^M$, in general. Therefore we have two solutions defined on the full set I.

If additionally $T2 = 0$, the second solution would require $n = m$ and then $T \equiv 0$. Thus only the first solution is possible, with $2 = T(2\,\text{Id})(x) = a(x)2^{n+1}x^n$, i.e., $a(x) = (2x)^{-n}$. Since $x = 0 \in I$ and $a \in C^\infty(I)$, it follows that $n = 0$ and $a \equiv 1$, i.e., $Tf = f'$ and $Af = f$ for all $f \in C^1(I)$. $\qquad\square$

Proof of Corollary 3.9. The operator T defined by (3.9) does not map arbitrary linear functions $f(x) = cx$, $c \in \mathbb{R}$ to polynomials, if $T \not\equiv 0$. In the case of (3.8), T again has the form

$$Tf = (bf \ln |f| + af')\, |f|^q\{\text{sgn } f\}.$$

This will not yield polynomials for all linear functions f unless $b \equiv 0$, $q = n \in \mathbb{N}_0$ and a is a polynomial function, i.e., $Tf = af'f^n$, $Af = f^{n+1}$ for all $f \in C^1(I)$.

Again, (3.10) yields the second solution with $p = n, q = m \in \mathbb{N}_0$.

If additionally $T2 = 0$, the second solution requires $n = m$, i.e., $T \equiv 0$. In the case of the first solution $T(2\,\text{Id}) = 2$ gives $2 = T(2\,\text{Id})(x) = a(x)2^{n+1}x^n$, i.e., $a(x) = (2x)^{-n}$. However, a is only a polynomial if $n = 0$, $a \equiv 1$. Then $Tf = f'$ and $Af = f$ for all $f \in C^1(I)$. $\qquad\square$

3.4 Notes and References

The basic result on the Leibniz rule equation, Theorem 3.1, is due to König, Milman [KM1]. The case $k = 0$ was shown before by Goldmann, Šemrl [GS].

Lemma 3.2 and Proposition 3.3 are taken from [KM1]. For $k = 1$, Theorems 3.5 and 3.7 were shown in [KM1], too.

The logarithm $F = \log$ satisfies $F(xy) = F(x) + F(y)$ for positive $x, y > 0$. However, there do not exist a function $F : \mathbb{R} \to \mathbb{R}$ and constants $c, d \in \mathbb{R}$ such that $F(xy) = cF(x) + dF(y)$ holds for all real numbers $x, y \in \mathbb{R}$. A function of this type sending products to sums requires replacing the constants c, d by functions, yielding in the simplest case the Leibniz rule in $\mathbb{R}$. On the real line $\mathbb{R}$ or the complex plane $\mathbb{C}$, there is the following version of the Leibniz rule:

Proposition 3.11. (a) *Let $F : \mathbb{R} \to \mathbb{R}$ be a measurable function satisfying*

$$F(xy) = F(x)y + xF(y), \quad x, y \in \mathbb{R}. \qquad (3.17)$$

Then there is $d \in \mathbb{R}$ such that $F(x) = d\,x \ln |x|$, $x \in \mathbb{R}$.

(b) *Let $F : \mathbb{C} \to \mathbb{C}$ be a measurable function satisfying*

$$F(zw) = F(z)w + zF(w), \quad z, w \in \mathbb{C}.$$

Then there is $d \in \mathbb{C}$ such that $F(z) = d z \ln |z|$, $z \in \mathbb{C}$.

Proof. (a) $F(1) = F(1^2) = 2F(1)$ implies $F(1) = 0$. Similarly $F(-1) = 0$, which implies $F(-x) = -F(x)$. For $xy \neq 0$,

$$\frac{F(xy)}{xy} = \frac{F(x)}{x} + \frac{F(y)}{y}.$$

Hence, $H(s) := F(e^s)/e^s$ is measurable and additive. By Proposition 2.1 there is $d \in \mathbb{R}$ with $H(s) = ds$. Then

$$F(x) = dx \ln |x|.$$

(b) We show by induction on n that for any $n \in \mathbb{N}$ and $z \in \mathbb{C}$, $F(z^n) = nz^{n-1}F(z)$: For $n = 2$ this is the assumption with $z = w$. Assuming this for n, we have $F(z^{n+1}) = F(z^n)z + z^n F(z) = (n+1)z^n F(z)$. Let $\zeta \in \mathbb{C}$ be an n-th root of unity. Then $0 = F(1) = F(\zeta^n) = n\zeta^{n-1}F(\zeta)$ implies that $F(\zeta) = 0$. Define $G(z) := \frac{F(z)}{z}$ for $z \in \mathbb{C} \setminus \{0\}$. Then $G(zw) = G(z) + G(w)$ for all $z, w \in \mathbb{C} \setminus \{0\}$. Hence $\phi : \mathbb{R} \to \mathbb{C}$ given by $\phi(t) := G(\exp(it))$, $t \in \mathbb{R}$, is additive and measurable. By Proposition 2.1 there is $c \in \mathbb{C}$ such that $\phi(t) = ct$ for all $t \in \mathbb{R}$. Since $F(\zeta) = 0$ for all roots of unity ζ, $c = 0$, i.e., $G|_{S^1} = 0$. The polar decomposition of $z \in \mathbb{C} \setminus \{0\}$, $z = |z| \exp(it)$ yields that $G(z) = G(|z|) + G(\exp(it)) = G(|z|)$ and for $z, w \in \mathbb{C} \setminus \{0\}$, $G(|zw|) = G(|z|) + G(|w|)$. Similarly as in part (a) we find $d \in \mathbb{C}$ such that $G(z) = G(|z|) = d \ln |z|$. Hence $F(z) = dz \ln |z|$ for all $z \in \mathbb{C} \setminus \{0\}$. Clearly $F(0) = 0$. □

Remark. The equation

$$F(xy) = F(x)B(y) + B(x)F(y), \quad x, y \in \mathbb{R} \tag{3.18}$$

for unknown functions $F, B : \mathbb{R} \to \mathbb{R}$ is a relaxation of equation (3.17). Proposition 2.13 gives the four (real) solutions of (3.18). The first of these, $B(x) = |x|^d \{\operatorname{sgn} x\}$, $F(x) = b \cdot \ln |x| \cdot B(x)$ has the property that B has a smaller order of growth as $|x| \to \infty$ than F. Comparing this with the operator functional equation (3.7),

$$T(f \cdot g) = Tf \cdot A_1 g + A_2 f \cdot Tg, \quad f, g \in C^k(I),$$

which has an algebraically similar form, the first solution of (3.7) has the property that $A = A_1 = A_2$ has a smaller order of differentiability than T.

We may also consider the Leibniz rule on real or complex spaces of polynomials or analytic functions. For $\mathbb{K} \in \{\mathbb{R}, \mathbb{C}\}$, let $\mathcal{P}(\mathbb{K})$ denote the space of polynomials with coefficients in $\mathbb{K}$ and $\mathcal{E}(\mathbb{K})$ be the space of real-analytic functions ($\mathbb{K} = \mathbb{R}$)

or entire functions ($\mathbb{K} = \mathbb{C}$) and $C(\mathbb{K})$ be the space of continuous functions on $\mathbb{K}$. Moreover, let $\mathcal{P}_n(\mathbb{K})$ be the subset of $\mathcal{P}(\mathbb{K})$ consisting of polynomials of degree $\leq n$.

On these spaces, there are different solutions of the Leibniz rule than those given in Theorem 3.1.

Example 1. Define $T : \mathcal{P}(\mathbb{K}) \to \mathcal{P}(\mathbb{K})$ by $Tf := \deg f \cdot f$, $f \in \mathcal{P}(\mathbb{K})$, where $\deg f$ denotes the degree of the polynomial f. Since $\deg(f \cdot g) = \deg f + \deg g$, T satisfies the Leibniz rule $T(f \cdot g) = Tf \cdot g + f \cdot Tg$ on $\mathcal{P}(\mathbb{K})$.

Example 2. Fix $x_0 \in \mathbb{K}$. For $f \in \mathcal{E}(\mathbb{K})$, let $n(f)$ denote the order of zero of f in x_0 (which may be zero if $f(x_0) \neq 0$). Define $T : \mathcal{E}(\mathbb{K}) \to \mathcal{E}(\mathbb{K})$ by $Tf := n(f) \cdot f$. Since $n(f \cdot g) = n(f) + n(g)$, T satisfies the Leibniz rule $T(f \cdot g) = Tf \cdot g + f \cdot Tg$ on $\mathcal{E}(\mathbb{K})$.

However, in both examples the operator T is not pointwise continuous in the sense that there are functions $f_m, f \in \mathcal{P}(\mathbb{K})$ or $\mathcal{E}(\mathbb{K})$ where $f_m \to f$ converges uniformly on compact sets but where $Tf_m(x)$ does not converge to $Tf(x)$ for some $x \in \mathbb{K}$, since the degree and the order of zero are not pointwise continuous operations. Let us therefore assume that $T : \mathcal{P}(\mathbb{K}) \to C(\mathbb{K})$ is pointwise continuous and satisfies the Leibniz rule. Does this guarantee that we have the same solutions as in Theorem 3.1? Again the answer is negative, as the following example due to Faifman [F3] shows:

Example 3 (Faifman).. If $T : \mathcal{P}(\mathbb{K}) \to C(\mathbb{K})$ satisfies the Leibniz rule $T(f \cdot g) = Tf \cdot g + f \cdot Tg$ for all $f, g \in \mathcal{P}(\mathbb{K})$, then for all $f_1, \ldots, f_n \in \mathcal{P}(\mathbb{K})$

$$T(\prod_{j=1}^{n} f_j) = \sum_{j=1}^{n} (\prod_{i=1, i \neq j}^{n} f_i) \, Tf_j. \tag{3.19}$$

Let us first consider the complex case $\mathbb{K} = \mathbb{C}$. Since any polynomial $f \in \mathcal{P}(\mathbb{C})$ factors as a product of linear terms, $f(z) = a\prod_{j=1}^{n}(z - z_j)$, with zeros $z_j \in \mathbb{C}$ and $a \in \mathbb{C} \setminus \{0\}$, it suffices to define $T(az + b)$, in order to define an operator $T : \mathcal{P}(\mathbb{C}) \to C(\mathbb{C})$ by applying (3.19), and then verify that this map T actually satisfies the Leibniz rule. Let $\phi : \mathbb{C} \to \mathbb{C}$ be given by $\phi(z) := z \ln |z|$, with $\phi(0) = 0$. Define

$$T(az + b) := \phi(a)z + \phi(b). \tag{3.20}$$

This map T satisfies the Leibniz rule on $\mathcal{P}_1(\mathbb{C})$ in the sense that $T(c(az + b)) = T(c)(az + b) + cT(az + b)$, since ϕ satisfies the Leibniz rule on $\mathbb{C}$. In terms of the elementary symmetric polynomials we have for $f \in \mathcal{P}_n(\mathbb{C})$

$$f(z) = a \prod_{j=1}^{n}(z - z_j) = \sum_{k=0}^{n}(-1)^k (\sum_{1 \leq j_1 < \cdots < j_k \leq n} az_{j_1} \cdots z_{j_k}) \, z^{n-k}. \tag{3.21}$$

Using (3.20) and requiring that (3.19) holds, yields the formula for $T : \mathcal{P}(\mathbb{C}) \to C(\mathbb{C})$

$$(Tf)(z) = \sum_{k=0}^{n} (-1)^k \left(\sum_{1 \leq j_1 < \cdots < j_k \leq n} \phi(a z_{j_1} \cdots z_{j_k}) \right) z^{n-k}, \qquad (3.22)$$

as induction on $n \in \mathbb{N}$ shows. Conversely, one checks that the operator T defined by (3.22) satisfies the Leibniz rule, using once more that ϕ satisfies it on $\mathbb{C}$. Moreover, this operator T is pointwise continuous on $\mathcal{P}(\mathbb{C})$, i.e., for any $f_m, f \in \mathcal{P}(\mathbb{C})$ with $f_m \to f$ uniformly on compact sets, we have $Tf_m(z) \to Tf(z)$ for any $z \in \mathbb{C}$, since the zeros depend continuously on the polynomials (in appropriate order) and ϕ is continuous. We remark that the pointwise continuity statement also holds, if $\deg f < \liminf_{m \to \infty} \deg f_m$.

Real polynomials $f \in \mathcal{P}(\mathbb{R})$ may be factored into linear and irreducible quadratic factors, the latter corresponding to two complex conjugate zeros. Applying the Leibniz rule (in $\mathbb{C}$) to such factors yields the real variable requirement for T

$$T(x^2 + px + q) = \frac{1}{2}(p \ln|q|)x + q \ln|q|, \quad p^2 < 4q.$$

Using this together with (3.20) and (3.19) then defines a pointwise continuous operator $T : \mathcal{P}(\mathbb{R}) \to C(\mathbb{R})$ satisfying the Leibniz rule. In both cases $\mathbb{K} \in \{\mathbb{R}, \mathbb{C}\}$, the image of T is actually again in $\mathcal{P}(\mathbb{K})$.

The question whether pointwise continuous operators $T : \mathcal{E}(\mathbb{K}) \to C(\mathbb{K})$ on the space of entire functions satisfying the Leibniz rule are of the same form as in Theorem 3.1 is open. The previous example does not extend to the space of entire functions $\mathcal{E}(\mathbb{K})$ since the (polynomial) functions given by $f_m(z) = (1 + \frac{z}{m})^m$ tend to $f(z) = \exp(z)$ uniformly on compact sets, but $Tf_m(z) = -z(1 + \frac{z}{m})^{m-1} \ln m$ for fixed $z \neq 0$ is a divergent sequence.

The extended Leibniz rule which was investigated in Theorem 3.7 in the space $C^k(I)$ may also be studied in the Schwartz space of complex-valued rapidly decreasing functions $\mathcal{S}(\mathbb{R}, \mathbb{C})$. The operator solutions Af are then expressed by integer powers of the functions f and their complex conjugates, and the images Tf are linear combinations of logarithmic derivatives of f and its complex conjugate or a difference of powers of f and its complex conjugate. We refer to König, Milman [KM13], where also criteria are given such that A is the identity and T the derivative. The extended Leibniz rule in $\mathcal{S}(\mathbb{R}, \mathbb{C})$ has applications to joint characterizations of the Fourier transform and the derivative [KM13].

Chapter 4

The Chain Rule

4.1 The chain rule on $C^k(\mathbb{R})$

The derivative $D : C^1(\mathbb{R}) \to C(\mathbb{R})$ satisfies the chain rule

$$D(f \circ g) = (Df) \circ g \cdot Dg$$

for all $f, g \in C^1(\mathbb{R})$. In this chapter, we study the question to which extent the chain rule formula characterizes the derivative. We consider general operators $T : C^1(\mathbb{R}) \to C(\mathbb{R})$ satisfying the *chain rule operator equation*

$$T(f \circ g) = (Tf) \circ g \cdot Tg, \quad f, g \in C^1(\mathbb{R}).$$

Due to the multiplicative structure of this equation, if T_1 and T_2 are operators satisfying the chain rule, so does the pointwise product $T_1 \cdot T_2$, and also do the positive powers of the pointwise modulus $|T_1|$. Suppose $H \in C(\mathbb{R})$ is a strictly positive continuous function. Then $Tf := H \circ f / H$ defines a map satisfying the chain rule as well. It is even defined on $C(\mathbb{R})$, not only on $C^1(\mathbb{R})$. Another example of a map $T : C^1(\mathbb{R}) \to C(\mathbb{R})$ verifying the chain rule is given by

$$Tf := \begin{cases} f', & f \in C^1(\mathbb{R}) \text{ is bijective}, \\ 0, & f \in C^1(\mathbb{R}) \text{ is not bijective}. \end{cases}$$

To avoid degenerate cases like this one, we impose the condition that T should not be identically zero on the half-bounded differentiable functions

$$C_b^1(\mathbb{R}) := \{f \in C^1(\mathbb{R}) \mid f \text{ is bounded from above or from below}\},$$

i.e., that there exist $f \in C_b^1(\mathbb{R})$ and $x \in \mathbb{R}$ such that $Tf(x) \neq 0$. For integers $k \in \mathbb{N}$, we also let $C_b^k(\mathbb{R}) := C^k(\mathbb{R}) \cap C_b^1(\mathbb{R})$.

Our main result states that a multiplicative combination of the previous examples, together with a possible factor $\operatorname{sgn} f'$, creates all possible solutions of

© Springer Nature Switzerland AG 2018
H. König, V. Milman, *Operator Relations Characterizing Derivatives*,
https://doi.org/10.1007/978-3-030-00241-1_4

the chain rule equation not only on $C^1(\mathbb{R})$ but also on $C^k(\mathbb{R})$ for any $k \in \mathbb{N}$. Again, all solutions operators are local, i.e., pointwise defined.

Theorem 4.1 (Chain rule). *Let $k \in \mathbb{N} \cup \{\infty\}$ and $T : C^k(\mathbb{R}) \to C(\mathbb{R})$ be an operator satisfying the* chain rule equation

$$T(f \circ g) = (Tf) \circ g \cdot Tg, \quad f, g \in C^k(\mathbb{R}). \tag{4.1}$$

Assume that $T|_{C_b^k(\mathbb{R})} \not\equiv 0$. Then there exist $p \geq 0$ and a positive continuous function $H \in C(\mathbb{R})$, $H > 0$, such that either

$$Tf = \frac{H \circ f}{H}|f'|^p \tag{4.2}$$

or

$$Tf = \frac{H \circ f}{H}|f'|^p \operatorname{sgn} f'. \tag{4.3}$$

In the second case we need $p > 0$ to guarantee that the image of T consists of continuous functions.

If $k = 0$ and $T : C(\mathbb{R}) \to C(\mathbb{R})$ satisfies (4.1), all solutions of T are given by $Tf = \frac{H \circ f}{H}$.

Conversely, the operators given by (4.2) or (4.3) satisfy the chain rule equation (4.1).

Under the additional initial condition $T(2\,\mathrm{Id}) = 2$ *(constant function), T has the form $Tf = f'$ or $Tf = |f'|$.*

If additionally to (4.1), $T(-2\,\mathrm{Id}) = -2$ holds, T is the derivative, $Tf = f'$.

In the formulation of similar results later we will combine statements like (4.2) and (4.3) by writing

$$Tf = \frac{H \circ f}{H}|f'|^p\{\operatorname{sgn} f'\},$$

the brackets $\{\cdot\}$ indicating that two possible solutions are given, one with the expression $\operatorname{sgn} f'$ and one without. Formulas (4.1), (4.2) and (4.3) are meant pointwise, e.g.,

$$T(f \circ g)(x) = (Tf)(g(x)) \cdot Tg(x), \quad x \in \mathbb{R},$$
$$Tf(x) = \frac{H(f(x))}{H(x)}|f'(x)|^p\{\operatorname{sgn} f'(x)\}, \quad x \in \mathbb{R}.$$

Remarks. (a) Note that we do not impose any a priori continuity condition on the operator T. A suitable level of continuity of T, however, is an a-posteriori consequence of the result.

(b) The proof will show that p and H are completely determined by the function $T(2\,\mathrm{Id}) \in C(\mathbb{R})$: we have $T(2\,\mathrm{Id}) > 0$, $p = \log_2 T(2\,\mathrm{Id})(0)$ and $H(x) = \prod_{n\in\mathbb{N}} \varphi(x/2^n)$ where φ is defined by $\varphi(x) = T(2\,\mathrm{Id})(x)/T(2\,\mathrm{Id})(0)$, and where the product converges uniformly on compact subsets of $\mathbb{R}$, with normalization $H(0) = 1$.

(c) For $C^\infty(\mathbb{R})$, there are not more solutions of the chain rule equation than for $C^1(\mathbb{R})$. Therefore, in the setup of spaces $C^k(\mathbb{R})$, the space $C^1(\mathbb{R})$ constitutes the natural domain of the chain rule. Of course, for $k = 0$, in $C(\mathbb{R})$ there is the non-surjective solution $Tf = \frac{H\circ f}{H}$ which does not depend on the derivative.

(d) For $p > 0$, let G be the antiderivative of $H^{1/p} > 0$. Then G is a strictly monotone $C^1(\mathbb{R})$-function and

$$Tf = \left|\frac{(G\circ f)'}{G'}\right|^p \{\mathrm{sgn}\, f'\} = \left|\frac{d(G\circ f)}{dG}\right|^p \left\{\mathrm{sgn}\left(\frac{d(G\circ f)}{dG}\right)\right\}.$$

In this sense, all solutions of (4.1) are p-th powers of some derivatives, up to signs.

As a consequence, the derivative is characterized by the Leibniz rule and the chain rule:

Corollary 4.2. *Let $k \in \mathbb{N}$ and suppose that $T : C^k(\mathbb{R}) \to C(\mathbb{R})$ satisfies the Leibniz rule and the chain rule,*

$$T(f\cdot g) = Tf\cdot g + f\cdot Tg, \quad T(f\circ g) = (Tf)\circ g\cdot Tg; \qquad f,g \in C^k(\mathbb{R}).$$

Then $T = 0$ or T is the derivative, $Tf = f'$ for all $f \in C^k(\mathbb{R})$.

Again, no continuity assumption on T is required here.

Proof of Corollary 4.2. By Theorem 3.1, T has the form $Tf = c\,f\ln|f| + d\,f'$ for suitable functions $c, d \in C(\mathbb{R})$. If $T \not\equiv 0$, c or d do not vanish identically and therefore T satisfies $T|_{C_b^k(\mathbb{R})} \not\equiv 0$. Hence, by Theorem 4.1, $Tf = \frac{H\circ f}{H}|f'|^p\{\mathrm{sgn}\, f'\}$ for some $p \geq 0$ and $H \in C(\mathbb{R})$, $H > 0$. Both forms of T can coincide only if $p = 1$, H is constant and $c = 0$, $d = 1$ and the $\mathrm{sgn}\, f'$-term occurs. Then $Tf = f'$, $f \in C(\mathbb{R})$. $\square$

Example. On suitable subsets of $C^k(I)$ or even $C(I)$, we may define operations T which satisfy the Leibniz rule and chain rule but are neither zero nor the derivative: Let $I = (1,\infty)$ and $C_+(I) := \{f : I \to I \mid f \text{ is continuous}\}$. Define $H \in C(I)$ by $H(x) = x\ln x$. Then the operator $T : C_+(I) \to C(I)$ given by $Tf = \frac{H\circ f}{H}$ is well defined and satisfies the Leibniz rule and the chain rule.

We now state a stronger version of Corollary 4.2: The derivative is also the only operator satisfying both the chain rule and the *extended* Leibniz rule studied in Theorem 3.7:

Corollary 4.3. *Suppose* $T, A : C^1(\mathbb{R}) \to C(\mathbb{R})$ *satisfy the chain rule and the extended Leibniz rule for all* $f, g \in C^1(\mathbb{R})$,

$$T(f \circ g) = Tf \circ g \cdot Tg \, ,$$
$$T(f \cdot g) = Tf \cdot Ag + Af \cdot Tg \, ,$$

and that T *does not vanish identically on the half-bounded functions and attains some negative values. Then* T *is the derivative,* $Tf = f'$, *and* $Af = f$ *for all* $f \in C^1(\mathbb{R})$.

Proof of Corollary 4.3. Theorem 4.1 yields that Tf is given by

$$Tf = \frac{H \circ f}{H} \, |f'|^p \, \operatorname{sgn} f'$$

for a suitable function $H \in C(\mathbb{R})$, $H > 0$ and $p > 0$. This form of Tf has to coincide with one of the solutions of the extended Leibniz rule (3.7) for $k = 1$, which were given by (3.8), (3.9) or (3.10) in Theorem 3.7. This is only possible for the first solution (3.8), and then only in the special case when $a(x) = d_1(x) = p(x) = 1$, $d_0(x) = 0$, and if the above function H satisfies $H = \mathbf{1}$ and $p = 1$, yielding $Tf = f'$, $Af = f$ for all $f \in C^1(\mathbb{R})$. $\qquad\square$

To prove Theorem 4.1 we first show, as in Chapter 3, that the operator T is localized. For this, we need that there are sufficiently many non-zero functions in the range of T.

Lemma 4.4. *Suppose the assumptions of Theorem 4.1 hold. Then for any open half-bounded interval* $I = (c, \infty)$ *or* $I = (-\infty, c)$ *with* $c \in \mathbb{R}$, *any* $y \in I$ *and any* $x \in \mathbb{R}$, *there exists* $g \in C^k(\mathbb{R})$ *such that* $g(x) = y$, $\operatorname{Im}(g) \subset I$ *and* $(Tg)(x) \neq 0$.

Proof. (i) Let $x \in \mathbb{R}$. We show that $(Tg)(x) \neq 0$ for a suitable function $g \in C_b^k(\mathbb{R})$: Since $T|_{C_b^k(\mathbb{R})} \neq 0$, there is $x_1 \in \mathbb{R}$ and a half-bounded function $h \in C_b^k(\mathbb{R})$ with $(Th)(x_1) \neq 0$. Define $\varphi, g \in C_b^k(\mathbb{R})$ by

$$\varphi(s) := s + x - x_1, \quad g(s) := h \circ \varphi^{-1}(s); \qquad s \in \mathbb{R}.$$

Then $h = g \circ \varphi$, $\varphi(x_1) = x$ and

$$0 \neq (Th)(x_1) = (Tg)(\varphi(x_1)) \cdot (T\varphi)(x_1) = (Tg)(x) \cdot (T\varphi)(x_1),$$

which implies $(Tg)(x) \neq 0$. Clearly $g \in C_b^k(\mathbb{R})$.

(ii) Suppose $I = (c, \infty)$ with $c \in \mathbb{R}$. Pick any $y \in I$ and $x \in \mathbb{R}$. By (i) there is $g \in C_b^k(\mathbb{R})$ with $(Tg)(x) \neq 0$. Let J be an open half-bounded interval with $\operatorname{Im}(g) \subset J$. Choose a bijective C^k-map $f : I \to J$ with $f(y) = g(x)$, noting that $g(x) \in J$. This may be done in such a way that f is extendable to a C^k-map $\tilde{f} : \mathbb{R} \to \mathbb{R}$ on $\mathbb{R}$, $\tilde{f}|_I = f$. Let

$$g_1 := f^{-1} \circ g : \mathbb{R} \longrightarrow I \subset \mathbb{R}.$$

Then $g_1 \in C^k(\mathbb{R})$, $g_1(x) = y$ and $\mathrm{Im}(g_1) \subset I$. Since $g = f \circ g_1 = \tilde{f} \circ g_1$, we find, using the chain rule equation (4.1),

$$0 \neq (Tg)(x) = (T\tilde{f})(y) \cdot (Tg_1)(x).$$

Hence $(Tg_1)(x) \neq 0$, $g_1(x) = y$ and $\mathrm{Im}(g_1) \subset I$. $\qquad \square$

Lemma 4.5. *Under the assumptions of Theorem 4.1, we have for any open, half-bounded interval I and any $f, f_1, f_2 \in C^k(\mathbb{R})$:*

(i) *If $f|_I = \mathrm{Id}$, then $(Tf)|_I = 1$.*

(ii) *If $f_1|_I = f_2|_I$, then $(Tf_1)|_I = (Tf_2)|_I$.*

Proof. (i) Assume $f|_I = \mathrm{Id}$. Take any $y \in I$, $x \in \mathbb{R}$. By Lemma 4.4, there is $g \in C^k(\mathbb{R})$ with $g(x) = y$, $\mathrm{Im}(g) \subset I$ and $(Tg)(x) \neq 0$. Then $f \circ g = g$ so that by (4.1)

$$0 \neq (Tg)(x) = T(f \circ g)(x) = (Tf)(y) \cdot (Tg)(x),$$

which implies that $(Tf)(y) = 1$. Since $y \in I$ was arbitrary, we conclude $(Tf)|_I = 1$.

(ii) Let $f_1|_I = f_2|_I$ and $x \in I$ be arbitrary. Choose a smaller open half-bounded interval $J \subset I$ and a function $g \in C^k(\mathbb{R})$ such that $x \in J$, $\mathrm{Im}(g) \subset I$ and $g|_J = \mathrm{Id}$. Then $f_1 \circ g = f_2 \circ g$ and $g(x) = x$. By part (i), $(Tg)|_J = 1$. Hence, again using the chain rule (4.1),

$$(Tf_1)(x) = (Tf_1)(g(x)) \cdot Tg(x) = T(f_1 \circ g)(x)$$
$$= T(f_2 \circ g)(x) = (Tf_2)(g(x)) \cdot Tg(x) = (Tf_2)(x),$$

which shows $(Tf_1)|_I = (Tf_2)|_I$. $\qquad \square$

Proposition 4.6. *Let $k \in \mathbb{N}_0 \cup \{\infty\}$ and $T : C^k(\mathbb{R}) \to C(\mathbb{R})$ satisfy the chain rule equation (4.1). Assume that $T|_{C_b^k(\mathbb{R})} \neq 0$. Then there is a function $F : \mathbb{R}^{k+2} \to \mathbb{R}$ such that for all $f \in C^k(\mathbb{R})$ and $x \in \mathbb{R}$*

$$Tf(x) = F\big(x, f(x), \dots, f^{(k)}(x)\big). \tag{4.4}$$

In the case $k = \infty$, this is supposed to mean that $Tf(x)$ depends on x and on all derivative values $f^{(j)}(x)$.

Proof. The result follows immediately from Proposition 3.3 for $I = \mathbb{R}$ and Lemma 4.5(ii). Note that (3.3) is used in the proof of Proposition 3.3 only for half-bounded intervals J. $\qquad \square$

Proof of Theorem 4.1. (i) Let $k \in \mathbb{N} \cup \{\infty\}$. We first show that $Tf(x)$ does not depend on any derivative values $f^{(j)}(x)$ of order $j \geq 2$. Let $x_0, y_0, z_0 \in \mathbb{R}$ and

$f, g \in C^k(\mathbb{R})$ satisfy $g(x_0) = y_0$, $f(y_0) = z_0$. Using the representation (4.4) of T, the chain rule equation (4.1) for T turns into a functional equation for F,

$$
\begin{aligned}
T(f \circ g)(x_0) &= F\big(x_0, z_0, f'(y_0)g'(x_0), (f \circ g)''(x_0), \dots\big) \\
&= (Tf)(y_0)Tg(x_0) \\
&= F\big(y_0, z_0, f'(y_0), f''(y_0), \dots\big)F\big(x_0, y_0, g'(x_0), g''(x_0), \dots\big). \quad (4.5)
\end{aligned}
$$

If $z_0 = x_0$, also $(g \circ f)(y_0)$ is defined and

$$
T(f \circ g)(x_0) = Tf(y_0)Tg(x_0) = Tg(x_0)Tf(y_0) = T(g \circ f)(y_0),
$$

i.e.,

$$
\begin{aligned}
F\big(x_0, x_0, f'(y_0)g'(x_0), (f \circ g)''(x_0), \dots\big) \\
= F\big(y_0, y_0, g'(x_0)f'(y_0), (g \circ f)''(y_0), \dots\big). \quad (4.6)
\end{aligned}
$$

By the Faà di Bruno formula, cf. Spindler [Sp], the derivatives of $(f \circ g)$ have the form

$$
(f \circ g)^{(j)} = f^{(j)} \circ g \cdot (g')^j + \varphi_j(f' \circ g, \dots, f^{(j-1)} \circ g, g', \dots g^{(j-1)}) + f' \circ g \cdot g^{(j)},
$$

for $2 \leq j \leq k$, where φ_j depends only on the lower-order derivatives of f and g, up to order $(j-1)$ (at y_0 and x_0). We have, e.g., $\varphi_2 = 0$, $\varphi_3(f' \circ g, f'' \circ g, g', g'') = 3f'' \circ g \cdot g' \cdot g''$.

Also, for any $x_0, y_0 \in \mathbb{R}$ and any sequence $(t_n)_{n \in \mathbb{N}}$ of real numbers, there is $g \in C^\infty(\mathbb{R})$ with $g(x_0) = y_0$ and $g^{(n)}(x_0) = t_n$ for any $n \in \mathbb{N}$, cf. Hörmander [Ho, p. 16]. This may be shown by adding infinitely many small bump functions. Similarly, given $(s_n)_{n \in \mathbb{N}}$, we may choose $f \in C^\infty(\mathbb{R})$ with $f(y_0) = x_0$ and $f^{(n)}(y_0) = s_n$, $n \in \mathbb{N}$.

Therefore, (4.6) implies, for all $x_0, y_0 \in \mathbb{R}$ and all $(s_n), (t_n)$,

$$
\begin{aligned}
F(x_0, x_0, s_1 t_1, t_1^2 s_2 + s_1 t_2, t_1^3 s_3 + s_1 t_3 + \varphi_{31}, \dots, t_1^j s_j + s_1 t_j + \varphi_{j1}, \dots) \\
= F(y_0, y_0, s_1 t_1, t_1 s_2 + s_1^2 t_2, t_1 s_3 + s_1^3 t_3 + \varphi_{32}, \dots, t_1 s_j + s_1^j t_j + \varphi_{j2}, \dots), \quad (4.7)
\end{aligned}
$$

where $\varphi_{j1}, \varphi_{j2} \in \mathbb{R}$ for $j \geq 3$ depend only on the values of $s_1, \dots, s_{j-1}$ and $t_1, \dots, t_{j-1}$, e.g., $\varphi_{31} = 3s_2 t_1 t_2$, $\varphi_{32} = 3t_2 s_1 s_2$. The last dots in (4.7) mean that the variables extend up to $j \leq k$ if $k \in \mathbb{N}$, or range over all j if $k = \infty$. Given $z_0 \in \mathbb{R}$, the functions g and f may be chosen with respect to (z_0, y_0) instead of (x_0, y_0) for the same sequences (t_n) and (s_n). Then (4.7) is also true with x_0 being replaced by z_0 which means that $F(x_0, x_0, s_1, \dots, s_j, \dots)$ is independent of x_0. We put

$$
K(s_1, \dots, s_j, \dots) := F(x_0, x_0, s_1, \dots, s_j, \dots).
$$

Assume that s_1, t_1 are such that $s_1 t_1 \notin \{0, 1, -1\}$. We claim that for arbitrary values (a_j) and (b_j)

$$
K(s_1 t_1, a_2, \dots, a_j, \dots) = K(s_1 t_1, b_2, \dots, b_j, \dots),
$$

i.e., that K only depends on the first variable $s_1 t_1$ if $s_1 t_1 \notin \{0, 1, -1\}$. To see this, first note that $\det \begin{pmatrix} t_1^j & s_1 \\ t_1 & s_1^j \end{pmatrix} = (s_1 t_1)((s_1 t_1)^{j-1} - 1) \neq 0$ for $j \geq 2$. Hence, we may solve successively and uniquely the sequence of (2×2)-linear equations for $(s_2, t_2), (s_3, t_3), \dots, (s_j, t_j)$

$$t_1^2 s_2 + s_1 t_2 = a_2, \qquad\qquad t_1 s_2 + s_1^2 t_2 = b_2,$$
$$t_1^3 s_3 + s_1 t_3 = a_3 - \varphi_{31}, \qquad t_1 s_3 + s_1^3 t_3 = b_3 - \varphi_{32},$$
$$\vdots \qquad\qquad\qquad\qquad \vdots$$
$$t_1^j s_j + s_1 t_j = a_j - \varphi_{j1}, \qquad t_1 s_j + s_1^j t_j = b_j - \varphi_{j2},$$

Here the values obtained for (s_2, t_2) are used to determine φ_{31} and φ_{32} according to the Faà di Bruno formula, and the values up to (s_{j-1}, t_{j-1}) to determine φ_{j1} and φ_{j2} accordingly. We then conclude from (4.7)

$$K(s_1 t_1, a_2, \dots, a_j, \dots) = K(s_1 t_1, b_2, \dots, b_j, \dots).$$

This means that $K(u_1, u_2, \dots, u_j, \dots)$ is independent of the variables $u_2, \dots, u_j, \dots$, if $u_1 \notin \{0, 1, -1\}$. We then put $\widetilde{K}(u_1) := K(u_1, u_2, \dots, u_j, \dots)$. If $u_1 = 1$ choose $t_1 = 2$, $s_1 = 1/2$, $u_1 = s_1 t_1 = 1$. Then by (4.5) and (4.6), we find that for any $s_2, \dots, s_j, \dots, t_2, \dots, t_j, \dots$ we have

$$K\left(1, 4s_2 + \tfrac{1}{2}t_2, \dots, 2^j s_j + \tfrac{1}{2^j}t_j + \varphi_j, \dots\right) = \widetilde{K}(2)\widetilde{K}\left(\tfrac{1}{2}\right).$$

Given arbitrary real numbers $u_2, \dots, u_j, \dots$, we find successively $s_2, t_2, s_3, t_3, \dots$ such that the left-hand side equals $K(1, u_2, u_3, \dots, u_j, \dots)$ and hence $\widetilde{K}(1) = K(1, u_2, \dots, u_j, \dots)$ is also independent of u_j for $j \geq 2$. A similar statement is true for $u_1 = -1$. To show that $K(0, u_2, \dots, u_j, \dots)$ is independent of the u_j for $j \geq 2$, too, choose $t_1 = a$, $s_1 = 0$ in (4.7) to find

$$K(0, a^2 s_2, \dots, a^j s_j + \varphi_{j1}, \dots) = K(0, a s_2, \dots, a s_j + \varphi_{j2}, \dots),$$

for all $a \in \mathbb{R}$, which again implies independence of further variables. We now write $K(u_1)$ for $\widetilde{K}(u_1)$. For values $y_0 \neq x_0 = z_0$, we then know by (4.5) that

$$F(x_0, y_0, t_1, t_2, \dots, t_j, \dots) = \frac{K(s_1 t_1)}{F(y_0, x_0, s_1, s_2, \dots, s_j, \dots)}.$$

Since the left-hand side is independent of $s_1, s_2, \dots, s_j, \dots$ and the right-hand side is independent of $t_2, \dots, t_j, \dots$, this equation has the form

$$F(x_0, y_0, t_1) = \frac{K(t_1)}{F(y_0, x_0, 1)}. \tag{4.8}$$

Note that $F(y_0, x_0, 1) \neq 0$ since, using Lemma 4.5(i),

$$F(y_0, x_0, 1)F(x_0, y_0, 1) = K(1) = T(\mathrm{Id})(x_0) = 1.$$

Define $G : \mathbb{R}^2 \to \mathbb{R}_{\neq 0}$ by $G(x_0, y_0) = 1/F(y_0, x_0, 1)$. Then by (4.8)

$$F(x_0, y_0, t_1) = G(x_0, y_0)K(t_1),$$

with $G(x_0, x_0) = 1$. Using the independence of the derivatives of order ≥ 2, (4.5) implies, for all $x_0, y_0, z_0 \in \mathbb{R}$, that

$$F(x_0, z_0, 1) = F(y_0, z_0, 1)F(x_0, y_0, 1),$$
$$G(x_0, z_0) = G(y_0, z_0)G(x_0, y_0).$$

Define $H : \mathbb{R} \to \mathbb{R}_{\neq 0}$ by $H(y) := G(0, y)$. Then

$$G(x, y) = G(x, 0)G(0, y) = G(0, y)/G(0, x) = H(y)/H(x).$$

Again using (4.8), we get

$$F(x_0, y_0, t_1) = \frac{H(y_0)}{H(x_0)}K(t_1), \tag{4.9}$$

and T has the form

$$Tf(x_0) = F\big(x_0, f(x_0), f'(x_0)\big) = \frac{H \circ f(x_0)}{H(x_0)}K(f'(x_0)), \quad f \in C^k(\mathbb{R}). \tag{4.10}$$

(ii) To identify the form of K, note that by (4.5) for $x_0 = y_0 = z_0$,

$$K(s_1 t_1) = K(s_1)K(t_1), \quad s_1, t_1 \in \mathbb{R},$$

i.e., K is multiplicative on $\mathbb{R}$. Let $b \neq 0$. Apply (4.10) to $f(x) = bx$, we get that $Tf(x) = \frac{H(bx)}{H(x)}K(b)$. Note that $K(b) \neq 0$ since otherwise, by multiplicativity, $K \equiv 0$. Since $Tf \in C(\mathbb{R})$, also $\frac{H(bx)}{H(x)}$ defines a continuous function in x which is strictly positive since H is never zero. We may assume that H is positive. Then for any $b \neq 0$, $\varphi(x) := \ln H(x) - \ln H(bx)$ defines a continuous function $\varphi \in C(\mathbb{R})$. By Proposition 2.8(a), $\ln H$ is measurable and hence also H is measurable. Choosing $f(x) = \frac{1}{2}x^2$ in (4.10), we conclude that

$$K(x) = Tf(x)\frac{H(x)}{H\big(\frac{1}{2}x^2\big)}.$$

Since Tf is continuous and H is measurable, also K is measurable. By Proposition 2.3, the multiplicative function K has the form $K(x) = |x|^p$ or $K(x) = |x|^p \operatorname{sgn} x$ for a suitable $p \in \mathbb{R}$, $x \neq 0$. Hence we conclude from (4.10) and the continuity of Tf that $\frac{H \circ f}{H}$ is continuous for any $f \in C^k(\mathbb{R})$ at any point $x \in \mathbb{R}$ such that $f'(x) \neq 0$.

(iii) We now show that H is continuous. For any $c \in \mathbb{R}$, let

$$b(c) := \overline{\lim_{y \to c}} \, H(y), \quad a(c) := \underline{\lim_{x \to c}} \, H(x).$$

We claim that $\frac{b(c)}{H(c)}$ and $\frac{a(c)}{H(c)}$ are constant functions of c. In the case that for some c_0, $b(c_0)$ or $a(c_0)$ are infinite or zero, this should mean that all other values $b(c)$ or $a(c)$ are also infinite or zero. Assume to the contrary that there are c_0 and c_1 such that $\frac{b(c_1)}{H(c_1)} < \frac{b(c_0)}{H(c_0)}$. Choose any maximizing sequence y_n, $\lim_{n\to\infty} y_n = c_0$ with $\lim_{n\to\infty} H(y_n) = b(c_0)$. Since for $f(t) = t + c_1 - c_0$, $\frac{H \circ f}{H}$ is continuous by part (ii), $\lim_{n\to\infty} \frac{H(y_n + c_1 - c_0)}{H(y_n)} = \frac{H(c_1)}{H(c_0)}$ exists and using $\overline{\lim}_{n\to\infty} H(y_n + c_1 - c_0) \le b(c_1)$, we arrive at the contradiction

$$\frac{b(c_0)}{H(c_0)} = \frac{H(c_1)}{H(c_0)} \frac{b(c_0)}{H(c_1)} = \lim_{n\to\infty} \frac{H(y_n + c_1 - c_0)}{H(y_n)} \frac{H(y_n)}{H(c_1)}$$

$$\le \overline{\lim}_{n\to\infty} \frac{H(y_n + c_1 - c_0)}{H(c_1)} \le \frac{b(c_1)}{H(c_1)} < \frac{b(c_0)}{H(c_0)}.$$

The argument is also valid assuming $b(c_1) < b(c_0) = \infty$. The proof for $a(c)$ is similar.

If H would be discontinuous at some point, it would be discontinuous anywhere since the functions $\frac{a}{H}$ and $\frac{b}{H}$ and hence $\frac{b}{a}$ are constant, under this assumption with $\frac{b}{a} > 1$. Assume that this is the case, and choose a sequence $(c_n)_{n\in\mathbb{N}}$ of pairwise disjoint numbers with $\lim_{n\to\infty} c_n = 0$. Let $\delta_n := \frac{1}{4} \min\{|c_n - c_m| \mid n \ne m\}$ and choose $0 < \epsilon_n < \delta_n$ such that $\sum_{n\in\mathbb{N}}(\epsilon_n/\delta_n)^k < \infty$ for all $k \in \mathbb{N}$, i.e., $(\epsilon_n)_{n\in\mathbb{N}}$ should decay much faster to zero than δ_n. Since H is discontinuous at any c_n, $\frac{b(c_n)}{a(c_n)} > 1$. By the above argument, this is independent of $n \in \mathbb{N}$, $1 < \frac{b}{a} := \frac{b(c_n)}{a(c_n)}$. By definition of $b(c_n)$ and $a(c_n)$, we may find $y_n, x_n \in \mathbb{R}$ with

$$|y_n - c_n| < \epsilon_n, \quad |x_n - c_n| < \epsilon_n, \quad \frac{H(y_n)}{H(x_n)} > \frac{b+a}{2a} > 1.$$

If $\frac{b}{a} = \infty$, choose them with $\frac{H(y_n)}{H(x_n)} > 2$. Let ψ be a C^∞-cutoff function like $\psi(x) = \exp\left(-\frac{x^2}{1-x^2}\right)$ for $|x| < 1$, and $\psi(x) = 0$ for $|x| \ge 1$, and put $g_n(x) = (y_n - x_n)\psi\left(\frac{x - x_n}{\delta_n}\right)$. The functions $(g_n)_{n\in\mathbb{N}}$ have disjoint support since for any $m \ne n$

$$|x_n - x_m| \ge |c_n - c_m| - 2\epsilon_n \ge 4\delta_n - 2\epsilon_n \ge 2\delta_n.$$

Hence $g_n(x_m) = (y_n - x_n)\delta_{nm}$. Since

$$\sum_{n\in\mathbb{N}} \|g_n^{(k)}\|_\infty \le \sum_{n\in\mathbb{N}} \left(\frac{|y_n - x_n|}{\delta_n}\right)^k \|\psi^{(k)}\|_\infty \le \sum_{n\in\mathbb{N}} \left(\frac{2\epsilon_n}{\delta_n}\right)^k \|\psi^{(k)}\|_\infty < \infty$$

holds for any $k \in \mathbb{N}$,

$$f(x) := x + \sum_{n\in\mathbb{N}} g_n(x), \quad x \in \mathbb{R}$$

defines a C^∞-function f with $f(x_n) = y_n$, $f(0) = 0$ and $f'(0) = 1 \neq 0$. Since $x_n \to 0$, $y_n = f(x_n) \to 0$, the continuity of $\frac{H \circ f}{H}$ yields the contradiction

$$1 = \frac{H(0)}{H(0)} = \lim_{n \to \infty} \frac{H(y_n)}{H(x_n)} > \frac{b+a}{2a} > 1.$$

This proves that H is continuous.

Now (4.10) implies that

$$Tf(x) = \frac{H \circ f(x)}{H(x)} |f'(x)|^p \{\operatorname{sgn} f'(x)\},$$

for any $f \in C^k(\mathbb{R})$, $x \in \mathbb{R}$. By assumption $Tf \in C(\mathbb{R})$ is continuous for any $f \in C^k(\mathbb{R})$. This requires $p \geq 0$, choosing functions f whose derivatives have zeros. In fact, if the term $\operatorname{sgn} f'(x)$ is present, $p > 0$ is needed to guarantee the continuity of all functions in the image of T.

(iv) If $T(2\operatorname{Id}) = 2$ is the constant function 2, then $\frac{H(2x)}{H(x)} 2^p = 2$ for all x, which for $x = 0$ yields $p = 1$. For $b = 1/2$, the function φ in part (ii) is constant,

$$\varphi(x) = \ln H(x) - \ln H(x/2) = 0.$$

Hence, the argument in the proof of Proposition 2.8(a) shows that $\ln H(x) = \ln H(1)$, $H(x) = H(1)$, taking $L = \ln H$ in Proposition 2.8(a). Hence, $\frac{H \circ f}{H} = 1$ and $Tf = f'$ or $Tf = |f'|$. If $T(-2\operatorname{Id}) = -2$, the only possible solution of Theorem 4.1 is $Tf = f'$.

Clearly, the operators T given by formulas (4.2) and (4.3) satisfy the chain rule (4.1). This proves Theorem 4.1. □

If the image of T consists of smooth functions, we have further restrictions on H and p:

Proposition 4.7. *Let $k \in \mathbb{N}$, $k \geq 2$ and suppose that $T : C^k(\mathbb{R}) \to C^{k-1}(\mathbb{R})$ satisfies the chain rule (4.1) with $T|_{C_b^k(\mathbb{R})} \neq 0$. Then there exists $H \in C^{k-1}(\mathbb{R})$, $H > 0$ and p with either*

$$p > k - 1 \quad and \quad Tf = \frac{H \circ f}{H} |f'|^p \{\operatorname{sgn} f'\}$$

or

$$p \in \{0, \ldots, k-1\} \quad and \quad Tf = \frac{H \circ f}{H} (f')^p, \quad f \in C^k(\mathbb{R}).$$

If the chain rule holds for $T : C^\infty(\mathbb{R}) \to C^\infty(\mathbb{R})$ with $T|_{C_b^\infty(\mathbb{R})} \neq 0$, there is $H \in C^\infty(\mathbb{R})$ and $p \in \mathbb{N}_0$ such that

$$Tf = \frac{H \circ f}{H} (f')^p, \quad f \in C^\infty(\mathbb{R}).$$

Proof. By Theorem 4.1, T is of the above form with $H \in C(\mathbb{R})$ and $p \geq 0$. Suppose T maps $C^k(\mathbb{R})$ into $C^{k-1}(\mathbb{R})$. Then the condition on p is needed to guarantee that Tf is in $C^{k-1}(\mathbb{R})$ for functions f whose derivatives have zeros.

We claim that H is smooth, i.e., $H \in C^{k-1}(\mathbb{R})$. Let $L := -\log H$. Obviously $L \in C^{k-1}(\mathbb{R})$ if and only if $H \in C^{k-1}(\mathbb{R})$. Take $f(x) = x/2$. By assumption $Tf \in C^{k-1}(\mathbb{R})$ and, hence,

$$\varphi(x) := L(x) - L(x/2)$$

defines a function $\varphi \in C^{k-1}(\mathbb{R})$. We prove by induction on $k \geq 2$ that $\varphi \in C^{k-1}(\mathbb{R})$ and $L \in C^{k-2}(\mathbb{R})$ imply that $L \in C^{k-1}(\mathbb{R})$.

For $k = 2$, $\varphi \in C^1(\mathbb{R})$ and $L \in C(\mathbb{R})$ since $H \in C(\mathbb{R})$. By Proposition 2.8(b) with $\psi = \varphi$ and $a = 1$, we get $L \in C^1(\mathbb{R})$.

To prove the induction step, assume $k \geq 3$, $\varphi \in C^{k-1}(\mathbb{R})$ and $L^{(k-2)} \in C(\mathbb{R})$. We have to show that $L \in C^{k-1}(\mathbb{R})$. Let $\psi(x) := \varphi^{(k-2)}(x) = L^{(k-2)}(x) - \frac{1}{2^{k-2}} L^{(k-2)}\left(\frac{x}{2}\right)$. Then $\psi \in C^1(\mathbb{R})$ and $L^{(k-2)} \in C(\mathbb{R})$. By Proposition 2.8(b) with $a = \frac{1}{2^{k-2}}$, $L^{(k-2)} \in C^1(\mathbb{R})$, i.e., $L \in C^{k-1}(\mathbb{R})$. This proves that $H \in C^{k-1}(\mathbb{R})$. $\qquad\square$

4.2 The chain rule on different domains

In the case of C^1-functions, there is an analogue of Theorem 4.1 for functions $f : \mathbb{R}^n \to \mathbb{R}^n$ on $\mathbb{R}^n$ when $n > 1$. For finite-dimensional Banach spaces X and Y and $k \in \mathbb{N}_0$, let

$$C^k(X, Y) = \{f : X \to Y \mid f \text{ is } k\text{-times continuously Fréchet differentiable}\},$$

with $C(X, Y) = C^0(X, Y)$. Let $L(X, Y) := \{f \in C(X, Y) \mid f \text{ is linear}\}$ and $C_b^k(X, \mathbb{R}^n) := \{f \in C^k(X, \mathbb{R}^n) \mid \text{Im}(f) \subset J \text{ for some open half-space } J \subset \mathbb{R}^n\}$. The derivative D is a map $D : C^1(\mathbb{R}^n, \mathbb{R}^n) \to C(\mathbb{R}^n, L(\mathbb{R}^n, \mathbb{R}^n))$ satisfying the chain rule

$$D(f \circ g)(x) = ((Df) \circ g)(x) \cdot (Dg)(x), \quad f, g \in C^1(\mathbb{R}^n, \mathbb{R}^n), \ x \in \mathbb{R}^n.$$

More generally, we consider operators $T : C^1(\mathbb{R}^n, \mathbb{R}^n) \to C(\mathbb{R}^n, L(\mathbb{R}^n, \mathbb{R}^n))$ satisfying the *chain rule equation*

$$T(f \circ g)(x) = ((Tf) \circ g)(x) \cdot (Tg)(x), \quad f, g \in C^1(\mathbb{R}^n, \mathbb{R}^n), \ x \in \mathbb{R}^n.$$

The multiplication on the right is the non-commutative composition of linear operators on $\mathbb{R}^n$. We do not write it with composition symbol $\circ$ to distinguish it from the composition of the non-linear functions f, g. In fact, in the following we will omit the symbol $\cdot$ for this composition. In stating the analogue of Theorem 4.1 for $n > 1$, we need another assumption on T.

An operator $T : C^1(\mathbb{R}^n, \mathbb{R}^n) \to C(\mathbb{R}^n, L(\mathbb{R}^n, \mathbb{R}^n))$ is *locally surjective* provided that there is $x \in \mathbb{R}^n$ so that

$$\{(Tf)(x) \mid f \in C^1(\mathbb{R}^n, \mathbb{R}^n), \ f(x) = x, \ \det f'(x) \neq 0\} \supseteq \mathrm{GL}(n, \mathbb{R}).$$

In the following result on the chain rule for maps of this type we use the notation $\det T|_{C_b^1}(\mathbb{R}^n, \mathbb{R}^n) \not\equiv 0$ to mean that there should be a function $f \in C_b^1(\mathbb{R}^n, \mathbb{R}^n)$ and a point $x \in \mathbb{R}^n$ such that $\det(Tf(x)) \neq 0$.

Theorem 4.8 (Multidimensional chain rule). *Let $n \geq 2$, and assume that $T : C^1(\mathbb{R}^n, \mathbb{R}^n) \to C(\mathbb{R}^n, L(\mathbb{R}^n, \mathbb{R}^n))$ satisfies the chain rule equation*

$$T(f \circ g)(x) = \big((Tf) \circ g\big)(x)\, Tg(x), \quad f, g \in C^1(\mathbb{R}^n, \mathbb{R}^n), \ x \in \mathbb{R}^n. \qquad (4.11)$$

Assume also that $\det T|_{C_b^1(\mathbb{R}^n, \mathbb{R}^n)} \not\equiv 0$ and that T is locally surjective. Then there are $p \geq 0$ and $H \in C(\mathbb{R}^n, \mathrm{GL}(n, \mathbb{R}))$ such that, if $n \in \mathbb{N}$ is odd, for all $f \in C^1(\mathbb{R}^n, \mathbb{R}^n)$ and $x \in \mathbb{R}^n$

$$(Tf)(x) = \big|\det f'(x)\big|^p (H \circ f)(x) f'(x) H(x)^{-1}.$$

If $n \in \mathbb{N}$ is even, T either has the same form or

$$Tf(x) = \mathrm{sgn}\big(\det f'(x)\big)\big|\det f'(x)\big|^p (H \circ f)(x) f'(x) H(x)^{-1},$$

the latter with $p > 0$.

Conversely, *these formulas define operators T which satisfy the chain rule and are locally surjective.*

If additionally to (4.11), $T(2\,\mathrm{Id})(x) = 2\,\mathrm{Id}$ *holds for all $x \in \mathbb{R}^n$, then $H = \mathrm{Id}$ and $Tf = f'$ or, if n is even, possibly $Tf = \mathrm{sgn}(\det f') f'$.*

Remarks. (a) Note that a priori we do not impose any continuity condition on T.

(b) For odd integers $n \in \mathbb{N}$, $p > 0$ and $H \in C(\mathbb{R}^n, \mathrm{GL}(n, \mathbb{R}))$,

$$(Tf)(x) := \mathrm{sgn}\big(\det f'(x)\big)\big|\det f'(x)\big|^p (H \circ f)(x) f'(x) H(x)^{-1}$$

also solves the chain rule equation, but is not locally surjective since in this case $\det((Tf)(x)) \geq 0$ for all $f \in C^1(\mathbb{R}^n, \mathbb{R}^n)$ with $f(x) = x$.

(c) If T is not assumed to be locally surjective, there are various other solutions of (4.11):

Take any continuous multiplicative homomorphism $\Phi : \mathbb{R} \to L(\mathbb{R}^n, \mathbb{R}^n)$ with $\Phi(0) = 0$ and $\Phi(1) = \mathrm{Id}$ and any continuous function $H \in C(\mathbb{R}^n, \mathrm{GL}(n, \mathbb{R}))$, and define

$$(Tf)(x) = (H \circ f)(x)\Phi\big(\det f'(x)\big) H(x)^{-1},$$

$x \in \mathbb{R}^n$, $f \in C^1(\mathbb{R}^n, \mathbb{R}^n)$. Then T satisfies (4.11). As for specific examples, take as Φ a one-parameter group like $\Phi(t) = \exp(\ln|t|A) = |t|^A$ for some fixed matrix $A \in L(\mathbb{R}^n, \mathbb{R}^n)$ and $t \in \mathbb{R}$. Here $\ln|t|$ might also be replaced by $(\mathrm{sgn}\,t) \cdot \ln|t|$.

(d) As in the case of one variable ($n = 1$), the function H is completely determined by the function $T(2\,\mathrm{Id})$. The inner automorphism defined by H, with additional composition by f, applied to the derivative, essentially yields T up to a character in terms of $\det f'$.

For the proof of Theorem 4.8 we refer to [KM2]. We will not reproduce it here since it is not in line with our main goals. We just mention a few steps of the proof.

The localization step for $n \geq 2$ is similar to the case $n = 1$, yielding

$$Tf(x) = F\big(x, f(x), f'(x)\big),$$

for a suitable function $F : \mathbb{R}^n \times \mathbb{R}^n \times L(\mathbb{R}^n, \mathbb{R}^n) \to L(\mathbb{R}^n, \mathbb{R}^n)$. The analysis of this representing function F is different from the case $n = 1$, due to the non-commutativity of the composition of linear maps in $L(\mathbb{R}^n, \mathbb{R}^n)$. However, again one may show that

$$K(v) := F(x, x, v) \in \mathrm{GL}(n, \mathbb{R}), \quad v \in \mathrm{GL}(n, \mathbb{R})$$

is independent of $x \in \mathbb{R}^n$ and multiplicative, $K(uv) = K(u)K(v)$ for all $u, v \in \mathrm{GL}(n, \mathbb{R})$, with $K(\mathrm{Id}) = \mathrm{Id}$, $K(v)^{-1} = K(v^{-1})$. The proof proceeds identifying these automorphisms K of $\mathrm{GL}(n, \mathbb{R})$ as inner automorphisms multiplied by characters in terms of $\det v$, i.e., powers of $|\det v|$, possibly multiplied by $\mathrm{sgn}(\det v)$. This result on the automorphisms of $\mathrm{GL}(n, \mathbb{R})$ replaces (the simpler) Proposition 2.3. Additional arguments are also needed to prove the continuity of H.

We may also consider the chain rule equation on real or complex spaces of polynomials or analytic functions. For $\mathbb{K} \in \{\mathbb{R}, \mathbb{C}\}$, let $\mathcal{P} := \mathcal{P}(\mathbb{K})$ denote the space of polynomials with coefficients in $\mathbb{K}$, $\mathcal{E} := \mathcal{E}(\mathbb{K})$ the space of real-analytic functions ($\mathbb{K} = \mathbb{R}$) or entire functions ($\mathbb{K} = \mathbb{C}$) and $C := C(\mathbb{K})$. Moreover, let $\mathcal{P}_n := \mathcal{P}_n(\mathbb{K})$ be the subset of $\mathcal{P}$ consisting of polynomials of degree $\leq n$. There are simple operators $T : \mathcal{P}(\mathbb{K}) \to C(\mathbb{K})$ satisfying the chain rule $T(f \circ g) = (Tf) \circ g \cdot Tg$ which have a different form than the solutions determined so far: For $f \in \mathcal{P}(\mathbb{K})$ and $c \in \mathbb{R}$, let $Tf := (\deg f)^c$, T mapping into the constant functions. Then T satisfies the chain rule on $\mathcal{P}$. More generally, if $\deg f = \prod_{j=1}^r p_j^{l_j}$ is the decomposition of $\deg f$ into prime powers and $c_j \in \mathbb{R}$, $Tf = \prod_{j=1}^r p_j^{c_j l_j}$ will satisfy the chain rule and also

$$Tf = \prod_{j=1}^r p_j^{c_j l_j} \frac{H \circ f}{H} |f'|^p \{\mathrm{sgn}\, f'\}^m$$

will define a map $T : \mathcal{P} \to C$ satisfying the chain rule, if $H \in C(\mathbb{K})$, $H \neq 0$, $p \geq 0$, $m \in \mathbb{N}_0$. We do not know whether this yields the general solution of the chain rule equation for $T : \mathcal{P} \to C$. However, we can give the general solution of the chain rule equation for such maps under a mild continuity assumption.

Let $X \in \{\mathcal{P}(\mathbb{K}), \mathcal{E}(\mathbb{K})\}$ and $Y \in \{\mathcal{P}(\mathbb{K}), \mathcal{E}(\mathbb{K}), C(\mathbb{K})\}$. An operator $T : X \to Y$ is *pointwise continuous at 0* provided that for any sequence $(f_n)_{n \in \mathbb{N}}$ of functions in X converging uniformly on all compact sets of $\mathbb{K}$ to a function $f \in X$, we have pointwise convergence of the images at zero, i.e., $\lim_{n \to \infty} (Tf_n)(0) = (Tf)(0)$. For $\xi \in \mathbb{K} \setminus \{0\}$, denote $\operatorname{sgn} \xi := \xi/|\xi|$. We then have the following two results for the chain rule.

Theorem 4.9. *Let $\mathbb{K} \in \{\mathbb{R}, \mathbb{C}\}$ and suppose that $T : \mathcal{P}(\mathbb{K}) \to C(\mathbb{K})$, $T \neq 0$, satisfies the chain rule equation*

$$T(f \circ g) = (Tf) \circ g \cdot Tg, \quad f, g \in \mathcal{P}(\mathbb{K}) \tag{4.12}$$

and is pointwise continuous at 0. Then there is a nowhere vanishing continuous function $H \in C(\mathbb{K})$ and there are $p \in \mathbb{K}$ with $\operatorname{Re}(p) \geq 0$ and $m \in \mathbb{Z}$ such that

$$Tf = \frac{H \circ f}{H} |f'|^p (\operatorname{sgn} f')^m. \tag{4.13}$$

For $\mathbb{K} = \mathbb{R}$, $m \in \{0, 1\}$ suffices and $H > 0$. For $p = 0$, only $m = 0$ yields a solution with range in $C(\mathbb{K})$. If T maps into the space $\mathcal{P}(\mathbb{K})$, H is constant and $p = m \in \mathbb{N}_0$ so that T has the form $Tf = f'^m$.

The result for entire functions is

Theorem 4.10. *Let $\mathbb{K} \in \{\mathbb{R}, \mathbb{C}\}$ and assume that $T : \mathcal{E}(\mathbb{K}) \to \mathcal{E}(\mathbb{K})$, $T \neq 0$, satisfies the chain rule equation*

$$T(f \circ g) = (Tf) \circ g \cdot Tg, \quad f, g \in \mathcal{E}(\mathbb{K})$$

and is pointwise continuous at 0. Then there is a function $h \in \mathcal{E}(\mathbb{K})$ and there is $m \in \mathbb{N}_0$ such that

$$Tf = \exp(h \circ f - f) \cdot f'^m.$$

Proof of Theorem 4.9. (a) Since $T \neq 0$, there are $n_0 \in \mathbb{N}$, $g \in \mathcal{P}_{n_0}(\mathbb{K})$ and $x_1 \in \mathbb{K}$ such that $Tg(x_1) \neq 0$. Let $n \in \mathbb{N}$, $n \geq n_0$. We restrict T to $\mathcal{P}_n(\mathbb{K}) =: \mathcal{P}_n$ and apply (4.12) for $f, g \in \mathcal{P}_n$ with $f \circ g \in \mathcal{P}_n$. For any $x_0 \in \mathbb{K}$, consider the shift $S(x) := x + x_1 - x_0$, $S \in \mathcal{P}_1 \subset \mathcal{P}_n$ and put $f := g \circ S$. Then by (4.12)

$$0 \neq (Tg)(x_1) = T(f \circ S^{-1})(x_1) = (Tf)(x_0)T(S^{-1})(x_1).$$

Hence, $Tf(x_0) \neq 0$. Moreover, $Th = T(h \circ \operatorname{Id}) = Th \cdot T(\operatorname{Id})$ for all $h \in \mathcal{P}_n$. Hence, $T(\operatorname{Id}) = 1$ is the constant function 1. For $x \in \mathbb{K}$, let $S_x(y) := x + y$, $S_x \in \mathcal{P}_1 \subset \mathcal{P}_n$. Again by (4.12)

$$1 = T(\operatorname{Id}) = T(S_{-x} \circ S_x) = T(S_{-x}) \circ S_x \cdot T(S_x).$$

Thus for all $y \in \mathbb{K}$, $T(S_x)(y) \neq 0$. In particular, $T(S_x)(0) \neq 0$ for all $x \in \mathbb{K}$. Again by (4.12)

$$T(f \circ S_x)(0) = (Tf)(x) \cdot T(S_x)(0),$$

so that for any $x \in \mathbb{K}$ and $f \in \mathcal{P}_n$

$$Tf(x) = \frac{T(f \circ S_x)(0)}{T(S_x)(0)}. \tag{4.14}$$

Let $f_j \in \mathcal{P}_n$ be a sequence converging uniformly on compacta to $f \in \mathcal{P}_n$. Then (4.14) and the pointwise continuity assumption at 0 imply that $\lim_{j \to \infty}(Tf_j)(x) = (Tf)(x)$ for all $x \in \mathbb{K}$, and not only for $x = 0$. By (4.14) it suffices to determine the form of $(Tf)(0)$ for any $f \in \mathcal{P}_n$. Since, for any $f \in \mathcal{P}_n$, $f(x) = \sum_{j=0}^{n} \frac{f^{(j)}(0)}{j!} x^j$ is determined by the sequence $(f^{(j)}(0))_{0 \le j \le n}$, $(Tf)(0)$ is a function of these values. Hence, there is $F_n : \mathbb{K}^{n+1} \to \mathbb{K}$ such that

$$(Tf)(0) = F_n\big(f(0), f'(0), \dots, f^{(n)}(0)\big), \quad f \in \mathcal{P}_n. \tag{4.15}$$

Since $(f \circ S_x)^{(j)} = f^{(j)} \circ S_x$, (4.14) and (4.15) imply

$$Tf(x) = \frac{F_n\big(f(x), f'(x), \dots, f^{(n)}(x)\big)}{F_n(x, 1, 0, \dots 0)}, \tag{4.16}$$

with $F_n(x, 1, 0, \dots, 0) = T(S_x)(0) \ne 0$ for any $x \in \mathbb{K}$.

(b) We now show that Tf does not depend on the higher derivatives $f^{(j)}$ for $j \ge 2$. Fix $x \in \mathbb{K}$ and define $G_n = G_{n,x} : \mathbb{K}^n \to \mathbb{K}$ by

$$G_n(\xi_1, \dots, \xi_n) := \frac{F_n(x, \xi_1, \dots, \xi_n)}{F_n(x, 1, 0, \dots, 0)}, \quad \xi_i \in \mathbb{K}. \tag{4.17}$$

For any $(\eta_1, \dots, \eta_n) \in \mathbb{K}^n$, there is a polynomial $g \in \mathcal{P}_n$, with $g(x) = x$ and $g^{(j)}(x) = \eta_j$ for $j = 1, \dots, n$. For $\xi_1 \in \mathbb{K}$, define $f \in \mathcal{P}_1 \subset \mathcal{P}_n$ by $f(y) := \xi_1(y - x) + x$. Then $f(x) = x$, $(f \circ g)^{(j)}(x) = \xi_1 \eta_j$ and $(g \circ f)^{(j)}(x) = \xi_1^j \eta_j$. Therefore, by (4.16) and (4.17),

$$\begin{aligned}
G_n(\xi_1 \eta_1, \dots, \xi_1 \eta_n) &= G_n\big((f \circ g)'(x), \dots, (f \circ g)^{(n)}(x)\big) \\
&= T(f \circ g)(x) = (Tf)(x)(Tg)(x) = (Tg)(x)(Tf)(x) \\
&= T(g \circ f)(x) = G_n(\xi_1 \eta_1, \dots, \xi_1^n \eta_n).
\end{aligned}$$

Given $(t_1, \dots, t_n) \in \mathbb{K}^n$ and $\alpha \in \mathbb{K}$, $\alpha \ne 0$, let $\eta_i = t_i/\alpha$.

Applying the previous equations with $\xi_1 = \alpha$, we conclude

$$G_n(t_1, t_2, \dots, t_n) = G_n(t_1, \alpha t_2, \dots, \alpha^{n-1} t_n). \tag{4.18}$$

Fix $t_1 \in \mathbb{K}$ and define $\widetilde{G}_n : \mathbb{K}^{n-1} \to \mathbb{K}$ by $\widetilde{G}_n(t_2, \dots, t_n) := G_n(t_1, t_2, \dots t_n)$. Then $\widetilde{G}_n$ is continuous at zero: if $t^{(m)} = (t_2^{(m)}, \dots, t_n^{(m)}) \to 0 \in \mathbb{K}^{n-1}$ for $m \to \infty$, choose polynomials $f_m \in \mathcal{P}_n$ with $f_m(x) = x$, $f_m'(x) = t_1$ and $f_m^{(j)}(x) = t_j^{(m)}$ for $2 \le j \le n$. Clearly, f_m converges uniformly on compact sets to f, where

$f(x) = t_1(y - x) + x$. By the assumption of pointwise continuity at 0 of T, (4.16) and (4.17),

$$\widetilde{G}_n(t_2^{(m)}, \ldots, t_n^{(m)}) = G_n(t_1, t_2^{(m)}, \ldots, t_n^{(m)}) = (Tf_m)(x)$$
$$\longrightarrow (Tf)(x) = G_n(t_1, 0, \ldots, 0) = \widetilde{G}_n(0, \ldots, 0).$$

Hence, $\widetilde{G}_n$ is continuous at 0. Letting $\alpha \to 0$ in (4.18), we find

$$G_n(t_1, t_2, \ldots, t_n) = \lim_{\alpha \to 0} G_n(t_1, \alpha t_2, \ldots, \alpha^{n-1} t_n) = G_n(t_1, 0, \ldots, 0), \qquad (4.19)$$

i.e., $G_n = G_{n,x}$ does not depend on the variables $(t_2, \ldots, t_n) \in \mathbb{K}^{n-1}$: Therefore Tf is independent of the higher derivatives of f.

(c) For any $f \in \mathcal{P}_n$ with $f(x) = x$ and $f'(x) = \xi_1$, we now know by (4.16), (4.17) and (4.19) that

$$(Tf)(x) = G_n\big(f'(x), \ldots f^{(n)}(x)\big) = G_n(\xi_1, 0, \ldots, 0)$$
$$= \frac{F_n(x, \xi_1, 0, \ldots, 0)}{F_n(x, 1, 0, \ldots, 0)} =: \phi(x, \xi_1). \qquad (4.20)$$

If $g \in \mathcal{P}_1$ satisfies $g(x) = x$, $g'(x) = \eta_1$, we have by (4.12) and (4.20)

$$\phi(x, \xi_1 \eta_1) = T(f \circ g)(x) = (Tf)(x)(Tg)(x) = \phi(x, \eta_1)\phi(x, \eta_1).$$

Therefore, $\phi(x, \cdot) : \mathbb{K} \to \mathbb{K}$ is multiplicative for every fixed $x \in \mathbb{K}$. It is also continuous: for $\xi_1^{(m)} \to \xi_1$ in $\mathbb{K}$, put $f_m(y) = \xi_1^{(m)}(y - x) + x$, $f(y) := \xi_1(y - x) + x$. Then $f_m \to f$ converges uniformly on compacta and hence

$$\phi(x, \xi_1^{(m)}) = (Tf_m)(x) \longrightarrow (Tf)(x) = \phi(x, \xi_1).$$

By Proposition 2.3 ($\mathbb{K} = \mathbb{R}$) and Proposition 2.4 ($\mathbb{K} = \mathbb{C}$) there are $p(x) \in \mathbb{K}$ with $\mathrm{Re}(p(x)) \geq 0$ and $m(x) \in \mathbb{Z}$ such that

$$\phi(x, \xi_1) = |\xi_1|^{p(x)}(\mathrm{sgn}\,\xi_1)^{m(x)}, \qquad (4.21)$$

$\mathrm{sgn}\,\xi_1 = \xi_1/|\xi_1|$ for $\xi \neq 0$ and $\phi(x, 0) = 0$, with $m(x) = 0$ if $\mathrm{Re}(p(x)) = 0$ and $m(x) \in \{0, 1\}$ if $\mathbb{K} = \mathbb{R}$.

(d) Let $H(x) = T(S_x)(0) = F_n(x, 1, 0, \ldots, 0)$. Then $H(x) \neq 0$ and by (4.16), (4.19), (4.20) and (4.21),

$$Tf(x) = \frac{F_x\big(f(x), f'(x), 0, \ldots, 0\big)}{F_n(x, 1, 0, \ldots, 0)} = \frac{H(f(x))}{H(x)}\phi\big(f(x), f'(x)\big)$$
$$= \frac{(H \circ f)(x)}{H(x)}|f'(x)|^{p(f(x))}\big(\mathrm{sgn}\,f'(x)\big)^{m(f(x))}. \qquad (4.22)$$

Choosing $f(x) = 2x$, we find that p is a continuous function since Tf and H are continuous. Actually, p is constant: Choosing arbitrary $x, y, z \in \mathbb{K}$ and functions $f, g \in \mathcal{P}_1$ with $g(x) = y$, $f(y) = z$, we have by (4.12) and (4.22),

$$\left|f'(y)g'(x)\right|^{p(yz)} \left(\operatorname{sgn} f'(y)g'(x)\right)^{m(yz)}$$

$$= |f'(y)|^{p(z)} \left(\operatorname{sgn} f'(y)\right)^{m(z)} |g'(x)|^{p(y)} \left(\operatorname{sgn} g'(x)\right)^{m(y)}.$$

Applying this first to polynomials with $f'(y) > 0$, $g'(x) > 0$, we find that $p(yz) = p(z) = p(y) =: p$ for all $y, z \in \mathbb{K}$, i.e., p is constant. Then, using functions with arbitrary sgn-values in S^1, we find that $m(yz) = m(z) = m(y) = m \in \mathbb{Z}$ may be taken constant. With $p = p(f(x))$ and $m = m(f(x))$, (4.22) gives the general solution for $T : \mathcal{P}_n \to C$, both for $\mathbb{K} = \mathbb{R}$ and $\mathbb{K} = \mathbb{C}$.

(e) Since (4.22) is independent of $n \in \mathbb{N}$, this is also the general solution for $T : \mathcal{P} \to C$. In the case that $T : \mathcal{P} \to \mathcal{P}$, i.e., that the range of T consists only of polynomials, all functions

$$Tf = \frac{H \circ f}{H} |f'|^p (\operatorname{sgn} f')^m, \quad f \in \mathcal{P},$$

have to be polynomials. Here $m \in \mathbb{Z}$, $p \in \mathbb{K}$, $\operatorname{Re}(p) \geq 0$. For $f(x) = \frac{1}{2}x^2$ this means that $\frac{H(\frac{1}{2}x^2)}{H(x)} |x|^p (\operatorname{sgn} x)^m$ is a polynomial. For $p = 0$ also $m = 0$ and $Tf = \frac{H \circ f}{H}$. For $p > 0$, Tf has a zero of order p in $x_0 = 0$. Since Tf is a polynomial, it follows that $p \in \mathbb{N}$ is a positive integer, and $Tf(x) = x^p g(x)$ with $g \in \mathcal{P}$, $g(0) \neq 0$. This implies that $m \in \mathbb{Z}$ has to be such that $x^p = |x|^p (\operatorname{sgn} x)^m$. Therefore $Tf = \frac{H \circ f}{H} f'^p \in \mathcal{P}$ for all $f \in \mathcal{P}$, with $p \in \mathbb{N}_0$. Applying this to linear functions $f(x) = ax + b$, $f^{-1}(y) = \frac{1}{a}y - \frac{b}{a} = x$, we find that $p(x) = \frac{H(ax+b)}{H(x)}$ and $\frac{H(x)}{H(ax+b)} = \frac{1}{p(x)}$ are polynomials in x. Therefore, $\frac{H(ax+b)}{H(x)} =: c_{a,b}$ is constant in $x \in \mathbb{K}$ for any fixed values $a, b \in \mathbb{K}$. In particular

$$\frac{H(2x)}{H(x)} = \frac{H(0)}{H(0)} = 1 =: c_{2,0}, \quad \frac{H(x+b)}{H(x)} =: c_{1,b}.$$

We find that

$$H(2x + 2b) = H(x + b) = c_{1,b} H(x) = c_{1,b} H(2x) = H(2x + b)$$

for all $x, b \in \mathbb{K}$. Therefore, $H(y + b) = H(y)$ for all $y, b \in \mathbb{K}$. Hence, H is constant and $\frac{H \circ f}{H} = 1$ for all $f \in \mathcal{P}$. We conclude that $Tf = f'^p$, $p \in \mathbb{N}_0$. $\qquad \square$

Proof of Theorem 4.10. Since $\mathcal{P}(\mathbb{K}) \subset \mathcal{E}(\mathbb{K})$, Theorem 4.9 yields that $T|_{\mathcal{P}(\mathbb{K})}$ has the form

$$Tf = \frac{H \circ f}{H} |f'|^p (\operatorname{sgn} f')^m, \quad f \in \mathcal{P}(\mathbb{K}), \tag{4.23}$$

with $m \in \mathbb{Z}$, $p \in \mathbb{K}$, $\operatorname{Re}(p) \geq 0$. We also know that H defined by $H(x) = T(S_x)(0)$ is continuous on $\mathbb{K}$. Let $c \in \mathbb{K}$, $c \neq 0$ be arbitrary. Applying (4.23) to $f(z) = cz$

and using that $Tf \in \mathcal{E}(\mathbb{K})$, we get that $z \mapsto \frac{H(cz)}{H(z)}$ is in $\mathcal{E}(\mathbb{K})$, i.e., real or complex analytic. Since H is nowhere zero, there exists an analytic function $k(c,\cdot) \in \mathcal{E}(\mathbb{K})$ such that $\frac{H(cz)}{H(z)} = \exp(k(c,z))$, with $k(c,0) = 0$. For $c, d \in \mathbb{K}$ we find

$$\exp\big(k(cd, z)\big) = \frac{H(cdz)}{H(z)} = \frac{H(cdz)}{H(dz)}\frac{H(dz)}{H(z)} = \exp\big(k(c, dz) + k(d, z)\big),$$

hence $k(cd, z) = k(c, dz) + k(d, z)$. In particular, for $z = 1$, $k(c,d) = k(cd, 1) - k(d, 1)$. Let $h(d) := k(d, 1)$ for $d \neq 0$. Then $k(c, d) = h(cd) - h(d)$, and with d replaced by z, $k(c, z) = h(cz) - h(z)$. Since H is continuous, k is continuous as a function of both variables. Therefore,

$$\lim_{c \to 0} k(c, z) = \lim_{c \to 0} h(cz) - h(z) := h(0) - h(z)$$

exists z-uniformly on compact subsets of $\mathbb{K}$. Since $k(c,\cdot) \in \mathcal{E}(\mathbb{K})$ for all $c \in \mathbb{K}$, we conclude that $h \in \mathcal{E}(\mathbb{K})$. For $w, z \in \mathbb{K} \setminus \{0\}$ define $c \in \mathbb{K}$ by $w = cz$. Then

$$\frac{H(w)}{H(z)} = \exp\big(k(c, z)\big) = \exp\big(h(w) - h(z)\big).$$

This extends by continuity to $w = 0$ or $z = 0$. Hence $\frac{H \circ f}{H} = \exp(h \circ f - h)$ for all $f \in \mathcal{P}(\mathbb{K})$. Since Tf, $\frac{H}{H \circ f}$ are in $\mathcal{E}(\mathbb{K})$, also $|f'|^p (\operatorname{sgn} f')^m$ has to be real-analytic ($\mathbb{K} = \mathbb{R}$) or analytic ($\mathbb{K} = \mathbb{C}$) for all polynomials f requiring that $p = m \in \mathbb{N}_0$, taking into account that $m \in \mathbb{Z}$, $\operatorname{Re}(p) \geq 0$. Therefore

$$Tf = \exp(h \circ f - h)f'^m, \quad f \in \mathcal{P}(\mathbb{K}), \tag{4.24}$$

$m \in \mathbb{N}_0$. Given any $f \in \mathcal{E}(\mathbb{K})$, its n-th order Taylor polynomials $p_n(f) \in \mathcal{P}(\mathbb{K})$ converge uniformly on compacta to f. By the assumption of pointwise continuity at 0 of T and (4.14), we have for any $z \in \mathbb{K}$, $Tf(z) = \lim_{n \to \infty} T(p_n(f))(z)$. Moreover, $\lim_{n \to \infty} h \circ p_n(f)(z) = h \circ f(z)$ and $\lim_{n \to \infty} p_n(f)'(z) = f'(z)$. Therefore, (4.24) holds for all $f \in \mathcal{E}(\mathbb{K})$. $\square$

Remark. Imposing the additional initial condition $T(-2\,\mathrm{Id}) = -2$ on T in Theorems 4.9 and 4.10 will imply that $p = m = 1$ and that H and h are constant so that $Tf = f'$, i.e., T is the derivative.

4.3   Notes and References

Theorem 4.1 on the solution of the chain rule operator equation was shown by Artstein-Avidan, König and Milman in [AKM].

The proof of the continuity of the function H in part (iii) of the proof of Theorem 4.1 uses similar arguments as in the proof of Theorem 2.6 and as in Step

12 of the proof of Theorem 2 of Alesker, Artstein-Avidan, Faifman and Milman [AAFM].

If the "compound" product $T(f \circ g) \cdot Tg$ on the right side of the chain rule is replaced by a simple product of Tf and Tg, the resulting equation essentially has only trivial solutions, since the right-hand side does not reflect the effects of the composition. We have the following result, cf. Proposition 8 of [KM3]:

Proposition 4.11. *Let $k \in \mathbb{N}_0$ and suppose that $T : C^k(\mathbb{R}) \to C(\mathbb{R})$ satisfies*

$$T(f \circ g) = Tf \cdot Tg, \quad f, g \in C^k(\mathbb{R}).$$

Assume also that for any $x \in \mathbb{R}$ and any open interval $J \subset \mathbb{R}$ there is $g \in C^k(\mathbb{R})$ with $Im(g) \subset J$ such that $Tg(x) \neq 0$. Then $Tf = \mathbf{1}$ for all $f \in C^k(\mathbb{R})$.

Theorem 4.1 admits a cohomological interpretation. The semigroup $G = (C^k(\mathbb{R}), \circ)$ with composition as operation acts on the abelian semigroup $M = (C(\mathbb{R}), \cdot)$ with pointwise multiplication as operation by composition from the right, $G \times M \to M$, $fH := H \circ f$. Thus, M is a module over G. Denote the functions from G^n to M by $F^n(G, M)$ and define the coboundary operators

$$d^n : F^n(G, M) \longrightarrow F^{n+1}(G, M), \quad n \in \mathbb{N}_0,$$

using the *additive* notation $+$ for the operation $\cdot$ on M, by

$$d^n \varphi(g_1, \ldots, g_{n+1}) = g_1 \varphi(g_2, \ldots, g_{n+1})$$
$$+ \sum_{i=1}^{n} (-1)^i \varphi(g_1, \ldots, g_{i-1}, g_i g_{i+1}, g_{i+2}, \ldots, g_{n+1}) + (-1)^{n+1} \varphi(g_1, \ldots, g_n),$$

for $\varphi \in F^n(G, M)$, $g_1, \ldots, g_{n+1} \in G$. Theorem 4.1 characterizes the cocycles in $\mathrm{Ker}(d^1)$ for $n = 1$. Then $\varphi = T : G = C^k(\mathbb{R}) \to M = C(\mathbb{R})$ has coboundaries

$$d^1 T(g_1, g_2) = g_1 T(g_2) - T(g_1 g_2) + T(g_2), \quad g_1, g_2 \in G.$$

As for cocycles T, $d^1 T = 0$ means in multiplicative notation

$$T(g_2 \circ g_1) = T(g_2) \circ g_1 \cdot Tg_1,$$

and these are just the solutions of the chain rule. For $n = 0$, $\varphi \in F^0(G, M)$ can be identified with $\varphi = H \in M = C(\mathbb{R})$ and we have in multiplicative notation $d^0 H(g) = \frac{H \circ g}{H}$ for $g \in G = C^k(\mathbb{R})$.

The cohomology group $H^1(G, M) = \mathrm{Ker}(d^1)/\mathrm{Im}(d^0)$ is hence, by Theorem 4.1, represented by the maps $g \mapsto |g'|^p \{\mathrm{sgn}\, g'\}$ from G to M.

We are grateful to L. Polterovich and S. Alesker for advising us on this cohomological interpretation of Theorem 4.1.

Theorem 4.8 on the chain rule equation in $\mathbb{R}^n$ was proved by König and Milman in [KM2]. The result on the inner automorphisms of $GL(n, \mathbb{R})$, which replaces Proposition 2.3 in the proof for $n > 1$, is taken from Dieudonné [D] and Hua [H]. We are grateful to J. Bernstein and R. Farnsteiner for discussions concerning the proof of Theorem 4.8.

Theorems 4.9 and 4.10 were shown in [KM11]. We would like to thank P. Domański for helpful discussions concerning these results.

Corollary 4.3 stated that the derivative is the only operator (not vanishing on the half-bounded functions) satisfying the chain rule and the extended Leibniz rule. It is interesting to note that on the complex plane there are different operators satisfying the chain rule and the extended Leibniz rule, though not with image in the continuous functions: By Aczél, Dhombres [AD], Theorem 7 in Chapter 5.2, there is a non-zero additive and multiplicative function $K : \mathbb{C} \to \mathbb{C}$ which is not the identity on $\mathbb{C}$. Let $C^1(\mathbb{C})$ denote the continuously differentiable (i.e., entire) functions from $\mathbb{C}$ to $\mathbb{C}$ and $F(\mathbb{C})$ denote all functions from $\mathbb{C}$ to $\mathbb{C}$. Define operators $T, A : C^1(\mathbb{C}) \to F(\mathbb{C})$ by $Tf := K(f')$ and $Af := K(f)$. Then (T, A) satisfy

$$T(f \circ g) = Tf \circ g \cdot Tg,$$
$$T(f \cdot g) = Tf \cdot Ag + Af \cdot Tg \; ; \; f, g \in C^1(\mathbb{C}),$$

but T is not the derivative and A is not the identity on $C^1(\mathbb{C})$.

The analogue of the chain rule in integration is the substitution formula. Let $c \in \mathbb{R}$ be fixed, $I : C(\mathbb{R}) \to C^1(\mathbb{R})$ denote the operator of definite integration from c to x and $D : C^1(\mathbb{R}) \to C(\mathbb{R})$ be the derivative. Then I is injective and

$$f \circ g - (f \circ g)(c) = I(Df \circ g \cdot Dg)$$

holds for all $f, g \in C^1(\mathbb{R})$. Modeling this, more generally we consider operators $T : C^1(\mathbb{R}) \to C(\mathbb{R})$ and $J : C(\mathbb{R}) \to C^1(\mathbb{R})$ such that for some fixed $c \in \mathbb{R}$ and all $f, g \in C^1(\mathbb{R})$

$$f \circ g - (f \circ g)(c) = J(Tf \circ g \cdot Tg).$$

The natural question then is whether T is closely connected to some derivative and J to some definite integral. Let us call $T : C^1(\mathbb{R}) \to C(\mathbb{R})$ non-degenerate if there is $y \in \mathbb{R}$ such that for all $x \in \mathbb{R}$ there is $f \in C_b^1(\mathbb{R})$ with $f(x) = y$ and $Tf(x) \neq 0$. Also $T(\mathrm{Id})(x) \neq 0$ is assumed for all $x \in \mathbb{R}$. We then have by König, Milman [KM12]:

Proposition 4.12. *Assume that $J : C(\mathbb{R}) \to C^1(\mathbb{R})$ and $T : C^1(\mathbb{R}) \to C(\mathbb{R})$ are operators such that for some fixed $c \in \mathbb{R}$*

$$f \circ g - (f \circ g)(c) = J(Tf \circ g \cdot Tg)$$

holds for all $f, g \in C^1(\mathbb{R})$. Suppose further that T is non-degenerate and that J is injective. Then there are constants $p > 0$, $d \neq 0$ such that for all $f \in C^1(\mathbb{R})$ and

$h \in C(\mathbb{R})$

$$Tf(x) = d \, |f'(x)|^p \, \operatorname{sgn} f'(x), \qquad (4.25)$$

$$Jh(x) = d^{-2/p} \int_c^x |h(s)|^{1/p} \, \operatorname{sgn} h(s) \, ds. \qquad (4.26)$$

If T additionally satisfies the initial condition $T(2\,\mathrm{Id}) = 2$, we have that $p = d = 1$ and

$$Tf(x) = f'(x), \quad Jf(s) = \int_c^x h(s) \, ds.$$

Hence T in (4.25) is a generalized derivative and J in (4.26) is a generalized definite integral. For the proof we refer to [KM12].

Chapter 5

Stability and Rigidity of the Leibniz and the Chain Rules

Equations modeling physical and mathematical phenomena should preferably be *stable*: reasonable perturbations of the equations should have solutions which are controlled perturbations of the solutions of the unperturbed equations. Even stronger, they may be *rigid*: this occurs if the perturbed equations turn out to have the same solutions as the unperturbed equations, so that these equations allow no reasonable perturbation.

In the previous chapter, we determined the solutions of the Leibniz rule $T(f \cdot g) = Tf \cdot g + f \cdot Tg$ and of the chain rule $T(f \circ g) = Tf \circ g \cdot Tg$, say for operators $T : C^k(\mathbb{R}) \to C(\mathbb{R})$. In this chapter, we show that these equations are stable under relaxations and perturbations and that the chain rule is even rigid in a certain setup. It is not too surprising that the chain rule is more stable than the Leibniz rule, since its operation also exchanges points $x \in \mathbb{R}$ in the domain of definition $\mathbb{R}$, whereas the Leibniz rule fixes these points $x \in \mathbb{R}$.

We consider relaxations and perturbations of two different types: firstly, we relax the equation by replacing the one operator T by three possibly different operators, $V, T_1, T_2 : C^k(\mathbb{R}) \to C(\mathbb{R})$, and study, e.g., in the case of the Leibniz rule, the solutions of

$$V(f \cdot g) = T_1 f \cdot g + f \cdot T_2 g, \quad f, g \in C^k(\mathbb{R}),$$

introducing additional freedom with a rule of a similar form, and secondly, we consider an additive perturbation

$$T(f \cdot g) = Tf \cdot g + f \cdot Tg + B(\cdot, f, g), \quad f, g \in C^k(\mathbb{R}),$$

where B is a given function of the independent and the function variables. The Leibniz rule and the chain rule turn out to be stable or even rigid in these situations. The convolution equation $T(f \cdot g) = Tf * Tg$ for bijective operators T on

© Springer Nature Switzerland AG 2018
H. König, V. Milman, *Operator Relations Characterizing Derivatives*,
https://doi.org/10.1007/978-3-030-00241-1_5

Schwartz space, characterizing the Fourier transform as mentioned in the introduction, may be relaxed as well, see Section 5.5.

5.1 Changing the operators

We start with the stability of the Leibniz rule under a change of operators. All function multiplications are meant to be defined pointwise.

Theorem 5.1 (Relaxed Leibniz rule). *Let $I \subset \mathbb{R}$ be an open interval, $k \in \mathbb{N}_0$ and $V, T_1, T_2 : C^k(I) \to C(I)$ be operators such that the relaxed Leibniz rule equation*

$$V(f \cdot g) = T_1 f \cdot g + f \cdot T_2 g \qquad (5.1)$$

holds for all $f, g \in C(I)$. Then there are continuous functions $c, d \in C(I)$ and $a_1, a_2 \in C(I)$ such that, with $T : C^k(I) \to C(I)$ defined by

$$Tf := c f \ln|f| + d f', \quad f \in C^k(I),$$

we have

$$
\begin{aligned}
Vf &= Tf + (a_1 + a_2)f, \\
T_1 f &= Tf + a_1 f, \\
T_2 f &= Tf + a_2 f, \qquad f \in C^k(I).
\end{aligned}
$$

For $k = 0$ we need $d = 0$. Conversely, these operators satisfy (5.1).

Remarks. The operator T satisfies the unperturbed Leibniz rule $T(f \cdot g) = Tf \cdot g + f \cdot Tg$, cf. Theorem 3.1. Adding $(a_1 + a_2) \operatorname{Id}, a_1 \operatorname{Id}, a_2 \operatorname{Id}$ to T obviously yields operators V, T_1, T_2 satisfying (5.1). The interesting fact is that these simple operations yield the general form of solutions of (5.1), so V, T_1, T_2 are very simple relaxations of T. Hence, as a consequence of (5.1), V, T_1 and T_2 are closely related. Actually, $a_i = T_i \mathbf{1}$.

Note that no continuity is imposed on any of the operators V, T_1 or T_2. Neither is linearity assumed; in fact, the operators are non-linear if $c \not\equiv 0$.

Proof. Exchanging $f, g \in C^k(I)$, we find

$$V(f \cdot g) = T_1 f \cdot g + f \cdot T_2 g = T_1 g \cdot f + g \cdot T_2 f.$$

Hence, $g \cdot (T_1 f - T_2 f) = f \cdot (T_1 g - T_2 g)$. Let $a_1 := T_1 \mathbf{1}$ and $a_2 := T_2 \mathbf{1}$. Since the ranges of T_1 and T_2 are in $C(I)$, a_1 and a_2 are continuous functions, $a_1, a_2 \in C(I)$. We get, for $g = \mathbf{1}$ and $f \in C^k(I)$,

$$T_1 f - a_1 f = T_2 f - a_2 f,$$

and

$$Vf = T_1 f + a_2 f = T_2 f + a_1 f.$$

Define $T : C^k(I) \to C(I)$ by $Tf := Vf - (a_1 + a_2)f$. Then $Tf = T_1 f - a_1 \cdot f = T_2 f - a_2 \cdot f$ and, using (5.1),

$$
\begin{aligned}
T(f \cdot g) &= V(f \cdot g) - (a_1 + a_2) \cdot f \cdot g \\
&= T_1 f \cdot g + f \cdot T_2 g - a_1 f \cdot g - f \cdot a_2 g \\
&= Tf \cdot g + f \cdot Tg, \qquad f, g \in C^k(I).
\end{aligned}
$$

Hence, T satisfies the (unperturbed) Leibniz rule (3.1), and by Theorem 3.1 there are continuous functions $c, d \in C(I)$ such that

$$
Tf = cf \ln|f| + d f', \quad f \in C^k(I).
$$

We then have $Vf = Tf + (a_1 + a_2)f$, $T_1 f = Tf + a_1 f$, $T_2 f = Tf + a_2 f$. $\qquad \square$

Corollary 5.2. *Assume that $V, T_1, T_2 : C^k(I) \to C(I)$ satisfy the relaxed Leibniz rule equation (5.1). Suppose further that $T_1 d_j = T_2 d_j = 0$ holds for two constant functions d_1, d_2 with values $d_1 \neq d_2$, $d_1 d_2 \neq 1$. Then there is $d \in C(I)$ such that*

$$
Vf = T_1 f = T_2 f = d f'.
$$

Proof. In Theorem 5.1 the assumption $T_1 d_j = T_2 d_j = 0$ for $j = 1, 2$ implies that $c(x) \ln|d_j| + a_i(x) = 0$ for all $i, j \in \{1, 2\}$, $x \in I$. Since $(1, \ln|d_j|) \in \mathbb{R}^2$ are linearly independent for $j = 1, 2$, we get $c = a_1 = a_2 = 0$. $\qquad \square$

We now turn to a relaxed form of the chain rule equation. For this, we need a weak condition of non-degeneration.

Definition. For $k \in \mathbb{N}$, an operator $V : C^k(\mathbb{R}) \to C(\mathbb{R})$ is *non-degenerate* provided that:

(i) for any $x \in \mathbb{R}$, there is $f \in C_b^k(\mathbb{R})$ such that $Vf(x) \neq 0$; and

(ii) for any $x \in \mathbb{R}$, there are $y \in \mathbb{R}$ and $f \in C^k(\mathbb{R})$ such that $f(y) = x$ and $Vf(y) \neq 0$.

Condition (i) means, in particular, that $V|_{C_b^k(\mathbb{R})} \not\equiv 0$.

We then have

Theorem 5.3 (Relaxed chain rule). *Let $k \in \mathbb{N}$ and $V, T_1, T_2 : C^k(\mathbb{R}) \to C(\mathbb{R})$ be operators such that the relaxed chain rule equation*

$$
V(f \circ g) = T_1 f \circ g \cdot T_2 g, \quad f, g \in C^k(\mathbb{R}), \tag{5.2}
$$

holds. Assume that V is non-degenerate. Then there is $p \geq 0$ and there are continuous functions $H, c_1, c_2 \in C(\mathbb{R})$, $H > 0$, such that with

$$
Tf := \frac{H \circ f}{H} |f'|^p \{\operatorname{sgn} f'\}, \quad f \in C^k(\mathbb{R}), \tag{5.3}
$$

we have

$$Vf = (c_1 \circ f) \cdot c_2 \cdot Tf,$$
$$T_1 f = (c_1 \circ f) \cdot Tf,$$
$$T_2 f = c_2 \cdot Tf.$$

Conversely, these operators V, T_1, T_2 *satisfy* (5.2).

Remarks. (a) As mentioned in Chapter 4, the notation $\{\operatorname{sgn} f'\}$ in equation (5.3) means that there are two possible operators T, one always with the term $\operatorname{sgn} f'$ and one without. If the term $\operatorname{sgn} f'$ appears, one needs $p > 0$ to ensure that the ranges of T and V, T_1 and T_2 consist of continuous functions. By Theorem 4.1, the operator T represents the general form of the solutions of the original chain rule $T(f \circ g) = Tf \circ g \cdot Tg$. Multiplying Tf by $c_1 \circ f \cdot c_2$, $c_1 \circ f$, c_2 yields operators $Vf, T_1 f, T_2 f$ which obviously satisfy (5.2). The interesting fact here is that these simple operations already provide the general solutions of (5.2). Again, no continuity assumption is imposed on any of the operators V, T_1 or T_2.

(b) To illustrate the result, suppose that in Theorem 5.3 we have $V = T_2$. Then $c_1 = \mathbf{1}$ and $T = T_1$.

Proof. Let $c_1 := T_1(\operatorname{Id})$, $c_2 := T_2(\operatorname{Id})$. Then $c_1, c_2 \in C(\mathbb{R})$. Choosing successively $g = \operatorname{Id}$ and $f = \operatorname{Id}$ in (5.2), we find that

$$Vf = c_2 \cdot T_1 f, \quad Vg = c_1 \circ g \cdot T_2 g, \qquad f, g \in C^k(\mathbb{R}).$$

Since V is non-degenerate, $c_1(x) \neq 0$ and $c_2(x) \neq 0$ for all $x \in \mathbb{R}$. Using these formulas for V and (5.2), we get

$$c_2 \cdot T_1(f \circ g) = V(f \circ g) = T_1 f \circ g \cdot T_2 g$$
$$= T_1 f \circ g \cdot \frac{c_2}{c_1 \circ g} \cdot T_1 g.$$

Put $Tf := \frac{1}{c_1 \circ f} \cdot T_1 f$. Then the last equalities mean that

$$T(f \circ g) = Tf \circ g \cdot Tg, \quad f, g \in C^k(\mathbb{R}).$$

Hence, T satisfies the chain rule equation and is non-degenerate as well. By Theorem 4.1, there are $p \geq 0$ and $H \in C(\mathbb{R})$, $H > 0$, such that

$$Tf = \frac{H \circ f}{H} |f'|^p \{\operatorname{sgn} f'\},$$

with $p > 0$ if the term $\operatorname{sgn} f'$ is present. Hence, by the definition of T,

$$T_1 f = c_1 \circ f \cdot Tf, \quad T_2 f = c_2 \cdot Tf, \quad Vf = c_1 \circ f \cdot c_2 \cdot Tf.$$

Conversely, the maps (V, T_1, T_2) defined by these formulas satisfy (5.2). $\square$

Corollary 5.4. *Assume that $V, T_1, T_2 : C^k(\mathbb{R}) \to C(\mathbb{R})$ satisfy the relaxed chain rule equation (5.2) and that V is non-degenerate. Suppose also that $V(2\,\mathrm{Id})$, $T_1(2\,\mathrm{Id})$ and $T_2(2\,\mathrm{Id})$ are constant functions. Then there are constants $c_1, c_2 \in \mathbb{R}$ and $p \geq 0$ such that, with $Tf = |f'|^p \{\mathrm{sgn}\, f'\}$, we have*

$$Vf = c_1 c_2 Tf, \quad T_1 f = c_1 Tf, \quad T_2 f = c_2 Tf.$$

If $V(2\,\mathrm{Id}) = T_1(2\,\mathrm{Id}) = T_2(2\,\mathrm{Id}) = 2$, either $Vf = T_1 f = T_2 f = f'$ or $Vf = T_1 f = T_2 f = |f'|$.

Proof. By Theorem 5.3, $c_2 = \frac{V(2\,\mathrm{Id})}{T_1(2\,\mathrm{Id})}$ and $c_1 \circ (2\,\mathrm{Id}) = \frac{V(2\,\mathrm{Id})}{T_2(2\,\mathrm{Id})}$. Thus by assumptions, c_1 and c_2 are constant functions. Therefore, also $H(2\,\mathrm{Id})/H$ is a constant function which, by the reasoning in part (iv) of the proof of Theorem 4.1, implies that H is constant. If $V(2\,\mathrm{Id}) = T_1(2\,\mathrm{Id}) = T_2(2\,\mathrm{Id}) = 2$, $c_1 = c_2 = 1$ and $p = 1$. $\square$

Corollary 5.5. *Let $k \in \mathbb{N}$ and assume that the operators $V, T_1, T_2, T_3, T_4 : C^k(\mathbb{R}) \to C(\mathbb{R})$ satisfy the relaxed Leibniz rule*

$$V(f \cdot g) = T_1 f \cdot g + f \cdot T_2 g$$

and the relaxed chain rule

$$V(f \circ g) = T_3 f \circ g \cdot T_4 g$$

for all $f, g \in C^k(\mathbb{R})$. If, for any $x \in \mathbb{R}$, there is $f \in C_b^k(\mathbb{R})$ with $Vf(x) \neq 0$, then

$$Vf = T_1 f = T_2 f = T_3 f = T_4 f = f'.$$

Proof. By Theorem 5.1, V has the form

$$Vf = c f \ln|f| + d f' + (a_1 + a_2)f.$$

Since for all $x \in I$ there is $f \in C_b^k(\mathbb{R})$ with $Tf(x) \neq 0$, at least one of the functions $c, d, a_1 + a_2$ is non-zero at any given point $x \in \mathbb{R}$. This implies that property (ii) of the condition of non-degeneration of V is satisfied, too. Theorem 5.3 then gives the general form of (V, T_3, T_4). The solutions in Theorems 5.1 and 5.3 will coincide only if $p = 1$, the $\{\mathrm{sgn}\, f'\}$-term appears, H is constant and $Vf = T_3 f = T_4 f = b f'$. Inserting this into (5.2) yields $b^2 = 1$, $b = 1$, i.e., $Vf = f'$. $\square$

This result should be compared with Corollary 4.3 where the derivative was characterized by the ordinary chain rule and the extended Leibniz rule.

5.2 Additive perturbations of the Leibniz rule

In this section we do not replace the operator T in its occurrences in the Leibniz or the chain rule by different operators. Instead, we allow perturbation terms and ask to what extent both equations are stable. We only consider perturbations that are local in the functions involved. We start with the Leibniz rule on C^1. Recall that a function $f : I \to \mathbb{R}$ is *locally bounded* if it is bounded on all compact subsets of I.

Theorem 5.6 (Stability of the Leibniz rule). *Let $I \subset \mathbb{R}$ be an open interval, $T : C^1(I) \to C(I)$ be an operator and $B : I \times \mathbb{R}^2 \to \mathbb{R}$ be a measurable function such that*

$$T(f \cdot g)(x) = Tf(x) \cdot g(x) + f(x) \cdot Tg(x) + B\big(x, f(x), g(x)\big) \qquad (5.4)$$

holds for all $f, g \in C^1(I)$ and any $x \in I$. Assume also that there exists a locally bounded function $M : I \to \mathbb{R}_{>0}$ such that for all $x \in I$, $(y, z) \in \mathbb{R}^2$

$$\big|B(x, y, z)\big| \leq M(x)|y||z|. \qquad (5.5)$$

Then there are continuous functions $c, d \in C(I)$, a locally bounded function $\widetilde{M} : I \to \mathbb{R}$ and a function $C : I \times \mathbb{R} \to \mathbb{R}$ with $|C(x, y)| \leq \widetilde{M}(x)|y|$ for $x \in I$, $y \in \mathbb{R}$, such that $C(x, f(x))$ is continuous in $x \in I$ for all $f \in C^1(I)$ and

$$Tf(x) = c(x)f(x) \ln|f(x)| + d(x)f'(x) + C\big(x, f(x)\big).$$

Remarks. (1) Theorem 5.6 implies that the solutions of the perturbed Leibniz rule (5.4) are perturbations of continuous solutions of the unperturbed Leibniz rule (cf. Theorem 3.1) by a continuous function $C(x, f(x))$ of controlled magnitude, $|C(x, f(x))| \leq \widetilde{M}(x)|f(x)|$. Note that the modulus of the entropy solution $|f(x)| \ln|f(x)|$ grows faster than $|C(x, f(x))|$ as $|f(x)| \to \infty$. Again, we do not impose any continuity assumptions on T.

(2) Let $C(x, y) = \frac{xy^2}{x^2 + y^4}$ for $(x, y) \neq (0, 0)$ and $C(0, 0) = 0$. Then, for any $f \in C^1(\mathbb{R})$, $C(x, f(x))$ is continuous in x, in particular, for those with $f(0) = 0$, but C is not continuous at $(0, 0)$. However, $C(x, f(x))$ cannot be continuous in x for any $f \in C(\mathbb{R})$, cf. [AFM, Lemma 3.1]. So Theorem 5.6 does not claim that C is continuous as a function of two variables.

(3) As a consequence of Theorem 5.6, any perturbation function B has to be of the form

$$B(x, y, z) = \big(C_1(x, y \cdot z) - C_1(x, y) - C_1(x, z)\big) \cdot y \cdot z,$$

where $C_1(x, y) = \frac{C(x, y)}{y}$. Conversely, if $C_1 : I \times \mathbb{R} \to \mathbb{R}$ is a continuous function and $\widetilde{M} : I \to \mathbb{R}$ is locally bounded with $|C_1(x, y)| \leq \widetilde{M}(x)$, define B by the above equation. Then the perturbed Leibniz rule (5.4) with this function B has a solution as given by Theorem 5.6.

Proof of Theorem 5.6. (a) We first show that T is a local operator. Let $J \subset I$ be an open subinterval and $f_1, f_2 \in C^1(\mathbb{R})$ be given with $f_1|_J = f_2|_J$. Choose any function $g \in C^1(I)$ with $\mathrm{supp}(g) \subset J$. Then $f_1 \cdot g = f_2 \cdot g$, hence $T(f_1 \cdot g) = T(f_2 \cdot g)$. Using the perturbed Leibniz rule formula (5.4), we get for any $x \in J$, after rearranging terms,

$$\big(Tf_1(x) - Tf_2(x)\big) \cdot g(x)$$
$$= \big(f_2(x) - f_2(x)\big) \cdot Tf(x) + B\big(x, f_2(x), g(x)\big) - B\big(x, f_1(x), g(x)\big) = 0,$$

since $f_1(x) = f_2(x)$. Choosing g with $g(x) \neq 0$, we find that $Tf_1(x) = Tf_2(x)$, so that $Tf_1|_J = Tf_2|_J$. Proposition 3.3 implies that there is a function $F : I \times \mathbb{R}^2 \to \mathbb{R}$ such that

$$Tf(x) = F\big(x, f(x), f'(x)\big)$$

holds for all $x \in I$ and $f \in C^1(I)$. Hence T is locally defined.

(b) For $g \in C^1(I)$, put $Sg(x) := \frac{T(\exp g)(x)}{\exp g(x)}$, $x \in I$. Then $S : C^1(I) \to C(I)$ is a local operator, too, and there is a function $G : I \times \mathbb{R}^2 \to \mathbb{R}$ with

$$Sg(x) = G\big(x, g(x), g'(x)\big), \quad g \in C^1(I), \ x \in I.$$

The operator equation (5.4) for T translates into the following one for S

$$S(g + h)(x) = Sg(x) + Sh(x) + \Psi\big(x, g(x), h(x)\big), \tag{5.6}$$

for $g, h \in C^1(I)$ where $\Psi(x, g(x), h(x)) := \frac{B(x, \exp(g(x)), \exp(h(x)))}{\exp(g(x) + h(x))}$. In view of assumption (5.5), we have, independently of g and h,

$$\big|\Psi(x, g(x), h(x))\big| \leq M(x), \quad x \in I,$$

where M is locally bounded. For any $x \in I$ and $\alpha = (\alpha_0, \alpha_1)$, $\beta = (\beta_0, \beta_1) \in \mathbb{R}^2$ choose $g, h \in C^1(I)$ with $g^{(j)}(x) = \alpha_j$, $h^{(j)}(x) = \beta_j$, $j = 0, 1$. Equation (5.6) then means, in terms of the function G representing S,

$$G(x, \alpha + \beta) = G(x, \alpha) + G(x, \beta) + \Psi(x, \alpha_0, \beta_0).$$

In particular, $G(x, 2\alpha) = 2G(x, \alpha) + \Psi(x, \alpha_0, \alpha_0)$. Iterating this equation, we find, for any $n \in \mathbb{N}$,

$$\frac{1}{2^n} G(x, 2^n \alpha) = G(x, \alpha) + \sum_{j=0}^{n-1} \frac{1}{2^{j+1}} \Psi(x, 2^j \alpha_0, 2^j \alpha_0). \tag{5.7}$$

Since $|\Psi(x, 2^j \alpha_0, 2^j \alpha_0)| \leq M(x)$, $\big|\sum_{j=n}^{\infty} \frac{1}{2^{j+1}} \Psi(x, 2^j \alpha_0, 2^j \alpha_0)\big| \leq \frac{M(x)}{2^n} \to 0$, as $n \to \infty$. Therefore, the series on the right-hand side of (5.7) converges, and the left-hand side has a limit as $n \to \infty$. Define

$$\widetilde{G}(x, \alpha) := \lim_{n \to \infty} \frac{1}{2^n} G(x, 2^n \alpha) =: G(x, \alpha) + \Phi(x, \alpha_0), \quad x \in I, \ \alpha \in \mathbb{R}^2$$

with $\Phi(x, \alpha_0) := \lim_{n \to \infty} \sum_{j=0}^{n-1} \frac{1}{2^{j+1}} \Psi(x, 2^j \alpha_0, 2^j \alpha_0)$. We have $|\Phi(x, \alpha_0)| \le M(x)$.
Since

$$\frac{1}{2^n} G(x, 2^n \alpha + 2^n \beta) = \frac{1}{2^n} G(x, 2^n \alpha) + \frac{1}{2^n} G(x, 2^n \beta) + \frac{1}{2^n} \Psi(x, 2^n \alpha_0, 2^n \beta_0),$$

$\widetilde{G}(x, \cdot) : \mathbb{R}^2 \to \mathbb{R}$ is additive,

$$\widetilde{G}(x, \alpha + \beta) = \widetilde{G}(x, \alpha) + \widetilde{G}(x, \beta), \quad x \in I, \ \alpha, \beta \in \mathbb{R}^2.$$

Note that $G(x, g(x), g'(x)) = Sg(x) = \frac{T(\exp(g))(x)}{\exp(g)(x)}$ defines a continuous function
of x since the images of T and S consist of continuous functions. Therefore, the
assumptions of Proposition 2.7 are satisfied for $(G, \widetilde{G}, \Psi)$, and we get that $\widetilde{G}(x, \cdot)$
is linear: for any $x \in I$ there are $c(x), d(x) \in \mathbb{R}$ such that

$$\widetilde{G}(x, \alpha) = c(x)\alpha_0 + d(x)\alpha_1.$$

By definition of S and G we have, for any $g \in C^1(I)$ and $x \in I$,

$$T(\exp(g))(x) = \exp(g(x))Sg(x) = \exp(g(x))G\big(x, g(x), g'(x)\big)$$
$$= \exp(g(x))\big(\widetilde{G}(x, g(x), g'(x)) - \Phi(x, g(x))\big).$$

This means that, for any $f \in C^1(I)$ with $f > 0$ and $g := \ln f$, we have

$$Tf(x) = f(x)\big(c(x) \ln |f|(x) + d(x)(\ln |f|)'(x)\big) - f(x)\Phi\big(x, \ln |f|(x)\big)$$
$$= c(x)f(x) \ln |f|(x) + d(x)f'(x) + C\big(x, f(x)\big), \tag{5.8}$$

with $C(x, f(x)) := -f(x)\Phi(x, \ln |f|(x))$, $|C(x, f(x))| \le M(x)|f(x)|$. We wrote $|f|$
in some places in (5.8) even though $f > 0$: we will show now that in this form the
formula is also true when $f' < 0$.

(c) Since T is a local operator, we may consider independently the points
where f is positive and where it is negative. Applying (5.4) to $f = g = \mathbf{1}$ and to
$f = g = -\mathbf{1}$, we find

$$T(\mathbf{1}) = -B(\cdot, 1, 1),$$
$$T(-\mathbf{1}) = \tfrac{1}{2}\big(-T(\mathbf{1}) + B(\cdot, -1, -1)\big) = \tfrac{1}{2}\big(B(\cdot, 1, 1) + B(\cdot, -1, -1)\big),$$

with $|T(\mathbf{1})| \le M$, $|T(-\mathbf{1})| \le M$. Now, let $f \in C^1(I)$, $x \in I$, be given with
$f(x) < 0$. Then $|f(x)| = -f(x)$, and applying (5.4), we find

$$Tf(x) = -T(-f)(x) - T(-\mathbf{1})(x)f(x) + B\big(x, -f(x), 1\big).$$

For the positive $|f(x)|$ we know by (5.8)

$$-T(|f|)(x) = -T(-f)(x) = c(x)f(x) \ln |f|(x) + d(x)f'(x) + C\big(x, -f(x)\big),$$

so that
$$Tf(x) = c(x)f(x)\ln|f(x)| + d(x)f'(x) + \widetilde{C}(x, f(x)),$$
where $\widetilde{C}(x, f(x)) := C(x, -f(x)) - T(-\mathbf{1})(x)f(x) + B(x, -f(x), 1)$ satisfies $|\widetilde{C}(x, f(x))| \le 3M(x)|f(x)|$. Take $\widetilde{M}(x) := 3M(x)$ in Theorem 5.6.

(d) We claim that c and d define continuous functions. If d would be discontinuous at some point $x_0 \in I$, choose a sequence $x_k \ne x_0$ in I such that the limits $\lim_{k\to\infty} x_k = x_0$, $\lim_{k\to\infty} d(x_k)$ exist and $\lim_{k\to\infty} d(x_k) \ne d(x_0)$. Recall that $|C(x, y)|/|y|$ is locally bounded. Choosing $f = 2$ in (5.8) and using the continuity of Tf, it follows that c is a locally bounded function, too. Thus we may find $M' > 0$ such that
$$\sup_{k\in\mathbb{N}} |c(x_k)| \le M', \quad \sup_{k\in\mathbb{N}} |C(x_k, y)| \le M'|y|, \quad y \in \mathbb{R}.$$
Choose $\epsilon > 0$ so small that $|\lim_{k\to\infty} d(x_k) - d(x_0)| \ge 3M'\epsilon(1 + \ln\frac{1}{\epsilon})$. Consider $f : I \to \mathbb{R}$, $f(x) := x - x_0 + \epsilon$. Since $Tf \in C(I)$, $\lim_{k\to\infty} Tf(x_k) = Tf(x_0)$. By (5.8)
$$Tf(x_k) - d(x_k) = c(x_k)(x_k - x_0 + \epsilon)\ln|x_k - x_0 + \epsilon| + C(x_k, x_k - x_0 + \epsilon)$$
$$=: \Delta(x_k)$$
and
$$Tf(x_0) - d(x_0) = c(x_0)\,\epsilon \ln|\epsilon| + C(x_0, \epsilon) =: \Delta(x_0).$$
This, however, leads to a contradiction, since
$$3M'\epsilon\left(1 + \ln\tfrac{1}{\epsilon}\right) \le \left|\lim_{k\to\infty} d(x_k) - d(x_0)\right|$$
$$\le \left|\lim_{k\to\infty} Tf(x_k) - Tf(x_0)\right| + \overline{\lim_{k\to\infty}} |\Delta(x_k)| + |\Delta(x_0)|$$
$$= \overline{\lim_{k\to\infty}} |\Delta(x_k)| + |\Delta(x_0)| \le 2M'\epsilon\left(1 + \ln\tfrac{1}{\epsilon}\right).$$
Hence, d is continuous. If c were discontinuous at some point $x_0 \in I$, choose again $x_k \ne x_0$ in I such that the limits $\lim_{k\to\infty} x_k = x_0$, $\lim_{k\to\infty} c(x_k)$ exist and $|\lim_{k\to\infty} c(x_k) - c(x_0)| =: \epsilon > 0$. With M' as above, choose N so large that $\epsilon \ln N > 2M'$. Let f be the constant function $f = N$. Since Tf is continuous, we get from (5.8)
$$Tf(x_k) = c(x_k)N\ln N + C(x_k, N) \longrightarrow Tf(x_0) = c(x_0)N\ln N + C(x_0, N).$$
This yields the contradiction
$$0 = \left|\lim_{k\to\infty} Tf(x_k) - Tf(x_0)\right|$$
$$\ge \left|\lim_{k\to\infty} c(x_k) - c(x_0)\right| N\ln N - \sup_k |C(x_k, N)| - |C(x_0, N)|$$
$$\ge (\epsilon \ln N - 2M')N > 0.$$

Hence c is continuous, too. Since c, d and Tf are continuous, for any $f \in C^1(I)$, so is $C(x, f(x))$ as a function of $x \in I$.

(e) If $f(x) = 0$ for some $f \in C^1(I)$, $x \in I$, we have $B(x, f(x), g(x)) = B(x, 0, g(x)) = 0$ for any $g \in C^1(I)$ by (5.5). If there is an open interval $J \subset I$ with $x \in J$ and $f|_J = 0$, we know already that $Tf|_J = 0$ and (5.8) holds. Otherwise, choose $x_k \to x$, $f(x_k) \neq 0$, $f(x) = 0$. Since Tf is continuous, $Tf(x_k) \to Tf(x)$. Applying (5.8) to $Tf(x_k)$ and using the continuity of c and d, we get that formula (5.8) also holds for $Tf(x)$,

$$Tf(x) = d(x)f'(x) + C(x, 0),$$

where we put $C(x_0, 0) := \lim_{k \to \infty} C(x_0, f(x_k))$, the limit being independent of the particular sequence x_k tending to x_0. Note that the limit of $f(x_k) \ln |f(x_k)|$ is zero. $\qquad \square$

5.3 Higher-order Leibniz rule

We next consider the Leibniz rule on $C^k(I)$-spaces for higher orders k of derivatives. For $f, g \in C^k(I)$,

$$(f \cdot g)^{(k)} = f^{(k)} \cdot g + f \cdot g^{(k)} + \sum_{j=1}^{k-1} \binom{k}{j} f^{(j)} \cdot g^{(k-j)}.$$

The last sum might be considered as an additive perturbation of the Leibniz rule by a function $B(\cdot, f, \ldots, f^{(k-1)}, g, \ldots, g^{(k-1)})$. Trying to characterize the k-th derivative by equations like this, note that the k-th derivative annihilates all polynomials of degree less that k, and then the perturbation term is zero. The following result states the converse.

Proposition 5.7 (Higher-order Leibniz rule). *Let $k \in \mathbb{N}$, $I \subset \mathbb{R}$ be an open interval and B be a function $B : I \times \mathbb{R}^{2k} \to \mathbb{R}$. Suppose that $T : C^k(I) \to C(I)$ is an operator satisfying*

$$T(f \cdot g)(x) = Tf(x) \cdot g(x) + f(x) \cdot Tg(x)$$
$$+ B\big(x, f(x), \ldots, f^{(k-1)}(x), g(x), \ldots, g^{(k-1)}(x)\big) \quad (5.9)$$

for all $f, g \in C^k(I)$ and $x \in I$. Suppose further that T annihilates all polynomials of degree $\leq k - 1$. Then T is a multiple of the k-th derivative, $Tf = d \cdot f^{(k)}$, for a suitable fixed function $d \in C(I)$, and B has the form

$$B\big(x, f(x), \ldots, g^{(k-1)}(x)\big) = d(x) \sum_{j=1}^{k-1} \binom{k}{j} f^{(j)}(x) g^{(k-j)}(x), \quad f, g \in C^k(I), \ x \in I.$$

Proof. (i) The proof of the localization is similar to the one of Theorem 5.6: If $f_1, f_2 \in C^k(I)$ satisfy $f_1|_J = f_2|_J$ for some open interval of $J \subset I$, and $g \in C^k(I)$ is a function with support in J, we have $f_1 \cdot g = f_2 \cdot g$. Hence, by (5.9) for all $x \in I$,

$$Tf_1(x) \cdot g(x) + f_1(x) \cdot Tg(x) + B\big(x, f_1(x), \ldots, g^{(k-1)}(x)\big)$$
$$= T(f_1 \cdot g)(x) = T(f_2 \cdot g)(x)$$
$$= Tf_2(x) \cdot g(x) + f_2(x) \cdot Tg(x) + B\big(x, f_2(x), \ldots, g^{(k-1)}(x)\big),$$

which yields for any $x \in J$ that $(Tf_1(x) - Tf_2(x)) \cdot g(x) = 0$. Choosing g with $g(x) \neq 0$, we conclude $Tf_1(x) = Tf_2(x)$, hence $Tf_1|_J = Tf_2|_J$. Therefore Proposition 3.3 implies that there is a function $F : I \times \mathbb{R}^{k+1} \to \mathbb{R}$ such that for all $f \in C^k(I)$, $x \in I$,

$$Tf(x) = F\big(x, f(x), \ldots, f^{(k)}(x)\big).$$

(ii) We claim that $Tf(x)$ only depends on x and the highest derivative $f^{(k)}(x)$. Let $\alpha_0, \ldots, \alpha_k, \beta_k \in \mathbb{R}$ and $x \in I$ be arbitrary. Choose $f \in C^k(I)$ with $f^{(j)}(x) = \alpha_j$ for all $j \in \{0, \ldots, k\}$ and $g \in C^k(I)$ with $g(x) = 1$, $g^{(j)}(x) = 0$ for all $j \in \{1, \ldots, k-1\}$ and $g^{(k)}(x) = \beta_k$. Then $(f \cdot g)^{(j)}(x) = \alpha_j$ for all $j \in \{0, \ldots, k-1\}$ and $(f \cdot g)^{(k)}(x) = \alpha_k + \alpha_0 \beta_k$. An application of (5.9) to f and g yields for the function F representing T that

$$F(x, \alpha_0, \ldots, \alpha_{k-1}, \alpha_k + \alpha_0 \beta_k) = F(x, \alpha_0, \ldots, \alpha_{k-1}, \alpha_k) + F(x, 1, 0, \ldots, 0, \beta_k) \alpha_0$$
$$+ B(x, \alpha_0, \ldots, \alpha_{k-1}, 1, 0, \ldots, 0). \quad (5.10)$$

Since T is zero on all polynomials of degree $\leq k - 1$, in particular on the constant functions, we have $F(x, 1, 0, \ldots, 0) = T\mathbf{1}(x) = 0$. Hence, choosing $\alpha_0 = 1$, $\alpha_1 = \cdots = \alpha_{k-1} = 0$ in (5.10), we get

$$F(x, 1, 0, \ldots, 0, \alpha_k + \beta_k) = F(x, 1, 0, \ldots, 0, \alpha_k) + F(x, 1, 0, \ldots, 0, \beta_k)$$
$$+ B(x, 1, 0, \ldots, 0, 1, 0, \ldots, 0).$$

For $\alpha_k = \beta_k = 0$ this yields $F(x, 1, 0, \ldots, 0) + B(x, 1, 0, \ldots, 0, 1, 0, \ldots, 0) = 0$, $B(x, 1, 0, \ldots, 0, 1, 0, \ldots, 0) = 0$. Therefore, $F(x, 1, 0, \ldots, 0, \cdot)$ is additive,

$$F(x, 1, 0, \ldots, 0, \alpha_k + \beta_k) = F(x, 1, 0, \ldots, 0, \alpha_k) + F(x, 1, 0, \ldots, 0, \beta_k).$$

Put $c(x, \alpha_k) := F(x, 1, 0, \ldots, 0, \alpha_k)$. Let $p_{k-1}(t) := \sum_{j=0}^{k-1} \frac{\alpha_j}{j!}(t - x)^j$. Since the degree of the polynomial p_{k-1} is $\leq k - 1$, we have $Tp_{k-1}(x) = 0$, i.e.,

$$F(x, \alpha_0, \ldots, \alpha_{k-1}, 0) = Tp_{k-1}(x) = 0.$$

Using this and putting $\alpha_k = 0$ in (5.10), we find

$$F(x, \alpha_0, \ldots, \alpha_{k-1}, \alpha_0 \beta_k) = F(x, \alpha_0, \ldots, \alpha_{k-1}, 0) + F(x, 1, 0, \ldots, 0, \beta_k) \alpha_0$$
$$+ B(x, \alpha_0, \ldots, \alpha_{k-1}, 1, 0, \ldots, 0)$$
$$= c(x, \beta_k) \alpha_0 + B(x, \alpha_0, \ldots, \alpha_{k-1}, 1, 0, \ldots, 0).$$

However, $c(x, 0) = 0$ by the additivity of $c(x, \cdot)$. Putting $\beta_k = 0$, we get that $B(x, \alpha_0, \ldots, \alpha_{k-1}, 1, 0, \ldots, 0) = 0$. Hence, renaming $\alpha_0 \beta_k$ as α_k, we get for $\alpha_0 \neq 0$

$$F(x, \alpha_0, \ldots, \alpha_{k-1}, \alpha_k) = c\left(x, \frac{\alpha_k}{\alpha_0}\right) \alpha_0,$$

and for all $f \in C^k(I)$, $x \in I$ with $f(x) \neq 0$,

$$Tf(x) = F\left(x, f(x), \ldots, f^{(k)}(x)\right) = c\left(x, \frac{f^{(k)}(x)}{f(x)}\right) f(x).$$

Since Tf is continuous, $c(x, g(x))$ is continuous for any continuous function $g \in C(I)$, just by taking a solution $f \in C^k(I)$ of $f^{(k)} = g \cdot f$ with $f(x) \neq 0$. Theorem 2.6, applied to the additive function $c(x, \cdot)$, yields that $c(x, \cdot)$ is linear and that there is a continuous function $d \in C(I)$ such that $c(x, \beta) = d(x) \cdot \beta$. Therefore, $Tf(x) = d(x)f^{(k)}(x)$, provided that $f(x) \neq 0$. This is true by continuity also if $f(x) = 0$:

Indeed, if $\alpha_0 = 0$, $(\alpha_j)_{j=1}^k \neq 0$ and fixing $x \in I$, consider the polynomial p_x given by $p_x(t) = \sum_{j=1}^k \frac{\alpha_j}{j!}(t - x)^j$. Since $p_x(t) \neq 0$ for $t \neq x$ close to x, we have $(Tp_x)(t) = d(t)p_x^{(k)}(t)$. However, for t tending to x, both sides have the limit

$$F(x, 0, \alpha_1, \cdots, \alpha_k) = (Tp_x)(x) = d(x)p_x^{(k)}(x) = d(x)\alpha_k,$$

and hence $Tf(x) = d(x)f^{(k)}(x)$ is also true for $f \in C^k(I)$ and $x \in I$ with $f(x) = 0$. $\qquad\square$

5.4   Additive perturbations of the chain rule

After studying additive perturbations of the Leibniz rule, we now turn to additive perturbations of the chain rule. The chain rule turns out to be rigid, under a weak condition of non-degeneration.

Definition. An operator $T : C^1(\mathbb{R}) \to C(\mathbb{R})$ is *locally non-degenerate* provided that, for any interval $J \subset \mathbb{R}$ and any $x \in J$, there are $g \in C^1(\mathbb{R})$ and $y \in \mathbb{R}$ with $g(y) = x$, $\mathrm{Im}(g) \subset J$ and $Tg(y) \neq 0$.

Additive perturbations of the chain rule for $T(f \circ g)$ naturally should involve functions of the values of $f \circ g$ and g since the information on f is coupled with g. This explains the setup in the following rigidity result for the chain rule.

Theorem 5.8 (Rigidity of the chain rule). *Assume that $T : C^1(\mathbb{R}) \to C(\mathbb{R})$ is an operator and $B : \mathbb{R}^3 \to \mathbb{R}$ is a function such that*

$$T(f \circ g)(x) = Tf \circ g(x) \cdot Tg(x) + B\big(x, f \circ g(x), g(x)\big) \qquad (5.11)$$

holds for all $f, g \in C^1(\mathbb{R})$, $x \in \mathbb{R}$. Assume also that T is locally non-degenerate and that Tf depends non-trivially on the derivative f'. Then $B = 0$ and T satisfies the chain rule. Hence, there are $p > 0$ and $H \in C(\mathbb{R})$, $H > 0$ such that

$$Tf(x) = \frac{H \circ f(x)}{H(x)} |f'(x)|^p \{\operatorname{sgn} f'(x)\}, \quad f \in C^1(\mathbb{R}), \ x \in \mathbb{R}.$$

Again, there are two types of solutions, one with the term $\operatorname{sgn} f'(x)$ and one without. The result means that the chain rule permits no additive perturbations of the above type, if Tf depends non-trivially on f'. If Tf does not depend on f', B might just be defined by

$$B\big(x, f \circ g(x), g(x)\big) = T(f \circ g)(x) - Tf \circ g(x) \cdot Tg(x),$$

the right-hand side of which we will show to be localized, i.e., being of the form $F(x, f \circ g(x)) - F(g(x), f \circ g(x))F(x, g(x))$.

Proof. (i) To prove that T is localized, let $J \subset \mathbb{R}$ be an open interval and $f_1, f_2 \in C^1(\mathbb{R})$ satisfy $f_1|_J = f_2|_J$. We claim that $Tf_1|_J = Tf_2|_J$ holds. Let $x \in J$. Since T is assumed to be locally non-degenerate, there is $g \in C^1(\mathbb{R})$ and $y \in \mathbb{R}$ such that $g(y) = x$, $\operatorname{Im}(g) \subset J$ and $Tg(y) \neq 0$. Then $f_1 \circ g = f_2 \circ g$ and, using (5.11)

$$Tf_1(x) \cdot Tg(y) + B\big(y, f_1(x), x\big) = T(f_1 \circ g)(x) = T(f_2 \circ g)(x)$$
$$= Tf_2(x) \cdot Tg(y) + B\big(y, f_2(x), x\big).$$

Since $f_1(x) = f_2(x)$, we get $(Tf_1(x) - Tf_2(x)) \cdot Tg(y) = 0$, and since $Tg(y) \neq 0$, $Tf_1(x) = Tf_2(x)$. Therefore, $Tf_1|_J = Tf_2|_J$. By Proposition 3.3, there is a function $F : \mathbb{R}^3 \to \mathbb{R}$ such that

$$Tf(x) = F\big(x, f(x), f'(x)\big)$$

holds for any $f \in C^1(\mathbb{R})$ and $x \in \mathbb{R}$.

(ii) We now analyze the form of F. For any $x, y, z, \alpha, \beta \in \mathbb{R}$, choose $f, g \in C^1(\mathbb{R})$ with $g(x) = y$, $f(y) = z$, $g'(x) = \alpha$, $f'(y) = \beta$. Then the operator equation (5.11) for T is equivalent to the functional equation for F,

$$F(x, z, \alpha\beta) = F(y, z, \beta)F(x, y, \alpha) + B(x, z, y). \tag{5.12}$$

Let $x = y = z$ and put $\phi_z := F(z, z, \cdot) : \mathbb{R} \to \mathbb{R}$, $\psi_z := B(z, z, z) \in \mathbb{R}$. Then

$$\phi_z(\alpha\beta) = \phi_z(\alpha)\phi_z(\beta) + \psi_z. \tag{5.13}$$

For $\alpha = 1$, we have $\phi_z(\beta)(1 - \phi_z(1)) = \psi_z$. If for some $z \in \mathbb{R}$, ψ_z were $\neq 0$, $\phi_z(1) \neq 1$ and hence $\phi_z(\beta) = \frac{\psi_z}{1-\phi_z(1)} =: a_z$ would be a constant function of β. Putting $\alpha = 1$ and $y = z$ in (5.12) would yield

$$F(x, z, \beta) = a_z F(x, z, 1) + B(x, z, z).$$

Interchanging y and z in (5.12), we get for $\beta = 1$

$$
\begin{aligned}
F(x, y, \alpha) &= F(z, y, 1)F(x, z, \alpha) + B(x, y, z) \\
&= a_z F(x, z, 1)F(z, y, 1) + F(z, y, 1)B(x, z, z) + B(x, y, z).
\end{aligned}
$$

The right-hand side is independent of the derivative variable α and hence Tf would not depend on the derivative f', contrary to the assumption in Theorem 5.8. Therefore, $\psi_z = B(z, z, z) = 0$ for all $z \in \mathbb{R}$.

(iii) We now know that $\phi_z = F(z, z, \cdot)$ is multiplicative by (5.13) and that $B(z, z, z) = 0$ for all $z \in \mathbb{R}$. Putting $g = \mathrm{Id}$ in (5.11), we find

$$
Tf(x) = Tf(x) \cdot T(\mathrm{Id})(x) + B\big(x, f(x), x\big),
$$

for all $f \in C^1(\mathbb{R})$, $x \in \mathbb{R}$. If $T(\mathrm{Id})(x)$ were $\neq 1$ for some $x \in \mathbb{R}$, $Tf(x) = B(x, f(x), x)/(1 - T(\mathrm{Id})(x))$ would be independent of the derivative $f'(x)$ at this point x, and by applying (5.11) to shift functions g, $Tf(z)$ would be independent of $f'(z)$ for all $z \in \mathbb{R}$, contradicting the assumption in Theorem 5.8. Hence $T(\mathrm{Id})(x) = 1$ for all $x \in \mathbb{R}$. We conclude for F and B that $F(x, x, 1) = T(\mathrm{Id})(x) = 1$ and $B(x, z, x) = 0$ for all $x, z \in \mathbb{R}$. Putting $f = \mathrm{Id}$ in (5.11) gives

$$
Tg(x) = T(\mathrm{Id})(g(x)) \cdot Tg(x) + B\big(x, g(x), g(x)\big) = Tg(x) + B\big(x, g(x), g(x)\big).
$$

Therefore also $B(x, z, z) = 0$ for all $x, z \in \mathbb{R}$.

Using this, we find, putting first $y = z$ and then $y = x$ in (5.12),

$$
F(x, z, \alpha\beta) = F(z, z, \alpha) \cdot F(x, z, \beta) = F(x, z, \beta) \cdot F(x, x, \alpha). \tag{5.14}
$$

We also used the symmetry in α and β on the left-hand side.

We claim that for any $x, z \in \mathbb{R}$, $F(x, z, 1) \neq 0$. If there were $x, z \in \mathbb{R}$ with $F(x, z, 1) = 0$, (5.14) would yield $F(x, z, \alpha) = 0$ for all $\alpha \in \mathbb{R}$, putting $\beta = 1$. Then for all $v \in \mathbb{R}$ by (5.12)

$$
F(x, v, \alpha) = F(z, v, \alpha)F(x, z, 1) + B(x, v, z) = B(x, v, z),
$$

and for all $u \in \mathbb{R}$, again using (5.12)

$$
\begin{aligned}
F(u, v, \alpha) &= F(x, v, \alpha)F(u, x, 1) + B(u, v, x) \\
&= B(x, v, z)F(u, x, 1) + B(u, v, x)
\end{aligned}
$$

would be independent of α for all $u, v \in \mathbb{R}$. This is impossible since Tf is assumed to depend non-trivially on f'. Hence, $F(x, z, 1) \neq 0$ for all $x, z \in \mathbb{R}$.

Therefore, putting $\beta = 1$ in (5.14), we get for any $x, z, \alpha \in \mathbb{R}$

$$
\frac{F(x, z, \alpha)}{F(x, z, 1)} = F(z, z, \alpha) = F(x, x, \alpha), \tag{5.15}
$$

which is independent of x and z and multiplicative in α. Put

$$\varphi(\alpha) := F(x, x, \alpha) = F(z, z, \alpha),$$

$\varphi(\alpha\beta) = \varphi(\alpha)\varphi(\beta)$ for all $\alpha, \beta \in \mathbb{R}$. By (5.15)

$$F(x, z, \alpha) = F(x, z, 1)\varphi(\alpha).$$

Note that for all $\alpha \neq 0$, $\varphi(\alpha) \neq 0$ since else by multiplicativity of φ, φ and F would be identically 0. Inserting this formula for F into (5.12), we find

$$F(x, z, 1)\varphi(\alpha\beta) = F(y, z, 1)\varphi(\beta) \cdot F(x, y, 1)\varphi(\alpha) + B(x, z, y).$$

Dividing this by $\varphi(\alpha\beta) = \varphi(\alpha)\varphi(\beta)$, we conclude that

$$F(x, z, 1) = F(y, z, 1) \cdot F(x, y, 1) + \frac{B(x, z, y)}{\varphi(\alpha\beta)},$$

for all $\alpha \neq 0 \neq \beta$. Comparing this with (5.12) for $\alpha = \beta = 1$, we get

$$B(x, z, y) = \frac{B(x, z, y)}{\varphi(\alpha\beta)},$$

for all $x, y, z, \alpha, \beta \in \mathbb{R}$. This implies that either B is identically zero or φ is identically 1. If $\varphi \equiv 1$, $F(x, z, \alpha) = F(x, z, 1)$ again would be independent of the derivative variable α, contrary to the assumption in Theorem 5.8. Therefore $B \equiv 0$, and T satisfies the unperturbed chain rule equation,

$$T(f \circ g)(x) = Tf \circ g(x) \cdot Tg(x), \quad f, g \in C^1(\mathbb{R}), \ x \in \mathbb{R}.$$

Hence, by Theorem 4.1, T has the form

$$Tf = \frac{H \circ f}{H}|f'|^p\{\operatorname{sgn} f'\},$$

with $p > 0$ and $H \in C(\mathbb{R})$, $H > 0$. Note that $p = 0$ is not allowed in Theorem 5.8 since then $Tf = \frac{H \circ f}{H}$ would be independent of the derivative f'. $\qquad\square$

5.5 Notes and References

Theorems 5.1 and 5.6 were shown by König, Milman [KM7] in the case $I = \mathbb{R}$. Theorems 5.3 and 5.8 are also taken from [KM7].

Proposition 5.7 was proven in [KM8].

As mentioned in the Introduction, the Fourier transform $\mathcal{F}$ on the Schwartz space $\mathcal{S}(\mathbb{R}^n)$ may be essentially characterized by the convolution equation

$$T(f \cdot g) = T(f) * T(g), \quad f, g \in \mathcal{S}(\mathbb{R}^n),$$

assuming that $T : \mathcal{S}(\mathbb{R}^n) \to \mathcal{S}(\mathbb{R}^n)$ is bijective, cf. [AAFM] and [AFM] . In the spirit of Section 5.1, we may also solve a relaxation of this equation: Suppose that $T, T_1, T_2 : \mathcal{S}(\mathbb{R}^n) \to \mathcal{S}(\mathbb{R}^n)$ are bijective operators satisfying

$$T(f \cdot g) = T_1(f) * T_2(g), \quad f, g \in \mathcal{S}(\mathbb{R}^n).$$

Then there are C^∞-functions $a_1, a_2 \in C^\infty(\mathbb{R}^n)$ which are never zero and a diffeomorphism $\omega : \mathbb{R}^n \to \mathbb{R}^n$ such that either for all $f \in \mathcal{S}(\mathbb{R}^n)$

$$Tf = a_1 * a_2 * \mathcal{F}(f \circ \omega), \ T_1 f = a_1 * \mathcal{F}(f \circ \omega), \ T_2 f = a_2 * \mathcal{F}(f \circ \omega),$$

or that for all $f \in \mathcal{S}(\mathbb{R}^n)$

$$Tf = a_1 * a_2 * \overline{\mathcal{F}(f \circ \omega)}, \ T_1 f = a_1 * \overline{\mathcal{F}(f \circ \omega)}, \ T_2 f = a_2 * \overline{\mathcal{F}(f \circ \omega)}.$$

The proof is based on the papers [AAFM] and [AFM] and the techniques of Section 5.1, reducing the convolution equation to a multiplicative equation by applying the Fourier transform.

Chapter 6

The Chain Rule Inequality and its Perturbations

In the previous chapter we showed that the chain rule operator equation shows a remarkable stability and rigidity, under modifications of operators or additive perturbations. In this chapter we study a different modification of the chain rule, replacing equalities by inequalities. Suppose $T : C^1(\mathbb{R}) \to C(\mathbb{R})$ is a map satisfying the *chain rule inequality*

$$T(f \circ g) \leq Tf \circ g \cdot Tg, \quad f, g \in C^1(\mathbb{R}). \tag{6.1}$$

Under mild assumptions on T, we determine the form of all operators T satisfying this inequality, provided that the image of T contains functions attaining negative values. There will be an assumption of non-degeneration of T which is a weak surjectivity type requirement. Moreover, we impose a weak continuity condition on T. In the case of the chain rule equation, the continuity of the operators was not assumed, but it was a consequence of the solution formulas. Here we have less information on T, and we require T to be pointwise continuous, as defined below. Remarkably, for functions f with positive derivative, the solutions Tf of the chain rule inequality (6.1) turn out to be the same as for the chain rule equation. For general functions the solutions of the chain rule inequality are bounded from above by corresponding solutions of the chain rule equality. This is a similar phenomenon as in Gronwall's inequality in its differential form, cf. Gronwall [G] or Hartman [H], where the solution of a differential inequality is bounded by the solution of the corresponding differential equation. We also state results for the opposite inequality $T(f \circ g) \geq Tf \circ g \cdot Tg$. The proofs are based in part on a result about submultiplicative functions on $\mathbb{R}$, which is of independent interest.

© Springer Nature Switzerland AG 2018
H. König, V. Milman, *Operator Relations Characterizing Derivatives*,
https://doi.org/10.1007/978-3-030-00241-1_6

6.1   The chain rule inequality

Studying the chain rule inequality, we will impose the following two conditions.

Definition. An operator $T : C^1(\mathbb{R}) \to C(\mathbb{R})$ is *non-degenerate* provided that, for any open interval $I \subset \mathbb{R}$ and any $x \in I$, there exists a function $g \in C^1(\mathbb{R})$ with $g(x) = x$, $\mathrm{Im}(g) \subset I$ and $Tg(x) > 1$. Let us call T *negatively non-degenerate* if there is $g \in C^1(\mathbb{R})$ with $g(x) = x$, $\mathrm{Im}(g) \subset I$ and $Tg(x) < -1$.

Definition. An operator $T : C^1(\mathbb{R}) \to C(\mathbb{R})$ is *pointwise continuous* if for any sequence of functions $f_n \in C^1(\mathbb{R})$ and $f \in C^1(\mathbb{R})$ with $f_n \to f$ and $f_n' \to f'$ converging uniformly on all compact subsets of $\mathbb{R}$, we have the pointwise convergence of $\lim_{n \to \infty} Tf_n(x) = Tf(x)$ for all $x \in \mathbb{R}$.

Theorem 6.1 (Chain rule inequality). *Let $T : C^1(\mathbb{R}) \to C(\mathbb{R})$ be an operator such that the chain rule inequality holds:*

$$T(f \circ g) \le Tf \circ g \cdot Tg, \quad f, g \in C^1(\mathbb{R}). \tag{6.1}$$

Assume in addition that T is non-degenerate and pointwise continuous. Suppose further that there exists $x \in \mathbb{R}$ with $T(-\,\mathrm{Id})(x) < 0$. Then there is a continuous function $H \in C(\mathbb{R})$, $H > 0$, and there are real numbers $p > 0$ and $A \ge 1$, such that T has the form

$$Tf = \begin{cases} \frac{H \circ f}{H} f'^p, & f' \ge 0, \\ -A\frac{H \circ f}{H}|f'|^p, & f' < 0, \end{cases} \quad f \in C^1(\mathbb{R}). \tag{6.2}$$

Remarks. (a) Let $Sf := \frac{H \circ f}{H}|f'|^p \,\mathrm{sgn}\, f'$. Then S satisfies the chain rule equation $S(f \circ g) = Sf \circ g \cdot Sg$. Equation (6.2) means that $Tf \le Sf$. Thus, the solutions of the chain rule inequality are bounded from above by solutions of the chain rule equation for which $A = 1$. Note that $-A = T(-\,\mathrm{Id})(0) \le -1$. Thus under the additional assumption $T(-\,\mathrm{Id})(0) = -1$ in Theorem 6.1, T satisfies the chain rule equation.

(b) Let $c > 0$. The modified operator inequality $T(f \circ g) \le c \cdot Tf \circ g \cdot Tg$ may be treated by considering $T_1 := c \cdot T$ which would satisfy $T_1(f \circ g) \le T_1 f \circ g \cdot T_1 g$.

(c) The condition $T(-\,\mathrm{Id})(x) < 0$ guarantees that there are sufficiently many negative functions in the range of T. If this is violated, there are many *positive* solution operators T of (6.1): Examples for non-negative solutions can be given by

$$Tf(x) = F\big(x, f(x), |f'(x)|\big),$$

where $F : \mathbb{R}^2 \times \mathbb{R}_{\ge 0} \to \mathbb{R}_{\ge 0}$ is a continuous function satisfying

$$F(x, z, \alpha\beta) \le F(y, z, \alpha)F(x, y, \beta) \tag{6.3}$$

for all $x, y, z, \alpha, \beta \in \mathbb{R}$. We might take, e.g.,

$$F(x, z, \alpha) = \exp(d(x, z)) \cdot K(\alpha),$$

where d is either a metric on $\mathbb{R}$ or $d(x, z) = z - x$, and $K : \mathbb{R}_{\geq 0} \to \mathbb{R}_{\geq 0}$ is continuous and *submultiplicative*, $K(\alpha\beta) \leq K(\alpha) \cdot K(\beta)$ for $\alpha, \beta \geq 0$. Non-trivial examples of such maps K besides power-type functions α^p, $p > 0$ and maxima of such functions are given, e.g., by $K(\alpha) = \ln(\alpha + c)$ with $c \geq e$, cf. Gustavsson, Maligranda, Peetre [GMP], and products of submultiplicative functions. Moreover, for any continuous submultiplicative function $K : \mathbb{R}_{\geq 0} \to \mathbb{R}_{\geq 0}$, $\widetilde{F} := K \circ F$ will also satisfy (6.3) if F does. There does not seem to be much hope of classifying the solutions of (6.1) without any negativity assumption like $T(-\operatorname{Id})(x) < 0$ for some $x \in \mathbb{R}$.

6.2   Submultiplicative functions

Let $K : \mathbb{R} \to \mathbb{R}$ be continuous and define $T : C^1(\mathbb{R}) \to C(\mathbb{R})$ by $Tf(x) := K(f'(x))$. This operator T will satisfy (6.1) if and only if K is submultiplicative, i.e., $K(\alpha\beta) \leq K(\alpha)K(\beta)$ for all $\alpha, \beta \in \mathbb{R}$. Hence, as a special case in the proof of Theorem 6.1, we have to classify submultiplicative functions on $\mathbb{R}$ attaining also negative values. This result is of independent interest and we formulate it as Theorem 6.2.

Theorem 6.2 (Submultiplicative functions). *Let $K : \mathbb{R} \to \mathbb{R}$ be a measurable function which is continuous in 0 and in 1 and submultiplicative, i.e.,*

$$K(\alpha\beta) \leq K(\alpha)K(\beta), \quad \alpha, \beta \in \mathbb{R}.$$

Assume further that $K(-1) < 0 < K(1)$. Then there exist real numbers $p > 0$ and $A \geq 1$ such that

$$K(\alpha) = \begin{cases} \alpha^p, & \alpha \geq 0, \\ -A|\alpha|^p, & \alpha < 0. \end{cases}$$

Hence, $K(-1) = -A \leq 1$. Note that $K|_{\mathbb{R}_{\geq 0}}$ is multiplicative, and if $K(-1) = -1$, K is multiplicative on $\mathbb{R}$, i.e., $K(\alpha) = |\alpha|^p \operatorname{sgn} \alpha$. As mentioned in Remark (b) above, there are many continuous submultiplicative functions $K : \mathbb{R}_{\geq 0} \to \mathbb{R}_{\geq 0}$ besides powers $K(\alpha) = \alpha^p$. However, these cannot be extended to continuous submultiplicative functions $K : \mathbb{R} \to \mathbb{R}$ with $K(-1) < 0$. There is a corresponding result for *supermultiplicative* functions on $\mathbb{R}$, $K(\alpha\beta) \geq K(\alpha)K(\beta)$, which gives the same form of K, except that then $0 < A \leq 1$.

Examples. (a) The measurability assumption in Theorem 6.2 is necessary. Otherwise, we may take a non-measurable additive function $f : \mathbb{R} \to \mathbb{R}$ as given in the comments following Proposition 2.1 and $A > 1$, and define $K(0) := 0$ and

$$K(\alpha) := \begin{cases} \exp(f(\ln \alpha)), & \alpha > 0, \\ -A \exp(f(\ln |\alpha|)), & \alpha < 0. \end{cases}$$

Then $K : \mathbb{R} \to \mathbb{R}$ is non-measurable and submultiplicative with $K(-1) < 0 < K(1)$.

(b) Let $d \geq 1$, $c \geq 0$, $c \neq d$, and put

$$K(\alpha) := \begin{cases} 1, & \alpha = 1, \\ -c, & \alpha = 0, \\ -d, & \alpha \notin \{0, 1\}. \end{cases}$$

Then K is measurable and submultiplicative with $K(-1) < 0 < K(1)$, but discontinuous at 0 and at 1.

The result corresponding to Theorem 6.1 in the supermultiplicative situation is

Theorem 6.3. *Let $T : C^1(\mathbb{R}) \to C(\mathbb{R})$ be an operator such that*

$$T(f \circ g) \geq Tf \circ g \cdot Tg, \quad f, g \in C^1(\mathbb{R}).$$

Assume also that T is negatively non-degenerate and pointwise continuous with $T(-\operatorname{Id})(x) < 0$ for some $x \in \mathbb{R}$. Then there exist numbers $p > 0$, $0 < B \leq 1$ and a function $H \in C(\mathbb{R})$, $H > 0$ such that

$$Tf = \begin{cases} \frac{H \circ f}{H} f'^p, & f' \geq 0, \\ -B \frac{H \circ f}{H} |f'|^p, & f' < 0, \end{cases} \quad f \in C^1(\mathbb{R}).$$

We first prove Theorem 6.2 which is used in the proof of Theorem 6.1. For this, we need two lemmas.

Lemma 6.4. *Let $K : \mathbb{R} \to \mathbb{R}$ be submultiplicative with $K(-1) < 0 < K(1)$. Assume that K is continuous in 0 and in 1. Then:*

(i) *$K(0) = 0$, $K(1) = 1$ and $K|_{\mathbb{R}_{<0}} < 0 < K|_{\mathbb{R}_{>0}}$.*

(ii) *There is $0 < \epsilon < 1$ such that $0 < K(\alpha) < 1$ for all $\alpha \in (0, \epsilon)$ and $1 < K(\alpha) < \infty$ for all $\alpha \in (1/\epsilon, \infty)$.*

Proof. Since $0 < K(1) = K(1^2) \leq K(1)^2$, $K(1) \geq 1$. Then $1 \leq K(1) = K((-1)^2) \leq K(-1)^2$, implying $K(-1) \leq -1$. By submultiplicativity $K(-1) \leq K(1)K(-1)$, $|K(-1)| \geq K(1)|K(-1)|$. Hence, $K(1) \leq 1$, $K(1) = 1$. Since K is continuous at 1, there is $\epsilon > 0$ such that $K|_{[1/(1+\epsilon), 1+\epsilon]} > 0$. For any $\alpha \in [1/(1+\epsilon), 1+\epsilon]$, $K(\alpha) > 0$ and $K(1/\alpha) > 0$. Hence, $0 < K(\alpha) \leq K(1/\alpha)K(\alpha^2)$, implying that $K(\alpha^2) > 0$, i.e., $K|_{[1/(1+\epsilon)^2, (1+\epsilon)^2]} > 0$. Inductively, we get that $K|_{\mathbb{R}_{>0}} > 0$, since $\mathbb{R}_{>0} = \bigcup_{n \in \mathbb{N}} [1/(1+\epsilon)^n, (1+\epsilon)^n]$. The inequality $K(0) = K((-1) \cdot 0) \leq K(-1) \cdot K(0)$ with $K(-1) < 0$ shows that $K(0) \leq 0$. Since $K|_{\mathbb{R}_{>0}} > 0$ and K is continuous in 0, we get $K(0) = 0$. Then there is $\epsilon > 0$ with $0 < K|_{(0,\epsilon)} < 1$. Since $1 \leq K(1) \leq K(\alpha) \cdot K(1/\alpha)$, it follows that $K|_{(1/\epsilon, \infty)} > 1$. Moreover, for any $\alpha > 0$, $K(-\alpha) \leq K(-1)K(\alpha) < 0$, i.e., $K|_{\mathbb{R}_{<0}} < 0$. $\square$

The second lemma is a well-known fact on subadditive functions on $\mathbb{R}$.

Lemma 6.5. *Assume that $f : \mathbb{R} \to \mathbb{R}$ is measurable and* subadditive, *i.e.,*

$$f(s + t) \leq f(s) + f(t), \quad s, t \in \mathbb{R}.$$

Define $p := \sup_{t<0} \frac{f(t)}{t}$ and $q := \inf_{t>0} \frac{f(t)}{t}$. Then f is bounded on compact intervals, $-\infty < p \leq q < \infty$ and $f(0) \geq 0$. Moreover, the limits $\lim_{t\to-\infty} \frac{f(t)}{t}$, $\lim_{t\to\infty} \frac{f(t)}{t}$ exist and $p = \lim_{t\to-\infty} \frac{f(t)}{t}$, $q = \lim_{t\to\infty} \frac{f(t)}{t}$.

Proof. (a) We first show that f is bounded from above on each compact subset of $(0, \infty)$. Fix $a > 0$ and put $A := f(a)$. Let $E := \{t \in (0, a) \mid f(t) \geq A/2\}$. Then E is measurable since f is measurable. Moreover, $(0, a) = E \cup (\{a\} - E)$, since $t_1, t_2 > 0$ with $a = t_1 + t_2$ implies that $t_1 \in E$ or $t_2 \in E$. Suppose there are $0 < \alpha < \beta < \infty$ such that $f|_{[\alpha,\beta]}$ is not bounded from above. There there is a sequence (t_n), $\alpha \leq t_n \leq \beta$ with $t_n \to t_0$, $\alpha \leq t_0 \leq \beta$ and $f(t_n) \geq 2n$. Let $E'_n := \{t \in (0, \beta) \mid f(t) \geq n\}$. For a fixed $n \in \mathbb{N}$, choose above $a = t_n$, $A = f(t_n) \geq 2n$. Then

$$|E'_n| \geq \left|\{t \in (0, t_n) \mid f(t) \geq n\}\right| =: |E_n| \geq \frac{t_n}{2} \geq \frac{\alpha}{2} > 0.$$

Since $E'_n \geq E'_{n+1}$, we get that $\left|\bigcap_{n\in\mathbb{N}} E'_n\right| \geq \frac{\alpha}{2}$. Therefore, $f|_{(0,\beta)}$ is infinite on a set of strictly positive measure, which is a contradiction. Therefore, f is bounded from above on any compact subset of $(0, \infty)$.

(b) Since $f(0) \leq f(0) + f(0)$, we have $f(0) \geq 0$. Also, for any $t \in \mathbb{R}$, $0 \leq f(0) \leq f(t) + f(-t)$. Hence, $\frac{f(-t)}{-t} \leq \frac{f(t)}{t}$ for any $t > 0$, and therefore $\overline{\lim}_{t\to-\infty} \frac{f(t)}{t} \leq \underline{\lim}_{t\to\infty} \frac{f(t)}{t}$.

Let $q := \inf_{t>0} \frac{f(t)}{t}$. We claim that the limit $\lim_{t\to\infty} \frac{f(t)}{t}$ exists and that $q = \lim_{t\to\infty} \frac{f(t)}{t}$. Assume that $q \in \mathbb{R}$; the case $q = -\infty$ is treated similarly. Let $\epsilon > 0$ and choose $b > 0$ with $\frac{f(b)}{b} \leq q + \epsilon$. For any $t \geq 3b$, there is $n \in \mathbb{N}$ with $t \in [(n+2)b, (n+3)b]$. Using the subadditivity of f, and the definition of q and b, we find

$$q \leq \frac{f(t)}{t} = \frac{f(nb + (t - nb))}{t} \leq \frac{n f(b) + f(t - nb)}{t}$$
$$= \frac{nb}{t}\frac{f(b)}{b} + \frac{f(t - nb)}{t} \leq \frac{nb}{t}(q + \epsilon) + \frac{f(t - nb)}{t}.$$

By part (a), f is bounded from above on $[2b, 3b]$. Let $M > 0$ be such that $f_{[2b,3b]} \leq M$. Since $t - nb \in [2b, 3b]$, we get

$$q \leq \frac{f(t)}{t} \leq \frac{nb}{t}(q + \epsilon) + \frac{M}{t}.$$

For $t \to \infty$, $\frac{nb}{t} \to 1$ and $\frac{M}{t} \to 0$. Therefore, for any $\epsilon > 0$,

$$\varlimsup_{t \to \infty} \frac{f(t)}{t} \leq q + \epsilon = \inf_{t>0} \frac{f(t)}{t} + \epsilon,$$

which shows that the limit $\lim_{t \to \infty} \frac{f(t)}{t}$ exists and is equal to q.

(c) Consider similarly $g(t) := f(-t)$, $t > 0$. Then by (b)

$$p := \sup_{t>0} \frac{f(-t)}{-t} = -\inf_{t>0} \frac{g(t)}{t} = -\lim_{t \to \infty} \frac{g(t)}{t} = \lim_{t \to -\infty} \frac{f(-t)}{-t}.$$

Now $\frac{f(-t)}{-t} \leq \frac{f(t)}{t}$ implies for $t \to \infty$ that $p \leq q$. Moreover, since $p > -\infty$, $q > -\infty$, and since $q < \infty$, $p < \infty$. Therefore, $-\infty < p \leq q < \infty$. $\square$

As a consequence of Lemma 6.5, we have that

$$f(t) = pt + a(t), \quad t < 0,$$
$$f(t) = qt + a(t), \quad t > 0,$$

where $a(t) \geq 0$ for all $t \neq 0$.

Proof of Theorem 6.2. (a) Let $K : \mathbb{R} \to \mathbb{R}$ be measurable and submultiplicative, continuous in 0 and in 1 with $K(-1) < 0 < K(1)$. By Lemma 6.4, $K(0) = 0$, $K(1) = 1$, $K|_{\mathbb{R}_{<0}} < 0 < K|_{\mathbb{R}_{>0}}$ and for a suitable $0 < \epsilon < 1$, $0 < K(\alpha) < 1$ for all $\alpha \in (0, \epsilon)$ and $1 < K(\alpha) < \infty$ for all $\alpha \in (1/\epsilon, \infty)$. Define $f(t) := \ln K(\exp(t))$, $t \in \mathbb{R}$. Then f is measurable and subadditive, and we have by Lemma 6.5

$$-\infty < p := \sup_{t<0} \frac{f(t)}{t} = \lim_{t \to -\infty} \frac{f(t)}{t} \leq q := \inf_{t>0} \frac{f(t)}{t} = \lim_{t \to \infty} \frac{f(t)}{t} < \infty.$$

Since f is negative for $t < 0$ and positive for $t > 0$, we have that $0 \leq p \leq q < \infty$, with

$$f(t) =: pt + a(t), \quad t < 0, \quad f(t) =: qt + a(t), \quad t > 0,$$

where $a(t) \geq 0$ for all t and $\lim_{t \to -\infty} \frac{a(t)}{t} = \lim_{t \to \infty} \frac{a(t)}{t} = 0$. This means, for all $0 < \alpha < 1$, that

$$K(\alpha) = \exp\big(f(\ln \alpha)\big) = \alpha^p \exp\big(a(\ln \alpha)\big) \geq \alpha^p,$$

and, for all $1 < \alpha < \infty$, that

$$K(\alpha) = \exp\big(f(\ln \alpha)\big) = \alpha^q \exp\big(a(\ln \alpha)\big) \geq \alpha^q.$$

(b) We claim that $p = q > 0$ holds. Using $K|_{\mathbb{R}_{<0}} < 0 < K|_{\mathbb{R}_{>0}}$, the submultiplicativity of K implies that for all $\beta < 0 < \alpha$

$$K(\alpha \beta) \leq K(\alpha) K(\beta), \quad |K(\alpha \beta)| \geq K(\alpha) |K(\beta)|.$$

Since $K(1) = 1$, $f(0) = 0$. Fix $t < 0$ and choose $\alpha < -1$, $0 < \beta < 1$, with $t = \alpha\beta$. Then, by submultiplicativity,

$$K(t) = K\big((-1)|\alpha|\beta\big) \leq K(-1)K(|\alpha|)K(\beta) \leq 0,$$
$$|K(t)| \geq |K(-1)|K(|\alpha|)K(\beta) \geq |\alpha|^q\beta^p = |t|^q\beta^{p-q},$$

using that $K(-1) \leq -1$ since $1 = K(1) \leq K(-1)^2$. Assuming $p \neq q$, i.e., $p < q$, and letting β tend to 0 (and α to $-\infty$), would yield the contradiction $|K(t)| = \infty$. Hence, $0 \leq p = q < \infty$. In fact, $0 < p = q$ since K is continuous at 0 with $K(0) = 0$ and $K(\beta) \geq \beta^p$ for $0 < \beta < 1$.

(c) Let $g(t) := \ln|K(-\exp(t))|$ for all $t \in \mathbb{R}$. Since, for any $s, t \in \mathbb{R}$,

$$K\big(-\exp(s)\exp(t)\big) \leq K\big(-\exp(s)\big)K\big(\exp(t)\big) \leq 0,$$

we get that

$$g(s+t) = \ln\big|K\big(-\exp(s)\exp(t)\big)\big| \geq \ln\big|K(-\exp(s))\big| + \ln K(\exp(t))$$
$$= g(s) + f(t) = g(s) + pt + a(t),$$

with $a(t) \geq 0$, for all t, and $\lim_{t\to\pm\infty}\frac{a(t)}{t} = 0$. Since $f(0) = 0$, $a(0) = 0$. Putting $s = 0$ yields

$$g(t) \geq g(0) + pt + a(t).$$

Putting $t = -s$ and renaming s by t gives

$$g(0) \geq g(t) - pt + a(-t).$$

Hence,

$$g(0) + pt + a(t) \leq g(t) \leq g(0) + pt - a(-t).$$

Since $a \geq 0$, this implies that $a = 0$ on $\mathbb{R}$. Therefore, for all $t \in \mathbb{R}$, $f(t) = pt$ and $g(t) = g(0) + pt$. We then find, for all $\beta < 0 < \alpha$,

$$K(\alpha) = \alpha^p, \quad |K(\beta)| = \exp\big(g(\ln|\beta|)\big) = \exp(g(0))|\beta|^p.$$

Since $\exp(g(0)) = |K(-1)| \geq 1$, $g(0) \geq 0$. Thus, $K(\beta) = K(-1)|\beta|^p$, proving Theorem 6.2 with $-A = K(-1) \leq -1$. □

6.3 Localization and Proof of Theorem 6.1

As the first step in the proof of Theorem 6.1 on the chain rule inequality , we show that T is locally defined. More precisely, $Tf(x)$ only depends on x, $f(x)$ and $f'(x)$.

Proposition 6.6. *Let $T : C^1(\mathbb{R}) \to C(\mathbb{R})$ be non-degenerate, pointwise continuous and satisfy the chain rule inequality (6.1). Assume also that there exists $x_0 \in \mathbb{R}$ such that $T(-\operatorname{Id})(x_0) < 0$. Then there is a function $F : \mathbb{R}^3 \to \mathbb{R}$ such that, for all $f \in C^1(\mathbb{R})$ and all $x \in \mathbb{R}$,*

$$Tf(x) = F\big(x, f(x), f'(x)\big). \tag{6.4}$$

To show this, we need a lemma.

Lemma 6.7. *Under the assumptions of Proposition 6.6 we have for any open interval $I \subset \mathbb{R}$:*

(a) *For all $x \in I$, there is $g \in C^1(I)$ with $g(x) = x$, $\operatorname{Im}(g) \subset I$ and $Tg(x) < -1$.*

(b) *For $c \in \mathbb{R}$, $f \in C^1(\mathbb{R})$ with $f|_I = c$, we have $Tf|_I = 0$.*

(c) *For $f \in C^1(\mathbb{R})$ with $f|_I = \operatorname{Id}|_I$, we have $Tf|_I = 1$.*

(d) *Take $f_1, f_2 \in C^1(\mathbb{R})$ with $f_1|_I = f_2|_I$ and assume that f_2 is invertible. Then $Tf_1|_I \leq Tf_2|_I$. Hence, if f_1 is invertible, too, $Tf_1|_I = Tf_2|_I$.*

Proof. (a) By (6.1), $T(\operatorname{Id})(x) \leq T(\operatorname{Id})(x)^2$ for all $x \in \mathbb{R}$. Hence, $T(\operatorname{Id})(x) \geq 1$ or $T(\operatorname{Id})(x) \leq 0$. If there would be $x_1 \in \mathbb{R}$ with $T(\operatorname{Id})(x_1) \leq 0$, use that by non-degeneration of T there is $g \in C^1(\mathbb{R})$, $g(x_1) = x_1$ and $Tg(x_1) > 1$. Then,

$$1 \leq Tg(x_1) = T(g \circ \operatorname{Id})(x_1) \leq Tg(x_1)T(\operatorname{Id})(x_1) \leq 0,$$

a contradiction. Hence $T(\operatorname{Id})(x) \geq 1$ for all $x \in I$. Also, $T(-\operatorname{Id})(x) < 0$: $1 \leq T(\operatorname{Id})(x) = T((-\operatorname{Id})^2)(x) \leq T(-\operatorname{Id})(-x)T(-\operatorname{Id})(x)$. By assumption, there is $x_0 \in \mathbb{R}$, with $T(-\operatorname{Id})(x_0) < 0$. If there would be $x_1 \in \mathbb{R}$ with $T(-\operatorname{Id})(x_1) > 0$, by continuity of the function $T(-\operatorname{Id})$ there would be $x_2 \in \mathbb{R}$ with $T(-\operatorname{Id})(x_2) = 0$, contradicting $1 \leq T(-\operatorname{Id})(-x_2)T(-\operatorname{Id})(x_2)$. Hence, $T(-\operatorname{Id})(x) < 0$ for all $x \in \mathbb{R}$. Also $1 \leq T(-\operatorname{Id})(0)^2$ yields $T(-\operatorname{Id})(0) \leq -1$.

Now let $I \subset \mathbb{R}$ be an open interval and $x_1 \in I$. Let $\epsilon > 0$ with $J = (x_1 - \epsilon, x_1 + \epsilon) \subset I$, $\tilde{J} := J - \{x_1\} = (-\epsilon, \epsilon)$. Since T is non-degenerate, there is $f \in C^1(\mathbb{R})$ with $f(0) = 0$, $\operatorname{Im}(f) \subset \tilde{J}$ and $Tf(0) > 1$. Then

$$T(-f)(0) \leq T(-\operatorname{Id})(0)Tf(0) < -1,$$

and $\operatorname{Im}(-f) \subset \tilde{J}$. We transport $-f$ back to J by conjugation with a shift. For $y \in \mathbb{R}$, let $S_y := \operatorname{Id} + y \in C^1(\mathbb{R})$ denote the shift by y. Since for $y_n \to y$, $S_{y_n} \to S_y$ and $S'_{y_n} \to S'_y$ converge uniformly on compacta, by the pointwise continuity of T, we have that $T(S_{y_n})(x) \to T(S_y)(x)$ for all $x \in \mathbb{R}$, i.e., $T(S_y)(x)$ is continuous in y for every fixed $x \in \mathbb{R}$. Since

$$1 \leq T(\operatorname{Id})(x_1) \leq T(S_{x_1})(0)T(S_{-x_1})(x_1),$$

we have $T(S_{x_1})(0) \neq 0$. Using $T(S_0)(0) = T(\text{Id})(0) \geq 1$, the continuity of $T(S_y)(0)$ in y implies that $T(S_{x_1})(0) > 0$. Let $g := S_{x_1} \circ (-f) \circ S_{-x_1}$. Then $g(x_1) = x_1$, $\text{Im}(g) \subset J \subset I$, and

$$Tg(x_1) \leq T(S_{x_1})(0)T(-f)(0)T(S_{-x_1})(x_1) < -1,$$

using $T(-f)(0) < -1$ and $1 \leq T(S_{x_1})(0)T(S_{-x_1})(x_1)$.

(b) For the constant function c, $c \circ g = c$, hence $Tc(x) \leq Tc(g(x))Tg(x)$ for all $g \in C^1(\mathbb{R})$. By non-degeneration of T and (a), there are $g_1, g_2 \in C^1(\mathbb{R})$ with $g_j(x) = x$, $\text{Im}(g_j) \subset I$ ($j \in \{1, 2\}$), and $Tg_2(x) < -1$, $Tg_1(x) > 1$. Applying the previous inequality to $g = g_1, g_2$, we find $Tc(x) = 0$.

Now suppose $f \in C^1(\mathbb{R})$ satisfies $f|_I = c$. Let $x \in I$ and g_1, g_2 be as before. Since $f \circ g_j = c$, for any $x \in I$, we have $0 = Tc(x) \leq Tf(x)Tg_j(x)$, yielding $Tf(x) = 0$. Hence $Tf|_I = 0$.

(c) Assume that $f \in C^1(\mathbb{R})$ satisfies $f|_I = \text{Id}|_I$. Let $x \in I$ and choose g_1, g_2 as in part (b). Then $f \circ g_j = g_j$ ($j = 1, 2$) and

$$Tg_j(x) = T(f \circ g_j)(x) \leq Tf(x)Tg_j(x).$$

This inequality for g_1 yields $Tf(x) \geq 1$, the one for g_2 that

$$|Tg_2(x)| \geq Tf(x)|Tg_2(x)|, \quad Tf(x) \leq 1.$$

Hence, $Tf(x) = 1$, $Tf|_I = 1$.

(d) Assume that $f_1|_I = f_2|_I$ and that f_2 is invertible. Let $g := f_2^{-1} \circ f_1$. Then $g \in C^1(\mathbb{R})$ with $f_1 = f_2 \circ g$ and $g|_I = \text{Id}|_I$. By (c), $Tg|_I = 1$. Hence, for any $x \in I$, we have $g(x) = x$ and

$$Tf_1(x) = T(f_2 \circ g)(x) \leq Tf_2(x)Tg(x) = Tf_2(x).$$

Therefore, $Tf_1|_I \leq Tf_2|_I$. $\qquad\square$

Proof of Proposition 6.6. (i) Let $\mathcal{C} := \{f \in C^1(\mathbb{R}) \mid f \text{ is invertible and } f'(x) \neq 0$ for all $x \in \mathbb{R}\}$. For any open interval $I \subset \mathbb{R}$ and $f_1, f_2 \in \mathcal{C}$ with $f_1|_I = f_2|_I$ we have by Lemma 6.7(d) that $Tf_1|_I = Tf_2|_I$, i.e., localization on intervals. Replacing a function $f \in C^1(\mathbb{R})$ by its tangent line approximation on the right side of a point x, and f on the left side of x is an operation inside $\mathcal{C}$. Therefore, the method of the proof of Proposition 3.3 yields that there is a function $F : \mathbb{R}^2 \times (\mathbb{R} \setminus \{0\}) \to \mathbb{R}$ such that for all $f \in \mathcal{C}$ and all $x \in \mathbb{R}$,

$$Tf(x) = F(x, f(x), f'(x)).$$

(ii) We now consider functions $f \in C^1(\mathbb{R})$ which are not invertible. Suppose $I := (y_0, y_1)$ is an interval where f is strictly increasing with $f'(x) > 0$, $x \in I$ and $f'(y_0) = f'(y_1) = 0$ (or $y_0 = -\infty$, $f'(y_1) = 0$ or $f'(y_0) = 0$, $y_1 = \infty$, with obvious modifications in the following). For $\epsilon > 0$ sufficiently small, $f'(y_0 + \epsilon) > 0$, $f'(y_1 - \epsilon) > 0$ for all $0 < \epsilon \leq \epsilon_0$. Define $\widetilde{f} \in C^1(\mathbb{R})$ by

$$\widetilde{f}(x) := \begin{cases} f(y_0), & x \leq y_0, \\ f(x), & x \in I, \\ f(y_1), & x \geq y_1. \end{cases}$$

Then $\widetilde{f}'(y_0) = \widetilde{f}'(y_1) = 0$ and $\widetilde{f}$ is the limit of functions $\widetilde{f}_\epsilon \in \mathcal{C}$ in the sense that $\widetilde{f}_\epsilon \to \widetilde{f}$ and $\widetilde{f}'_\epsilon \to \widetilde{f}'$ converge uniformly on compacta. One may choose

$$\widetilde{f}_\epsilon(x) := \begin{cases} f(y_0 + \epsilon) + f'(y_0 + \epsilon)\big(x - (y_0 + \epsilon)\big), & x \leq y_0 + \epsilon, \\ f(x), & x \in (y_0 + \epsilon, y_1 - \epsilon), \\ f(y_1 - \epsilon) + f'(y_1 - \epsilon)\big(x - (y_1 - \epsilon)\big), & x \geq y_1 - \epsilon. \end{cases}$$

Note that $\widetilde{f}_\epsilon \in \mathcal{C}$ for any $0 < \epsilon \leq \epsilon_0$ since $\widetilde{f}_\epsilon$ is invertible with $\widetilde{f}'_\epsilon(x) > 0$ for all $x \in \mathbb{R}$. By (i) for any $x \in I_\epsilon := (y_0 + \epsilon, y_1 - \epsilon)$

$$T\widetilde{f}_\epsilon(x) = F\big(x, \widetilde{f}_\epsilon(x), \widetilde{f}'_\epsilon(x)\big) = F\big(x, f(x), f'(x)\big).$$

Since T is pointwise continuous, for any $x \in (y_0, y_1)$

$$T\widetilde{f}(x) = \lim_{\epsilon \to 0} T\widetilde{f}_\epsilon(x) = F\big(x, f(x), f'(x)\big).$$

By definition of $\widetilde{f}_\epsilon$, $f|_{I_\epsilon} = \widetilde{f}_\epsilon|_{I_\epsilon}$. Since $\widetilde{f}_\epsilon \in \mathcal{C}$, we have by Lemma 6.7(d) that $Tf(x) \leq T\widetilde{f}_\epsilon(x) = F(x, f(x), f'(x))$ for any $x \in I_\epsilon$. For $\epsilon \to 0$ this shows that

$$Tf(x) \leq T\widetilde{f}(x) = F\big(x, f(x), f'(x)\big), \quad x \in (y_0, y_1).$$

(iii) We now show the converse inequality $T\widetilde{f}(x) \leq Tf(x)$ for $x \in (y_0, y_1)$. We may write $\widetilde{f} = f \circ g$ where

$$g(x) = \begin{cases} y_0, & x \leq y_0, \\ x, & x \in (y_0, y_1), \\ y_1, & x \geq y_1. \end{cases}$$

If g were in $C^1(\mathbb{R})$, $g|_{(y_0, y_1)} = \text{Id}$, $Tg|_{(y_0, y_1)} = 1$ so that

$$T\widetilde{f}(x) \leq Tf(x)Tg(x) = Tf(x),$$

which would prove the claim. However, $g \notin C^1(\mathbb{R})$. Therefore, we approximate g

by smooth functions $g_\epsilon \in C^1(\mathbb{R})$. Let

$$
g_\epsilon(x) := \begin{cases}
y_0 + \frac{\epsilon}{2}, & x < y_0, \\
y_0 + \frac{\epsilon^2 + (x - y_0)^2}{2\epsilon}, & y_0 \leq x \leq y_0 + \epsilon, \\
x, & y_0 + \epsilon \leq x \leq y_1 - \epsilon, \\
y_1 - \frac{\epsilon^2 + (x - y_1)^2}{2\epsilon}, & y_1 - \epsilon \leq x \leq y_1, \\
y_1 - \frac{\epsilon}{2}, & x \geq y_1.
\end{cases}
$$

Then $g_\epsilon(y_1) = y_1 - \frac{\epsilon}{2}$, $g'_\epsilon(y_1) = 0$, $g_\epsilon(y_1 - \epsilon) = y_1 - \epsilon$, $g'_\epsilon(y_1 - \epsilon) = 1$, and similar equations hold for y_0 and $y_0 + \epsilon$ so that $g_\epsilon \in C^1(\mathbb{R})$ for any $\epsilon > 0$. Note that $f \circ g_\epsilon \to \widetilde{f}$, $(f \circ g_\epsilon)' \to \widetilde{f}'$ uniformly on compacta, with $f \circ g_\epsilon$, $\widetilde{f} \in C^1(\mathbb{R})$: Namely, we have $g'_\epsilon = 1$ in $(y_0 + \epsilon, y_1 - \epsilon)$ and $0 \leq g'_\epsilon \leq 1$ in $(y_1 - \epsilon, y_1)$, $g'_\epsilon = 0$ in (y_1, ∞). Since $g_\epsilon|_{I_\epsilon} = \mathrm{Id}\,|_{I_\epsilon}$, we have $Tg_\epsilon|_{I_\epsilon} = 1$ by Lemma 6.7(c). Thus by (6.1) for all $x \in I_\epsilon$

$$
T(f \circ g_\epsilon)(x) \leq Tf(g_\epsilon(x)) Tg_\epsilon(x) = Tf(x).
$$

Now the pointwise continuity of T implies for all $x \in (y_0, y_1)$

$$
T\widetilde{f}(x) = \lim_{\epsilon \to 0} T(f \circ g_\epsilon)(x) \leq Tf(x).
$$

Together with part (ii), we get

$$
T\widetilde{f}(x) = Tf(x) = F\big(x, f(x), f'(x)\big), \quad x \in (y_0, y_1).
$$

(iv) We now know that (6.4) holds for all $f \in C^1(\mathbb{R})$ and all open intervals (y_0, y_1) of strict monotonicity of f. On intervals J where f is constant, $Tf|_J = 0$ by Lemma 6.7(b), and $F(x, y, 0) = 0$ is a result of continuity arguments like $\lim_{\epsilon \to 0} T\widetilde{f}_\epsilon(x) = T\widetilde{f}(x)$ for boundary points of J together with $Tf|_J = 0$. Equation (6.4) then means $0 = 0$. Equation (6.4) similarly extends to limit points of intervals of monotonicity of f or to limit points of intervals of constancy of f. Hence (6.4) holds for all $f \in C^1(\mathbb{R})$ and all $x \in I$. □

Proof of Theorem 6.1. (a) By Proposition 6.6 there is $F : \mathbb{R}^3 \to \mathbb{R}$ such that for all $f \in C^1(\mathbb{R})$, $x \in \mathbb{R}$,

$$
Tf(x) = F\big(x, f(x), f'(x)\big).
$$

The chain rule inequality is equivalent to the functional inequality for F,

$$
F(x, z, \alpha\beta) \leq F(y, z, \alpha) F(x, y, \beta) \tag{6.5}
$$

for all $x, y, z, \alpha, \beta \in \mathbb{R}$. Just choose $f, g \in C^1(\mathbb{R})$ with $g(x) = y$, $f(y) = z$, $g'(x) = \beta$, $f'(y) = \alpha$. The equations $Tc = 0$, $T(\mathrm{Id}) = 1$ imply that

$$
F(x, y, 0) = 0, \quad F(x, x, 1) = 1. \tag{6.6}
$$

Note that $F(x, y, 1) = T(S_{y-x})(x)$ where $S_{y-x} = \mathrm{Id} + (y - x)$ is the shift by $y - x$. Since $T(S_{y-x})(x)$ depends continuously on $y - x$, cf. the proof of (a) of Lemma 6.7, and since by (6.5) and (6.6)

$$1 = F(x, x, 1) \leq F(y, x, 1)F(x, y, 1),$$

we first get that $F(x, y, 1) \neq 0$ and then $F(x, y, 1) > 0$ for all $x, y \in \mathbb{R}$. We showed in the proof of (a) of Lemma 6.7 that $T(-\mathrm{Id})(0) \leq -1$. Hence, $F(0, 0, -1) = T(-\mathrm{Id})(0) \leq -1$ and for any $x \in \mathbb{R}$

$$F(x, x, -1) \leq F(0, x, 1)F(0, 0, -1)F(x, 0, 1) \leq -1$$

using $1 = F(0, 0, 1) \leq F(0, x, 1)F(x, 0, 1)$.

(b) Fix $x_0 \in \mathbb{R}$ and put $K(\alpha) := F(x_0, x_0, \alpha)$ for $\alpha \in \mathbb{R}$. By (6.5) for $x = y = z = x_0$, K is submultiplicative on $\mathbb{R}$ with $K(-1) < 0 < K(1)$. Further K is continuous as implied by the pointwise continuity of T: Assume $\alpha_n \to \alpha$ in $\mathbb{R}$. Consider $f_n(x) := \alpha_n(x - x_0) + x_0$, $f(x) := \alpha(x - x_0) + x_0$. Then $f_n(x_0) = f(x_0) = x_0$ and $f'_n(x) = \alpha_n \to \alpha = f'(x)$. Hence, $f_n \to f$, $f'_n \to f'$ converge uniformly on compacta and therefore $Tf_n(x_0) \to Tf(x_0)$, which means

$$K(\alpha_n) = F(x_0, x_0, \alpha_n) = Tf_n(x_0) \to Tf(x_0) = F(x_0, x_0, \alpha) = K(\alpha).$$

Theorem 6.2 yields that there are $p(x_0) > 0$ and $A(x_0) = |F(x_0, x_0, -1)| \geq 1$ such that

$$K(\alpha) = \begin{cases} \alpha^{p(x_0)}, & \alpha \geq 0, \\ -A(x_0)|\alpha|^{p(x_0)}, & \alpha < 0. \end{cases} \tag{6.7}$$

For any $x, y, z \in \mathbb{R}$ by (6.5)

$$F(x, x, \alpha) \leq F(z, x, 1)F(z, z, \alpha)F(x, z, 1) = d(x, z)F(z, z, \alpha),$$

where $d(x, z) := F(z, x, 1)F(x, z, 1) \geq 1$ is a number independent of α. Fixing x, z with $x \neq z$, we have for all $\alpha > 0$ that $\alpha^{p(x)-p(z)} \leq d(x, z)$. If $p(x) \neq p(z)$, we would get a contradiction for either $\alpha \to 0$ or for $\alpha \to \infty$. Hence, the exponent $p := p(x)$ is independent of $x \in \mathbb{R}$.

(c) We next analyze the form of $F(x, z, \alpha)$ for $x \neq z$. Let $x, z \in \mathbb{R}$, $x \neq z$. By (6.5) and (6.7) for all $\alpha > 0$, $\beta \in \mathbb{R}$,

$$F(x, z, \alpha\beta) \leq F(x, z, \beta)F(x, x, \alpha) = \alpha^p F(x, z, \beta)$$

and

$$F(x, z, \beta) \leq F(x, z, \alpha\beta)F\left(x, x, \frac{1}{\alpha}\right) = \frac{1}{\alpha^p}F(x, z, \alpha\beta).$$

Therefore,

$$F(x, z, \alpha\beta) \leq \alpha^p F(x, z, \beta) \leq F(x, z, \alpha\beta),$$

and we have equality $F(x, z, \alpha\beta) = \alpha^p F(x, z, \beta)$. Putting here $\beta = 1$ and $\beta = -1$, we find that

$$F(x, z, \alpha) = \begin{cases} F(x, z, 1)\alpha^p, & \alpha \geq 0, \\ F(x, z, -1)|\alpha|^p, & \alpha < 0. \end{cases} \tag{6.8}$$

We know that $F(x, z, 1) > 0$. On the other hand,

$$F(x, z, -1) \leq F(0, z, 1)F(0, 0, -1)F(x, 0, 1) < 0.$$

Let $c_\pm(x, z) := F(x, z, \pm 1)$ and $a(x, z) := |c_-(x, z)|/c_+(x, z)$. Since

$$c_-(x, z) = F(x, z, -1) \leq F(x, z, 1)F(x, x, -1) \leq -F(x, z, 1) = -c_+(x, z),$$

we have $a(x, z) \geq 1$ for all $x, z \in \mathbb{R}$. Choose $\alpha, \beta \in \{+1, -1\}$ in (6.5) to find that

$$c_+(x, z) \leq c_+(y, z)c_+(x, y),$$
$$c_-(x, z) \leq c_-(y, z)c_+(x, y) \quad \text{and}$$
$$c_-(x, z) \leq c_+(y, z)c_-(x, y).$$

Using these inequalities and $c_-(x, z) < 0$, we get

$$c_+(x, z) \max\big(a(y, z), a(x, y)\big) \leq c_+(y, z)c_+(x, y) \max\big(a(y, z), a(x, y)\big)$$
$$= \max\big(|c_-(y, z)|c_+(x, y), c_+(y, z)|c_-(x, y)|\big)$$
$$\leq |c_-(x, z)| = c_+(x, z)a(x, z). \tag{6.9}$$

Since $c_+(x, z) > 0$, this implies for all $x, y, z \in \mathbb{R}$ that $\max(a(y, z), a(x, y)) \leq a(x, z)$, which yields $a(x, y) \leq a(x, 0) \leq a(0, 0)$ and $a(0, 0) \leq a(x, 0) \leq a(x, y)$. Therefore, a is constant, $a(x, y) = a(0, 0)$ for all $x, y \in \mathbb{R}$. Let $A := a(0, 0)$. Then $A \geq 1$ and $c_-(x, z) = -Ac_+(x, z)$. Since we now have equalities everywhere in (6.9), we conclude $c_+(x, z) = c_+(y, z)c_+(x, y)$. For $y = 0$, $c_+(x, z) = c_+(0, z)c_+(x, 0)$, $1 = c_+(x, x) = c_+(0, x)c_+(x, 0)$. Put $H(x) := c_+(0, x)$. Then $H > 0$ and $c_+(x, z) = \frac{H(z)}{H(x)}$. Hence, by (6.8),

$$F(x, z, \alpha) = \begin{cases} \frac{H(z)}{H(x)}\alpha^p, & \alpha \geq 0, \\ -A\frac{H(z)}{H(x)}|\alpha|^p, & \alpha < 0. \end{cases}$$

Note that $H(z) = F(0, z, 1) = T(S_z)(0)$ depends continuously on z. Finally, using (6.4), we have

$$Tf(x) = \begin{cases} \frac{H \circ f(x)}{H(x)} f'(x)^p & f'(x) \geq 0 \\ -A\frac{H \circ f(x)}{H(x)} |f'(x)|^p & f'(x) < 0 \end{cases}; \qquad f \in C^1(\mathbb{R}), \ x \in \mathbb{R}.$$

This ends the proof of Theorem 6.1. $\qquad\qquad\qquad\qquad\qquad\qquad\qquad\qquad\qquad\quad$ □

The proof of Theorem 6.3 is similar to the one of Theorem 6.1.

6.4   Rigidity of the chain rule

In Theorem 5.8 we showed that the chain rule is rigid: the perturbed chain rule equation

$$T(f \circ g)(x) - Tf \circ g(x) \cdot Tg(x) = B\big(x, f \circ g(x), g(x)\big) \qquad (6.10)$$

under weak conditions implies that $B \equiv 0$ and that (6.10) has the same solutions as the unperturbed chain rule. We now consider an extension of (6.10) and study the more general inequality

$$\big|T(f \circ g)(x) - Tf \circ g(x) \cdot Tg(x)\big| \le B\big(x, f \circ g(x), g(x)\big). \qquad (6.11)$$

Theorem 5.8 required no continuity assumption on T. Since (6.11) allows more freedom than (6.10), we need a stronger condition of non-degeneration of T to solve (6.11). We also assume that T is pointwise continuous.

Definition. An operator $T : C^1(\mathbb{R}) \to C(\mathbb{R})$ is *strongly non-degenerate* provided that, for all open intervals $I \subset \mathbb{R}$, all $x \in I$ and all $t > 0$, there are functions $f_1, f_2 \in C^1(\mathbb{R})$ with $f_1(x) = f_2(x) = x$, $\mathrm{Im}(f_1) \subset I$, $\mathrm{Im}(f_2) \subset I$, and $Tf_1(x) > t$, $Tf_2(x) < -t$.

Note that the model chain rule equality has derivative-type solutions, and then these assumptions are clearly satisfied.

We then have the following rigidity result for the chain rule.

Theorem 6.8 (Strong rigidity of the chain rule). *Assume that $T : C^1(\mathbb{R}) \to C(\mathbb{R})$ is strongly non-degenerate and pointwise continuous. Suppose there is a function $B : \mathbb{R}^3 \to \mathbb{R}$ such that T satisfies*

$$\big|T(f \circ g)(x) - Tf \circ g(x) \cdot Tg(x)\big| \le B\big(x, f \circ g(x), g(x)\big). \qquad (6.11)$$

for all $f, g \in C^1(\mathbb{R})$, $x \in \mathbb{R}$. Assume also that there is $x_0 \in \mathbb{R}$ such that $T(-\mathrm{Id})(x_0) < 0$. Then (6.11) has the same solutions as the unperturbed chain rule, i.e., B can be chosen to be zero: There is $p > 0$ and a function $H \in C(\mathbb{R})$, $H > 0$, such that

$$Tf(x) = \frac{H \circ f(x)}{H(x)} |f'(x)|^p \operatorname{sgn} f'(x), \quad f \in C^1(\mathbb{R}), \ x \in \mathbb{R}.$$

The proof of this theorem relies on the follow localization result.

Proposition 6.9. *Under the assumptions of Theorem 6.8, there is a function $F : \mathbb{R}^3 \to \mathbb{R}$ such that, for all $f \in C^1(\mathbb{R})$ and all $x \in \mathbb{R}$,*

$$Tf(x) = F\big(x, f(x), f'(x)\big).$$

Proof. Using Proposition 3.3, it suffices to show that for any open interval $I \subset \mathbb{R}$ and $f_1, f_2 \in C^1(\mathbb{R})$ with $f_1|_I = f_2|_I$ we have $Tf_1|_I = Tf_2|_I$. Let $x \in I$. Since T

is strongly non-degenerate, we may choose functions $g_n \in C^1(\mathbb{R})$ with $g_n(x) = x$, $\text{Im}(g_n) \subset I$ and $\lim_{n \to \infty} Tg_n(x) = \infty$. Then by (6.11)

$$-B\big(x, f_1(x), x\big) \leq T(f_1 \circ g_n)(x) - Tf_1(x) \cdot Tg_n(x) \leq B\big(x, f_1(x), x\big).$$

Since $\lim_{n \to \infty} \frac{B(x, f_1(x), x)}{Tg_n(x)} = 0$, we get by dividing the previous inequalities by $Tg_n(x)$ that

$$Tf_1(x) = \lim_{n \to \infty} \frac{T(f_1 \circ g_n)(x)}{Tg_n(x)},$$

where the limit exists. Note that $f_1 \circ g_n = f_2 \circ g_n$ since $f_1|_I = f_2|_I$. Therefore, $Tf_1(x) = Tf_2(x)$ and consequently $Tf_1|_I = Tf_2|_I$. □

Using Proposition 6.9, the operator chain rule inequality (6.11) for T is equivalent to the functional inequality for F:

$$\big|F(x, z, \alpha\beta) - F(y, z, \alpha)F(x, y, \beta)\big| \leq B(x, z, y), \tag{6.12}$$

for all $x, y, z, \alpha, \beta \in \mathbb{R}$. For $x = y = z$ and $\phi_x := F(x, x, \cdot)$, $d_x := B(x, x, x)$, this means

$$\big|\phi_x(\alpha\beta) - \phi_x(\alpha)\phi_x(\beta)\big| \leq d_x. \tag{6.13}$$

Since T is strongly non-degenerate, $\overline{\lim}_{\alpha \in \mathbb{R}} \phi_x(\alpha) = \infty$, $\underline{\lim}_{\alpha \in \mathbb{R}} \phi_x(\alpha) = -\infty$. Actually, we can show that $\overline{\lim}_{\alpha \to \infty} \phi_x(\alpha) = \infty$, $\underline{\lim}_{\alpha \to -\infty} \phi_x(\alpha) = -\infty$, cf. [KM10]. The pointwise continuity of T implies that $\phi_x : \mathbb{R} \to \mathbb{R}$ is continuous. These facts suffice to show that the *nearly multiplicative* function ϕ_x is actually multiplicative:

Proposition 6.10. *Suppose that* $\phi : \mathbb{R} \to \mathbb{R}$ *is continuous with* $\overline{\lim}_{\alpha \to \infty} \phi_x(\alpha) = \infty$ *and* $\underline{\lim}_{\alpha \to -\infty} \phi_x(\alpha) = -\infty$. *Assume also that there is* $d \in \mathbb{R}$ *such that for all* $\alpha, \beta \in \mathbb{R}$

$$\big|\phi(\alpha\beta) - \phi(\alpha)\phi(\beta)\big| \leq d. \tag{6.14}$$

Then ϕ *is multiplicative, i.e.,* d *may be chosen zero, and there is* $p > 0$ *such that*

$$\phi(\alpha) = |\alpha|^p \operatorname{sgn} \alpha.$$

Proof. Choose $\beta_n \in \mathbb{R}$ such that $0 < \phi(\beta_n) \to \infty$. Then by (6.14)

$$\left| \frac{\phi(\alpha\beta_n)}{\phi(\beta_n)} - \phi(\alpha) \right| \leq \frac{d}{\phi(\beta_n)} \to 0,$$

and hence $\phi(\alpha) = \lim_{n \to \infty} \frac{\phi(\alpha\beta_n)}{\phi(\beta_n)}$, where the limit exists for all $\alpha \in \mathbb{R}$. In particular, $\phi(0) = 0$, $\phi(1) = 1$. We conclude that for any $\alpha, \gamma \in \mathbb{R}$

$$\phi(\alpha)\phi(\gamma) = \lim_{n \to \infty} \frac{\phi(\alpha\beta_n)}{\phi(\beta_n)} \frac{\phi(\gamma\beta_n)}{\phi(\beta_n)}.$$

Now $\phi(\alpha\,\beta_n)\phi(\gamma\,\beta_n) \leq \phi(\alpha\,\gamma\,\beta_n^2) + d$ and $\phi(\beta_n)\phi(\beta_n) \geq \phi(\beta_n^2) - d$. Hence

$$\phi(\alpha)\phi(\gamma) \leq \lim_{n\to\infty} \frac{\phi(\alpha\gamma\beta_n^2) + d}{\phi(\beta_n^2) - d} = \lim_{n\to\infty} \frac{\phi(\alpha\gamma\beta_n^2)}{\phi(\beta_n^2)} = \phi(\alpha\gamma),$$

since $\phi(\beta_n^2) \to \infty$, too, in view of $|\phi(\beta_n)^2 - \phi(\beta_n^2)| \leq d$. Similarly $\phi(\alpha\gamma) \geq \phi(\alpha)\phi(\gamma)$. Therefore ϕ is multiplicative and continuous, with negative values for $\alpha \to -\infty$. Proposition 2.3 implies that there is $p > 0$ such that $\phi(\alpha) = |\alpha|^p \, \mathrm{sgn}\,\alpha$. $\qquad\square$

Proof of Theorem 6.8. By Proposition 6.9, $Tf(x) = F(x, f(x), f'(x))$, where F satisfies (6.12). We analyze the form of F. By Proposition 6.10 and (6.13), there is $p(x) > 0$ such that $F(x, x, \alpha) = \phi_x(\alpha) = \alpha^{p(x)}$ for any $\alpha > 0$. For $x \neq z$, by choosing successively $y = x$ and $y = z$ in (6.12), we find

$$\left| F(x, z, \alpha\beta) - F(x, z, \alpha)\beta^{p(x)} \right| \leq B(x, z, x), \quad \beta > 0, \ \alpha \in \mathbb{R},$$

and

$$\left| F(x, z, \alpha\beta) - \alpha^{p(z)} F(x, z, \beta) \right| \leq B(x, z, z), \quad \alpha > 0, \ \beta \in \mathbb{R}.$$

Exchange α and β in the first inequality. Then the triangle inequality yields $|\alpha^{p(x)} - \alpha^{p(z)}||F(x, z, \beta)| \leq B(x, z, x) + B(x, z, z)$ for any $\alpha > 0$, $\beta \in \mathbb{R}$. This obviously implies $p(x) = p(z)$ for $\beta = 1$, $\alpha \to \infty$, since $F(x, z, 1) \neq 0$, which is an easy consequence of (6.12). Thus for any $x, \alpha \in \mathbb{R}$

$$F(x, x, \alpha) = |\alpha|^p \, \mathrm{sgn}\,\alpha$$

with $p := p(x) = p(z) > 0$. Since $\frac{B(x,z,x)}{\beta^p} \to 0$ for $\beta \to \infty$, we also conclude for all $\alpha \in \mathbb{R}$

$$F(x, z, \alpha) = \lim_{\beta\to\infty} \frac{F(x, z, \alpha\beta)}{\beta^p}. \qquad (6.15)$$

For any $\alpha > 0$, $\alpha\beta \to \infty$ if $\beta \to \infty$, and therefore

$$F(x, z, \alpha) = \lim_{\beta\to\infty} \frac{F(x, z, \alpha\beta)}{\beta^p} = \alpha^p \lim_{\alpha\beta\to\infty} \frac{F(x, z, \alpha\beta)}{(\alpha\beta)^p} = \alpha^p F(x, z, 1)$$

for any $x, z \in \mathbb{R}$. For $\alpha < 0$ we have

$$F(x, z, \alpha) = \lim_{\beta\to\infty} \frac{F(x, z, \alpha\beta)}{\beta^p} = |\alpha|^p \lim_{|\alpha|\beta\to\infty} \frac{F(x, z, -|\alpha|\beta)}{|\alpha|^p\beta^p} = |\alpha|^p F(x, z, -1).$$

Dividing (6.12) by $(\alpha\beta)^p$ for $\alpha, \beta > 0$, we get

$$\left| \frac{F(x, z, \alpha\beta)}{\alpha^p\beta^p} - \frac{F(y, z, \alpha)}{\alpha^p} \frac{F(x, y, \beta)}{\beta^p} \right| \leq \frac{B(x, z, y)}{\alpha^p\beta^p}.$$

By (6.15), this implies $F(x, z, 1) = F(y, z, 1)F(x, y, 1)$ for all $x, y, z \in \mathbb{R}$. For $\alpha \to \pm\infty$, $\beta \to \mp\infty$, a similar argument yields that for all $x, y, z \in \mathbb{R}$, $F(x, z, -1) =$

$F(y, z, 1)F(x, y, -1) = F(y, z, -1)F(x, y, 1)$. Since $F(x, x, 1) = \phi_x(1) = 1^p = 1$ for all $x \in \mathbb{R}$, $1 = F(y, x, 1)F(x, y, 1)$ for all $x, y \in \mathbb{R}$. Let $H(y) := F(0, y, 1)$. Then $F(y, 0, 1) = \frac{1}{H(y)}$ and

$$F(x, z, 1) = F(0, z, 1)F(x, 0, 1) = \frac{H(z)}{H(y)},$$

$$F(x, z, -1) = F(z, z, -1)F(x, z, 1) = -F(x, z, 1) = -\frac{H(z)}{H(y)},$$

using that $F(z, z, -1) = \phi_z(-1) = -1$. We conclude that

$$F(x, z, \alpha) = \frac{H(z)}{H(x)}|\alpha|^p \operatorname{sgn}\alpha, \quad x, z, \alpha \in \mathbb{R}.$$

Note that $H(y) = F(0, y, 1) = T(S_y)(0)$ depends continuously on y, where S_y denotes as before the shift by y. Using Proposition 6.9, we get

$$Tf(x) = F\big(x, f(x), f'(x)\big) = \frac{H \circ f(x)}{H(x)}|f'(x)|^p \operatorname{sgn} f'(x),$$

for any $f \in C^1(\mathbb{R})$, $x \in \mathbb{R}$. This solves the chain rule operator equation, so that B in (6.11) can be chosen to be zero, and proves Theorem 6.8. $\qquad\square$

We now turn to a further extension of the chain rule, the one-sided perturbed chain rule inequality. Let $B : \mathbb{R}^3 \to \mathbb{R}$ be a function and $T : C^1(\mathbb{R}) \to C(\mathbb{R})$ be an operator satisfying

$$T(f \circ g)(x) - Tf \circ g(x) \cdot Tg(x) \le B\big(x, f \circ g(x), g(x)\big), \qquad (6.16)$$

for all $f \in C^1(\mathbb{R})$, $x \in \mathbb{R}$. This is more general than the two-sided inequality considered in Theorem 6.8, and also more general than the one-sided chain rule inequality considered in Theorem 6.1.

In the results proved so far, the operator T was localized. The operator inequality (6.16), however, is too general that localization could always be shown, even under strong non-degeneration and continuity assumptions on T. We provide an example.

Example. Let $H \in C(\mathbb{R})$ be a non-constant function with $4 \le H \le 5$. For $f \in C^1(\mathbb{R})$, $x \in \mathbb{R}$, with $f'(x) \in (-1, 0)$, let $I_{f,x}$ denote the interval $I_{f,x} := [x + f'(x)(1 + f'(x)), x]$. Then $0 < |I_{f,x}| \le 1/4$. Let $Jf(x) := \frac{1}{|I_{f,x}|}\int_{I_{f,x}} f(y)dy$ denote the average of f in $I_{f,x}$. Define an operator $T : C^1(\mathbb{R}) \to C(\mathbb{R})$ by putting, for any $f \in C^1(\mathbb{R})$, $x \in \mathbb{R}$,

$$Tf(x) := \begin{cases} \frac{H \circ f(x)}{H(x)}f'(x), & f'(x) \ge 0, \\[2mm] \frac{H \circ f(x)}{H(x)}4f'(x), & f'(x) \le -2, \\[2mm] \frac{H \circ f(x)}{H(x)}\left(7 + \frac{15}{2}f'(x)\right), & -2 < f'(x) \le -1, \\[2mm] \frac{H \circ Jf(x)}{H(x)}\frac{1}{2}f'(x), & -1 < f'(x) < 0. \end{cases}$$

Then T satisfies, for all $f, g \in C^1(\mathbb{R})$, $x \in \mathbb{R}$,

$$T(f \circ g)(x) - Tf \circ g(x) \cdot Tg(x) \leq 5. \qquad (6.17)$$

Obviously T is not localized since it depends on the integral average Jf if $f'(x) \in (-1, 0)$. Note here that for $f'(x) \geq 0$ or $f'(x) \leq -2$, $Tf(x)$ has the form given in Theorem 6.1 for $B = 0$, with $p = 1$, $A = 4$. For $-2 < \alpha = f'(x) < 0$ there is a continuous perturbation of the line 4α by $\frac{1}{2}\alpha$ if $\alpha \in (-1, 0)$ and by $7 + \frac{15}{2}\alpha$ if $\alpha \in (-2, -1]$. Note that Tf is continuous for all $f \in C^1(\mathbb{R})$: if $x_n \in \mathbb{R}$ are such that $f'(x_n) \in (-1, 0)$ and $x_n \to x$ with $f'(x_n) \to -1$ or $f'(x_n) \to 0$, then $Jf(x_n) \to f(x)$ since $|I_{f, x_n}| \to 0$.

To prove (6.17), use $\frac{4}{5} \leq \frac{H(z)}{H(y)} \leq \frac{5}{4}$ for all $y, z \in \mathbb{R}$, and distinguish the following cases: (1) $\alpha, \beta \geq 0$; (2) $\alpha, \beta \leq -2$; (3) $\alpha, \beta \in (-2, 0)$; (4) $\alpha \leq -2$, $\beta \in (-2, -1]$; (5) $\alpha \leq -2$, $\beta \in (-1, 0)$; (6) $\alpha > 0$, $\alpha\beta \leq -2$; (7) $\alpha > 0$, $\alpha\beta \in (-2, -1]$; (8) $\alpha > 0$, $\alpha\beta \in (-1, 0)$. The estimates to show (6.17) are easy in each case but a bit tedious. They can be found in detail in [KM10]. $\qquad \square$

Assuming localization in addition to (6.16), i.e., that there is a function $F : \mathbb{R}^3 \to \mathbb{R}$ such that for all $f \in C^1(\mathbb{R})$, $x \in \mathbb{R}$,

$$Tf(x) = F\big(x, f(x), f'(x)\big)$$

holds, the operator inequality (6.16) is equivalent to the functional inequality

$$F(x, z, \alpha\beta) \leq F(y, z, \alpha)F(x, z, \beta) + B(x, z, y) \qquad (6.18)$$

for all $x, y, z, \alpha, \beta \in \mathbb{R}$. Similar to the two-sided case in (6.12), the most important special case to solve is the one of $x = y = z$ which means

$$\phi_x(\alpha\beta) \leq \phi_x(\alpha)\phi_x(\beta) + d_x,$$

where $\phi_x := F(x, x, \cdot)$ and $d_x := B(x, x, x)$. We have the following result on these *nearly submultiplicative functions.*

Theorem 6.11. *Let $\phi : \mathbb{R} \to \mathbb{R}$ be continuous with $\overline{\lim}_{\alpha \to \infty} \phi(\alpha) = \infty$. Suppose that there is $\alpha_0 < 0$ with $\phi(\alpha_0) < 0$ and that there is $d \in \mathbb{R}$ such that we have for all $\alpha, \beta \in \mathbb{R}$*

$$\phi(\alpha\beta) \leq \phi(\alpha)\phi(\beta) + d.$$

Then $d \geq 0$ and there are $p > 0$ and $A \geq 1$ such that for all $\alpha > 0$

$$\phi(\alpha) = \alpha^p, \quad -A\alpha^p \leq \phi(-\alpha) \leq \min\left(-\tfrac{1}{A}\alpha^p, -A\alpha^p + d\right).$$

Moreover, the limit $\lim_{\alpha \to \infty} \frac{\phi(-\alpha)}{-\alpha^p}$ exists and $A = \lim_{\alpha \to \infty} \frac{\phi(-\alpha)}{-\alpha^p}$.

Remarks. For $d \neq 0$, $\phi|_{\mathbb{R}_{<0}}$ is not of power type, but is close to the power-type function $-A|\alpha|^p$ for large $|\alpha|$, $\alpha < 0$. Interestingly enough, $\phi|_{\mathbb{R}_{\geq 0}}$ is of power type α^p. For $p = 1$, $A = 2$, the function

$$\phi(\alpha) := \begin{cases} \alpha, & \alpha \geq 0, \\ \frac{1}{2}\alpha, & \alpha \in [-1,0), \\ 3 + \frac{7}{2}\alpha, & \alpha \in [-2,-1), \\ 2\alpha, & \alpha < -2 \end{cases}$$

provides an explicit example satisfying the assumptions of Theorem 6.11 with $d = \frac{3}{2}$, $\phi(\alpha\beta) \leq \phi(\alpha)\phi(\beta) + \frac{3}{2}$, which is not of power type on $\mathbb{R}_{<0}$.

The proof of Theorem 6.11 is an asymptotic modification of the one of Theorem 6.2 for submultiplicative functions when $d = 0$. We only provide the essential steps of the proof, which are:

(a) Show $\lim_{\alpha \to \infty} \phi(\alpha) = \infty$, $\lim_{\alpha \to -\infty} \phi(\alpha) = -\infty$.

(b) Choose $b > 1$ close to 1. Let $\phi_1 := b\phi$, $\phi_2 := \frac{1}{b}\phi$. Then for large $\gamma_0 = \gamma_0(b)$, $\phi_1(\alpha\beta) \leq \phi_1(\alpha)\phi_1(\beta)$ if $\alpha\beta \geq \gamma_0$ and $\phi_2(\alpha\beta) \leq \phi_2(\alpha)\phi_2(\beta)$ if $\alpha\beta \leq -\gamma_0$: ϕ_1 and ϕ_2 are submultiplicative for large $\alpha\beta$ in the positive, respectively negative range.

(c) Define $f(t) := \ln \phi_1(\exp(t))$, $g(t) := \ln |\phi_2(-\exp(t))|$ for $t \in \mathbb{R}$. Then, for $t_0 := \ln \gamma_0$,

$$0 \leq p := \inf_{t \geq t_0} \frac{f(t)}{t} = \lim_{t \to \infty} \frac{f(t)}{t} < \infty, \quad f(t) = pt + a(t).$$

(d) For $t, s \in \mathbb{R}$ with $t + s \geq t_0$,

$$g(t+s) \geq g(t) + f(s) - 2\ln b.$$

Using $0 \leq f(t) + f(-t)$ for all $t \in \mathbb{R}$, show that

$$\big| g(t) - [c + pt + a(t - t_0)] \big| \leq 2\ln b,$$

for all $t \geq t_0$, where $c := g(t_0) - pt_0$, and a satisfies $\lim_{t \to \infty} \frac{a(t)}{t} = 0$ with $a(t) \geq 0$ for $t \geq t_0$.

(e) Improve the bound for a to $a(t) \leq 6\ln b$ for all $t \geq t_1$, for a suitable $t_1 \geq t_0$. Then $\phi_1(\alpha) = \alpha^p \exp(a(\ln \alpha))$, $\alpha > 0$ is asymptotically α^p for large α, if b is close to 1 and thus $\ln b$ is close to 0. Further $\phi_2(\alpha) \simeq -A|\alpha|^p \exp\big(a(\ln|\frac{\alpha}{\alpha_1}|)\big)$, $\alpha_1 := \exp(t_1)$, for large negative α.

(f) Use $\phi = \frac{1}{b}\phi_1 = b\phi_2$, take the limit as $b \to 1$ and prove that $\phi(\alpha) = \alpha^p$ for $\alpha > 0$ and $\lim_{\alpha \to -\infty} \frac{\phi(\alpha)}{-|\alpha|^p} = A$.

We do not give the details here, but refer to [KM10].

Using Theorem 6.11, we may prove the following result on the one-sided perturbed chain rule inequality, assuming localization which cannot be guaranteed otherwise, as shown by the previous example.

Theorem 6.12. *Assume that $T : C^1(\mathbb{R}) \to C(\mathbb{R})$ is strongly non-degenerate, pointwise continuous, and that there is $x_0 \in \mathbb{R}$ with $T(-\mathrm{Id})(x_0) < 0$. Suppose that there is a function $B : \mathbb{R}^3 \to \mathbb{R}$ such that the perturbed chain rule inequality*

$$T(f \circ g)(x) \le Tf \circ g(x) \cdot Tg(x) + B\big(x, f \circ g(x), g(x)\big)$$

holds for all $f, g \in C^1(\mathbb{R})$, $x \in \mathbb{R}$. Assume also that there is $F : \mathbb{R}^3 \to \mathbb{R}$, so that

$$Tf(x) = F\big(x, f(x), f'(x)\big), \quad f \in C^1(\mathbb{R}), \ x \in \mathbb{R}.$$

Then there are $p > 0$, $A \ge 1$, $H \in C(\mathbb{R})$, $H > 0$ and a function $K : \mathbb{R}^2 \times \mathbb{R}_{<0} \to \mathbb{R}_{<0}$ which is continuous in the second and the third variable satisfying

$$-A\alpha^p \le K(x, z, -\alpha)$$

$$\le \min\left(-\frac{1}{A}\alpha^p, -A\alpha^p + \frac{H(x)}{H(z)}\min\big[B(x, z, x), B(x, z, z)\big]\right)$$

for all $x, z \in \mathbb{R}$, $\alpha > 0$, and for which $A = \lim_{\beta \to \infty} \frac{K(x, z, -\beta)}{-\beta^p}$ exists for all $x, z \in \mathbb{R}$, the limit A being independent of x and z, such that for all $f \in C^1(\mathbb{R})$ and $x \in \mathbb{R}$

$$Tf(x) = \begin{cases} \frac{H \circ f(x)}{H(x)} f'(x)^p, & f'(x) \ge 0, \\ \frac{H \circ f(x)}{H(x)} K\big(x, f(x), f'(x)\big), & f'(x) < 0. \end{cases}$$

The property of K means that for negative values of $f'(x)$, $Tf(x)$ is reasonably close to $-A\frac{H \circ f(x)}{H(x)}|f'(x)|^p$, deviating from this value by at most

$$\min[B(x, f(x), x), B(x, f(x), f(x))],$$

i.e., deviating by at most this amount from the solution in Theorem 6.1 for $B = 0$.

For the proof of Theorem 6.12 we refer to [KM10]. Theorem 6.11 has an analogue for nearly supermultiplicative functions $\phi(\alpha\beta) \ge \phi(\alpha)\phi(\beta) - d$ and Theorem 6.12 has an analogue for the perturbed supermultiplicative operator inequality

$$T(f \circ g)(x) \ge Tf \circ g(x) \cdot Tg(x) - B\big(x, f \circ g(x), g(x)\big).$$

6.5   Notes and References

The result on the chain rule inequality, Theorem 6.1 and Proposition 6.6 on the localization of the operator T were shown by König and Milman in [KM9]. Theorem 6.2 on submultiplicative functions on $\mathbb{R}$ is also found in [KM9].

The proof of Lemma 6.5 on subadditive functions on $\mathbb{R}$ follows Hille, Phillips, [HP, Chap. VII]. If additionally in this lemma f is continuous at 0 and $f(0) = 0$ holds, f is continuous at any $t \in \mathbb{R}$, cf. also Hille, Phillips, [HP]. A simpler variant of Lemma 6.5 for sequences goes back to Fekete [Fe], p. 233, cf. also Pólya, Szegö [PS], Problem 98.

The rigidity result Theorem 6.8 for the is taken from [KM10]. Theorems 6.11 and 6.12 as well as the example before Theorem 6.11 are shown in [KM10], too.

Submultiplicative maps may not only be considered on the real line as in Theorem 6.2, but also on function spaces like $C^k(I)$. But, as in the case of multiplicative operators, cf. [M], [MS], [AAFM] or [AFM], one assumes that the mapping is bijective. Let us call $T : C^k(I) \to C^k(I)$ C^k-*pointwise continuous* provided that for all sequences $f_n \in C^k(I)$ and all $f \in C^k(I)$ with $f_n^{(j)} \to f^{(j)}$ converging uniformly on compact subsets of I for all $j \in \{0, \dots, k\}$ we have that $Tf_n(x) \to Tf(x)$ converges for all $x \in I$. Then the following result holds for submultiplicative operators, cf. Faifman, König and Milman [FKM]:

Proposition 6.13. *Let $I \subset \mathbb{R}$ be open and $k \in \mathbb{N}_0$. Suppose that $T : C^k(I) \to C^k(I)$ is bijective, C^k-pointwise continuous and submultiplicative, i.e.,*

$$T(f \cdot g)(x) \leq Tf(x) \cdot Tg(x), \quad f, g \in C^k(I), \ x \in I. \tag{6.19}$$

Assume also that $T(-1) < 0$ and that $Tf \geq 0$ holds if and only if $f \geq 0$ for all $f \in C^k(I)$. Then there exist a homeomorphism $u : I \to I$ and two continuous functions $p, A \in C(I)$ with $A \geq 1$, $p > 0$ such that

$$(Tf)(u(x)) = \begin{cases} f(x)^{p(x)}, & f(x) \geq 0, \\ -A(x)\,|f(x)|^{p(x)}, & f(x) < 0. \end{cases}$$

Conversely, T defined this way satisfies (6.19).

For $k \in \mathbb{N}$, we have that $A = p = 1$ and that u is a C^k-diffeomorphism, so that

$$Tf(u(x)) = f(x).$$

Thus, for $k \in \mathbb{N}$, the operator is even *multiplicative* and *linear*.

We indicate some steps of the proof.

Step 1. For $x \in I$, an approximate indicator at x is a function $f \in C^k(I)$ with $f \geq 0$ such that there are open neighborhoods $x \in J_1 \subset J_2$ of x with $f|_{J_1} = \mathbf{1}$ and $f|_{I \setminus J_2} = 0$. Let AI_x denote the set of all approximate indicators at x. Define a set-valued map from I to the subsets of I by $u(x) := \bigcap_{f \in AI_x} \operatorname{supp}(Tf)$, where $\operatorname{supp}(Tf)$ denotes the support of Tf. One shows that $u(x)$ is either empty or consists of only one point and that for $f \in AI_x$, $Tf|_{u(x)} = 1$. Also for any $f \in C^k(I)$ and $x \in I$, $\operatorname{sgn} Tf|_{u(x)} = \operatorname{sgn} f(x)$. Here the fact that $f \geq 0$ implies $Tf \geq 0$ is used.

Step 2. Let G denote the set of all $x \in I$ for which there is an approximate indicator $f \in AI_x$ with compact support. Then obviously, $u(x)$ is not empty and hence consists of one point, and $u : G \to u(G) \subset I$ can be considered as a point map. Using among other things that $Tf \geq 0$ implies $f \geq 0$, one proves that $u(G)$ and G are dense in I and that $u : G \to u(G) \subset I$ is continuous and injective.

Step 3. One shows that for any open subset $J \subset I$ and any $f_1, f_2 \in C^k(I)$ with $f_1|_J = f_2|_J$ we have that $Tf_1|_{u(J)} = Tf_2|_{u(J)}$, after proving for any $h \in C^k(I)$ that $h|_J = \mathbf{1}$ implies $Th|_{u(J)} = \mathbf{1}$ and that $h|_J = -\mathbf{1}$ implies $Th|_{u(J)} = T(-\mathbf{1})|_{u(J)}$. This yields the localization of T on G: There is F such that

$$Tf(u(x)) = F(x, f(x), \ldots, f^{(k)}(x))$$

for any $f \in C^k(I)$ and $x \in G$. Moreover, $\operatorname{sgn} F(x, \alpha_0, \ldots, \alpha_k) = \operatorname{sgn} \alpha_0$.

Step 4. The operator inequality for T translates into a functional inequality for F. One proves that F does not depend on the variables $(\alpha_1, \ldots, \alpha_k)$. Theorem 6.2 then yields that for any $f \in C^k(I)$ and all $x \in G$

$$(Tf)(u(x)) = \begin{cases} f(x)^{p(x)}, & f(x) \geq 0, \\ -A(x) \, |f(x)|^{p(x)}, & f(x) < 0, \end{cases}$$

where $A \geq 1$ and $p \geq 0$ are continuous functions on G. The functions and operators are then extended by continuity to all of I, with $u : I \to I$ being a homeomorphism. For $k \in \mathbb{N}$, considering the inverse operator expressed with powers $\frac{1}{p(x)}$, shows that $A = p = 1$ and that u is a C^k-diffeomorphism. □

Chapter 7

The Second-Order Leibniz Rule

In the previous chapters we investigated the solutions and the stability properties of the Leibniz and the chain rule operator equations. These equations formalized properties of the first derivative of a function. In this and the next chapter we study equations which are motivated by identities for the second derivative. One of our goals is to find simple properties which characterize the Laplacian. In the setup of Leibniz type equations, this will be done in Sections 7.1 and 7.2.

Let $I \subset \mathbb{R}$ be an open interval. Then for $f, g \in C^2(I)$

$$D^2(f \cdot g) = D^2 f \cdot g + f \cdot D^2 g + 2Df \cdot Dg,$$

where D^2 and D denote the second and first derivative, respectively. This is a very particular setting of the operator functional equation

$$T(f \cdot g) = Tf \cdot g + f \cdot Tg + Af \cdot Ag, \quad f, g \in C^k(I), \tag{7.1}$$

for operators $T, A : C^k(I) \to C(I)$, namely for $k = 2$, $T = D^2$ and $A = \sqrt{2}D$. In this chapter we will study the general form of the solutions of (7.1) under mild additional assumptions. Note that for $A = 0$, (7.1) is just the Leibniz rule equation, so its solutions may be added to any solution of (7.1), and they can be considered as the "homogeneous" solution. Actually, the operators T and A are strongly coupled by (7.1) and there are fewer solutions (T, A) than one might at first imagine.

To characterize the Laplacian, we also consider functions in $C^2(I) = C^2(I, \mathbb{R})$, where $I \subset \mathbb{R}^n$ is an open set. Then for the Laplacian $\Delta := \sum_{i=1}^n \frac{\partial^2}{\partial x_i^2}$,

$$\Delta(f \cdot g) = \Delta f \cdot g + f \cdot \Delta g + 2\langle Df, Dg \rangle, \quad f, g \in C^2(I, \mathbb{R}).$$

For $x \in I \subset \mathbb{R}^n$, $Df(x), Dg(x) \in \mathbb{R}^n$ and $\langle \cdot, \cdot \rangle$ denotes the standard scalar product on $\mathbb{R}^n$. Formalizing this, we will also investigate the solutions of the operator

© Springer Nature Switzerland AG 2018
H. König, V. Milman, *Operator Relations Characterizing Derivatives*,
https://doi.org/10.1007/978-3-030-00241-1_7

equation

$$T(f \cdot g) = Tf \cdot g + f \cdot Tg + \langle Af, Ag \rangle, \quad f, g, \in C^k(I, \mathbb{R}), \tag{7.2}$$

for operators $T : C^k(I, \mathbb{R}) \to C(I, \mathbb{R})$ and $A : C^k(I, \mathbb{R}) \to C(I, \mathbb{R}^n)$. For clarity, we include in the notation $C^k(I, X)$ the space $X \in \{\mathbb{R}, \mathbb{R}^n\}$, into which C^k-smooth functions $f : I \to X$ are mapped, unless evidently $X = \mathbb{R}$.

We characterize the Laplacian by this equation, orthogonal invariance and the annihilation of affine functions. But we also determine the general solution of (7.2) under weak assumptions.

For $f \in C^1(I, \mathbb{R})$, $I \subset \mathbb{R}^n$, we use the notation $Df = f'$ instead of grad f and $D^2 f = f''$ instead of the Hessian Hess f.

The second-order chain rule for $f, g \in C^2(\mathbb{R})$,

$$D^2(f \circ g) = D^2 f \circ g \cdot (Dg)^2 + Df \circ g \cdot D^2 g,$$

may be better understood if we ignore the form of the specific operator D and study the "second-order-type" operator equation

$$T(f \circ g) = Tf \circ g \cdot A_1 g + A_2 f \circ g \cdot Tg, \quad f, g \in C^k(\mathbb{R})$$

for a priori arbitrary operators T, A_1 and A_2. We will investigate the solutions of this equation in Chapter 9.

7.1   Second-order Leibniz rule equation

If the operators T and A satisfying (7.1) would be localized, as most of the operators T in the previous chapters, equation (7.1) would turn into a functional equation for two unknown functions which then could be analyzed. However, without further assumptions, T and A will not be localized as the following simple example shows.

Example. Define $T, A : C^2(\mathbb{R}) \to C(\mathbb{R})$ by

$$Tf(x) := -f(x) + f(x+1), \quad Af(x) := f(x) - f(x+1),$$

for $f \in C^2(\mathbb{R})$, $x \in \mathbb{R}$. Then T and A satisfy (7.1) but $Tf(x)$ is not a function of $(x, f(x), f'(x), f''(x))$, but depends also on the values of f at $x+1$. Here, $x+1$ might be replaced by $x + \varphi(x)$ for any continuous function $\varphi \in C(\mathbb{R})$.

In the example, for functions $f \in C^2(\mathbb{R})$ supported in a small neighborhood of x, $Af(x) = f(x)$. We have to exclude this possibility to prove the localization of T and A. To do so, we introduce a condition of non-degeneration of A, already in the more general setup of functions on open subsets of $\mathbb{R}^n$.

Definition. Let $k \in \mathbb{N}_0$, $n \in \mathbb{N}$ and $I \subset \mathbb{R}^n$ be an open set. An operator $A :$ $C^k(I, \mathbb{R}) \to C(I, \mathbb{R}^m)$ is *non-degenerate* provided that for all open subsets $J \subset I$ and all $x \in J$ there exist $(m + 1)$ functions $g_i \in C^k(I, \mathbb{R})$ with support in J such that the $(m + 1)$ vectors $(g_i(x), Ag_i(x)) \in \mathbb{R}^{m+1}$, $i \in \{1, \ldots, m + 1\}$, are linearly independent in $\mathbb{R}^{m+1}$.

In the case of (7.1), $m = 1$, $(g_i(x), Ag_i(x)) \in \mathbb{R}^2$ should be linearly independent for $i = 1, 2$: Locally near x, A should not be proportional to the identity operator. This assumption of non-degeneration excludes a type of "resonance" situation between two involved operators, namely A and the identity.

Under the assumption of non-degeneration of A, we determine the general solution of (7.1). We do it slightly more generally for functions $f : I \to \mathbb{R}$ on domains I in $\mathbb{R}^n$, to prepare for the solution of the operator equation (7.2) which will be based on the following two theorems which are the central results of this chapter.

The first theorem is, in fact, a special case of the second. However, it is much easier to state. It is the most interesting special case of the second theorem.

Definition. Let $k \geq 2$, $n \in \mathbb{N}$ and $I \subset \mathbb{R}^n$ be an open set. An operator $A :$ $C^k(I, \mathbb{R}) \to C(I, \mathbb{R})$ *depends non-trivially on the derivative* if there is $x \in I$ and there are functions $f_1, f_2 \in C^k(I, \mathbb{R})$ with $f_1(x) = f_2(x)$ but $Af_1(x) \neq Af_2(x)$.

Theorem 7.1. *Let $n \in \mathbb{N}$, $k \in \mathbb{N}_0$ and $I \subset \mathbb{R}^n$ be open and connected. Assume that $T, A : C^k(I, \mathbb{R}) \to C(I, \mathbb{R})$ satisfy the* second-order Leibniz rule equation

$$T(f \cdot g) = Tf \cdot g + f \cdot Tg + Af \cdot Ag, \quad f, g \in C^k(I, \mathbb{R}),$$

and that A is non-degenerate and depends non-trivially on the derivative. Then there are continuous functions $a \in C(I, \mathbb{R})$ and $b, c \in C(I, \mathbb{R}^n)$ such that we have for all $f \in C^k(I, \mathbb{R})$ and all $x \in I$

$$Tf(x) = \frac{1}{2}\langle f''(x)c(x), c(x)\rangle + Rf(x),$$
$$Af(x) = \langle f'(x), c(x)\rangle.$$

where $Rf(x) = \langle f'(x), b(x)\rangle + a(x)f(x) \ln|f(x)|$ is an additive "homogeneous" solution, i.e., a solution of the ordinary Leibniz rule.

Conversely, these operators satisfy the second-order Leibniz rule.

Hence, up to the additive homogeneous term, the solution for T is just a second directional derivative. We now state the main result when no assumption on the dependence on the derivative is imposed.

Theorem 7.2 (Second-order Leibniz rule). *Let $n \in \mathbb{N}$, $k \in \mathbb{N}_0$ and $I \subset \mathbb{R}^n$ be open and connected. Assume that $T, A : C^k(I, \mathbb{R}) \to C(I, \mathbb{R})$ satisfy the* second-order Leibniz rule equation

$$T(f \cdot g) = Tf \cdot g + f \cdot Tg + Af \cdot Ag, \quad f, g \in C^k(I, \mathbb{R}), \tag{7.3}$$

and that A is non-degenerate. Then there are three possible families of solutions,
two of which possibly might be defined on subsets partitioning I and joined to form
solutions on the full set I. More precisely, there are two disjoint subsets $I_1, I_2 \subset I$,
one of them possibly empty, with $I = I_1 \cup I_2$, I_2 open, and there are functions
$b, c : I \to \mathbb{R}^n$ and $a, d : I \to \mathbb{R}$ which are continuous except possibly on ∂I_2, and
there is $p \in C(I_2, \mathbb{R})$ with $p > -1$, such that after subtracting from T the solution
R of the homogeneous equation given by

$$Rf(x) = a(x)f(x)\ln|f(x)| + \langle f'(x), b(x)\rangle, \quad f \in C^k(I, \mathbb{R}), \ x \in I,$$

the operators $T_1 := T - R$ and A have one of the following three forms:
either

- $T_1 f(x) = \frac{1}{2}\langle f''(x)c(x), c(x)\rangle$, $Af(x) = \langle f'(x), c(x)\rangle$, $x \in I$, $k \geq 2$;
  or

- $T_1 f(x) = \frac{1}{2}d(x)^2 f(x)\big(\ln|f(x)|\big)^2$, $Af(x) = d(x)f(x)\ln|f(x)|$, $x \in I_1$;
  or

- $T_1 f(x) = d(x)Af(x)$, $Af(x) = d(x)f(x)\big(\{\operatorname{sgn} f(x)\}|f(x)|^{p(x)} - 1\big)$, $x \in I_2$.

In the first case, for $k \geq 2$, the functions a, b, c are continuous on I, whereas in
the second and third case, for $k \geq 0$, the functions a, b, d may have discontinuities
in points of ∂I_2. In the last formula, there are two solutions on I_2, one with
the $\{\operatorname{sgn} f(x)\}$-term present and the second without it. If the $\{\operatorname{sgn} f(x)\}$-term is
present, $p = -1$ is allowed, too.

If $I_1 \neq I$ and $I_2 \neq I$, the last two solutions should be combined to form
a solution on $I = I_1 \cup I_2$ where the images of T and A need to be contained
in the continuous functions $C(I)$. This is possible by appropriate choices of the
parametric functions a, b, c and d, as an example below shows.

Conversely, the operators (T, A) defined by these formulas satisfy equation
(7.3).

Remarks. (a) The coupled solutions on I_1 and I_2 only depend on x and the function
values $f(x)$, but not on the derivatives of f at x. Therefore Theorem 7.1 is a special
case of Theorem 7.2.

(b) We do not impose any continuity conditions on T or A. We also do not
require T or A to be linear, which, in fact, is only fulfilled in the case of the first
solution when $a = 0$, i.e., when T is essentially the second derivative and A is
essentially the first derivative.

(c) Note that for all solutions, the operator A can be extended from $C^k(I, \mathbb{R})$
to $C^1(I, \mathbb{R})$ if $k \geq 2$, and the operator T from $C^k(I, \mathbb{R})$ to $C^2(I, \mathbb{R})$, if $k \geq 3$, by the
same formulas. Therefore, on $C^k(I, \mathbb{R})$ with $k \geq 3$, there are not more solutions
than on $C^2(I, \mathbb{R})$. In the case of the second and third solution, A can even be
extended to $C(I, \mathbb{R})$ and T to $C^1(I, \mathbb{R})$ or $C(I, \mathbb{R})$, depending on whether $b \neq 0$ or
$b = 0$. Therefore,

$$\big(C^2(I, \mathbb{R}), C^1(I, \mathbb{R})\big), \quad \big(C^1(I, \mathbb{R}), C(I, \mathbb{R})\big), \quad \text{and} \quad \big(C(I, \mathbb{R}), C(I, \mathbb{R})\big)$$

are the natural domain spaces for (T, A).

(d) In the first solution, the second derivative part in T is the second directional derivative of f at x in the direction of $v(x) = \frac{c(x)}{\|c(x)\|}$, multiplied by $\frac{1}{2}\|c(x)\|^2$. However, the direction $v(x)$ changes continuously with $x \in I$. Similarly, $Af(x)$ is a multiple of the directional derivative $\frac{\partial f}{\partial v(x)}(x)$.

Starting with the derivations $Rf = f'$ and $Rf = f \ln|f|$ solving the Leibniz rule by Theorem 3.1, the main parts of the first two solutions might be considered as "second iterated derivations". In particular, on I_1, $\frac{1}{2}f(\ln|f|)^2$ plays the role of the second iterated derivation, when starting with the entropy function $f \ln|f|$ as first derivation, with natural domain $C(I_1, \mathbb{R})$. Nevertheless, the solutions of (7.3) are slightly different from the second iterations of the solutions of the first-order equation (3.1): Iterating $Rf = f \ln|f|$ gives

$$R^2 f = f \, (\ln|f|)^2 + f \, (\ln|\ln|f||).$$

In the last solution on I_2, when the $\{\mathrm{sgn}\, f(x)\}$-term does not appear, actually $p > -1$ is required to guarantee that the ranges of T and A consist of continuous functions. In the third solution on I_2, different from the first two formulas on I and on I_1, the operators T_1 and A are proportional.

(e) Only very few tuning operators A are possible when solving (7.3), and then they determine the main operator T to a large degree. Choosing, e.g., $A = \sqrt{2}\, D$, D being the derivative, yields that T_1 is given by the second derivative.

(f) As in the case of the extended Leibniz rule, considered in Theorem 3.7, the last couple of solutions for (T, A) in Theorem 7.2 on I_1 and I_2 could and should be joined to provide solution operators (T, A) with ranges in $C(I)$, $I = I_1 \cup I_2$. To indicate that this is possible, we adjust the example following Theorem 3.7 to the second-order Leibniz rule.

Example. For $n = 1$, $k \in \mathbb{N}_0$, define operators $T, A : C^k(-1, 1) \to C(-1, 1)$ by

$$Tf(x) = \begin{cases} \frac{1}{2}f(x)\big(\ln|f(x)|\big)^2, & x \in (-1, 0], \\ \frac{1}{x^2}f(x)\big(|f(x)|^x - 1 - x\ln|f(x)|\big), & x \in (0, 1), \end{cases}$$

$$Af(x) = \begin{cases} f(x)\ln|f(x)|, & x \in (-1, 0], \\ \frac{1}{x}f(x)\big(|f(x)|^x - 1\big), & x \in (0, 1). \end{cases}$$

On $I_1 = (-1, 0]$, (T, A) is a solution of the second type, on $I_2 = (0, 1)$ one of the third type, with $-\frac{1}{x}f(x)\ln|f(x)|$ being the homogenous part of the solution for T on $(0, 1)$. For any $f \in C(-1, 1)$, Af and Tf are continuous at $x = 0$ since $\lim_{x \to 0} \frac{1}{x}(|f(x)|^x - 1) = \ln|f(0)|$ and

$$\lim_{x \to 0} \frac{1}{x^2}\big(|f(x)|^x - 1 - x\ln|f(x)|\big) = \frac{1}{2}(\ln|f(0)|)^2.$$

The operator A is non-degenerate: For $0 \in J \subset (-1, 1)$ open, just choose functions $g_1, g_2 \in C^k(I)$ with $\operatorname{supp}(g_i) \subset J$ and $g_1(0) = 2$, $g_2(0) = 3$. Then $(g_i(0), g_i(0) \ln g_i(0)) \in \mathbb{R}^2$ are linearly independent for $i = 1, 2$.   □

(g) Since there is no continuous approximation of derivative values $f'(x)$ by only functions of x and $f(x)$, the first solution cannot be continuously approximated by one of the other two solutions. Hence, the first solution has the full domain I if it occurs.

Corollary 7.3. *Suppose that the assumptions of Theorem 7.2 are satisfied. Assume, in addition, that T annihilates all constant functions. Then $k \geq 2$ is required and there are $b, c \in C(I, \mathbb{R}^n)$ such that for all $f \in C^k(I, \mathbb{R})$ and all $x \in I$*

$$Tf(x) = \frac{1}{2}\langle f''(x)c(x), c(x)\rangle + \langle f'(x), b(x)\rangle, \quad Af(x) = \langle f'(x), c(x)\rangle.$$

If T also annihilates the linear functions, the function b is zero.

If $I = \mathbb{R}^n$ and T or A are *isotropic*, i.e., commute with all shifts S_y, $y \in \mathbb{R}^n$, $AS_y = S_y A$, where $S_y f(x) := f(x + y)$, $x \in \mathbb{R}^n$, the vector functions c, b are constant. Then T is a multiple of the second directional derivative in the fixed direction of c at x plus a multiple of the first directional derivative in the direction of b at x.

Proof of Corollary 7.3. Choosing different constant functions, the assumption that T annihilates the constant functions shows that the parameter function d in the second and third solution for T_1 in Theorem 7.2 has to be zero, and also the function a in R. This leaves only the first solution for T_1. Hence, under the assumptions of Corollary 7.3, the solution of (7.3) has the form

$$Tf(x) = \frac{1}{2}\langle f''(x)c(x), c(x)\rangle + \langle f'(x), b(x)\rangle, \quad Af(x) = \langle f'(x), c(x)\rangle.$$

If T also annihilates the linear functions, the function b has to be zero, too. Note that $c \neq 0$ since otherwise A would be degenerate. Therefore also $T \neq 0$.   □

To prove Theorem 7.2, we first show that the operators T and A are localized. In the following we again represent the ℓ-th derivative $f^{(\ell)}(x)$ of f at x by the $M(n, \ell)$ independent ℓ-th order iterated partial derivatives

$$\left(\frac{\partial^\ell f(x)}{\partial x_{i_1} \cdots \partial x_{i_\ell}}\right)_{1 \leq i_1 \leq \cdots \leq i_\ell \leq n},$$

as done in Proposition 3.6.

Proposition 7.4. *Let $k \in \mathbb{N}_0$, $n, m \in \mathbb{N}$, $I \subset \mathbb{R}^n$ be open, and $T : C^k(I, \mathbb{R}) \to C(I, \mathbb{R})$ and $A : C^k(I, \mathbb{R}) \to C(I, \mathbb{R}^m)$ be operators such that*

$$T(f \cdot g)(x) = Tf(x) \cdot g(x) + f(x) \cdot Tg(x) + \langle Af(x), Ag(x)\rangle, \qquad (7.4)$$

$f, g \in C^k(I, \mathbb{R})$, $x \in I$. *Assume that A is non-degenerate. Let $M(n, \ell) := \binom{n+\ell-1}{\ell}$ and $N(n, k) := \sum_{\ell=0}^{k} M(n, \ell) = \binom{n+k}{k}$. Then there are functions $F : I \times \mathbb{R}^{N(n,k)} \to \mathbb{R}$ and $E : I \times \mathbb{R}^{N(n,k)} \to \mathbb{R}^m$ such that for all $f \in C^k(I, \mathbb{R})$ and $x \in I$*

$$Tf(x) = F\big(x, f(x), f'(x), \dots, f^{(k)}(x)\big)$$

and

$$Af(x) = E\big(x, f(x), f'(x), \dots, f^{(k)}(x)\big).$$

In the case of equation (7.3) we need this result for $m = 1$, but in (7.2) – to be considered later – we will use it for $m = n$.

Proof. Let $J \subset I \subset \mathbb{R}^n$ be open and $f_1, f_2 \in C^k(I, \mathbb{R})$ satisfy $f_1|_J = f_2|_J$. For any $g \in C^k(I, \mathbb{R})$ with support in J we have $f_1 \cdot g = f_2 \cdot g$ and hence by (7.4)

$$Tf_1 \cdot g + f_1 \cdot Tg + \langle Af_1, Ag \rangle = T(f_1 \cdot g) = T(f_2 \cdot g)$$
$$= Tf_2 \cdot g + f_2 \cdot Tg + \langle Af_2, Ag \rangle.$$

Therefore, for any $x \in J$, with $f_1(x) = f_2(x)$,

$$\big(Tf_1(x) - Tf_2(x)\big) \cdot g(x) + \langle Af_1(x) - Af_2(x), Ag(x) \rangle = 0. \tag{7.5}$$

Since A is non-degenerate, we may find $(m+1)$ functions $g_1, \dots, g_{m+1} \in C^k(I, \mathbb{R})$ with support in J such that the $(m+1)$ vectors $(g_i(x), Ag_i(x)) \in \mathbb{R}^{m+1}$ for $i \in \{1, \dots, m+1\}$ are linearly independent. Applying (7.5) to these functions g_i instead of g, we conclude that $Tf_1(x) - Tf_2(x) = 0$ and $Af_1(x) - Af_2(x) = 0$. This yields that $Tf_1|_J = Tf_2|_J$ and $Af_1|_J = Af_2|_J$. Note here that Tf_j is $\mathbb{R}$-valued and Af_j is $\mathbb{R}^m$-valued, for $j \in \{1, 2\}$. Hence by Proposition 3.6 there are functions $F : I \times \mathbb{R}^{N(n,k)} \to \mathbb{R}$ and $E : I \times \mathbb{R}^{N(n,k)} \to \mathbb{R}^m$ such that

$$Tf(x) = F\big(x, f(x), \dots, f^{(k)}(x)\big)$$

and

$$Af(x) = E\big(x, f(x), \dots, f^{(k)}(x)\big), \quad f \in C^k(I, \mathbb{R}), \ x \in I. \qquad \square$$

Proof of Theorem 7.2. (a) We use Proposition 7.4 for $m = 1$ to conclude that there are functions $F, E : I \times \mathbb{R}^{N(n,k)} \to \mathbb{R}$ such that

$$Tf(x) = F\big(x, f(x), f'(x), \dots, f^{(k)}(x)\big)$$

and

$$Af(x) = E\big(x, f(x), f'(x), \dots, f^{(k)}(x)\big), \quad f \in C^k(I, \mathbb{R}), \ x \in I.$$

Putting $f = \mathbf{1}$ in (7.3) yields that, for all $g \in C^k(I, \mathbb{R})$ and $x \in I$,

$$g(x) \cdot T\mathbf{1}(x) + Ag(x) \cdot A\mathbf{1}(x) = 0.$$

By non-degeneracy of A, we may choose $g_1, g_2 \in C^k(I, \mathbb{R})$ such that $(g_i(x), Ag_i(x)) \in \mathbb{R}^2$ are linearly independent for $i = 1, 2$. Applying the previous equality to $g = g_1$ and $g = g_2$ yields $T\mathbf{1}(x) = 0$, $A\mathbf{1}(x) = 0$, i.e., $T\mathbf{1} = A\mathbf{1} = 0$. This means that $F(x, 1, 0, \dots, 0) = E(x, 1, 0, \dots, 0) = 0$ for all $x \in I$.

For $g \in C^k(I, \mathbb{R})$, put $f := \exp(g)$. Then $f > 0$ and $f \in C^k(I, \mathbb{R})$. In the following, we will analyze the form of the solutions Tf and Af of (7.3) for strictly positive functions $f > 0$. Only in part (e) of the proof we turn to general functions.

With $f = \exp(g)$ we may define operators $S, B : C^k(I, \mathbb{R}) \to C(I, \mathbb{R})$ by

$$Sg := T(\exp(g))/\exp(g), \quad Bg := A(\exp(g))/\exp(g), \quad g \in C^k(I, \mathbb{R}).$$

Then, for any $g_1, g_2 \in C^k(I, \mathbb{R})$, by (7.3)

$$
\begin{aligned}
S(g_1 + g_2) &= \frac{T\big(\exp(g_1) \cdot \exp(g_2)\big)}{\exp(g_1) \cdot \exp(g_2)} \\
&= \frac{T(\exp(g_1))}{\exp(g_1)} + \frac{T(\exp(g_2))}{\exp(g_2)} + \frac{A(\exp(g_1))}{\exp(g_1)} \cdot \frac{A(\exp(g_2))}{\exp(g_2)} \\
&= Sg_1 + Sg_2 + Bg_1 \cdot Bg_2.
\end{aligned}
\tag{7.6}
$$

Since the derivatives of $f = \exp(g)$ can be expressed in terms of functions of g and the derivatives of g, the operators S and B are localized, too. Hence, there are functions $G, H : I \times \mathbb{R}^{N(n,k)} \to \mathbb{R}$ such that, for all $g \in C^k(I, \mathbb{R})$,

$$Sg(x) = \frac{T(\exp(g))(x)}{\exp(g)(x)} = G\big(x, g(x), \dots, g^{(k)}(x)\big)$$

and

$$Bg(x) = \frac{A(\exp(g))(x)}{\exp(g)(x)} = H\big(x, g(x), \dots, g^{(k)}(x)\big), \tag{7.7}$$

$G(x, 0, \dots, 0) = H(x, 0, \dots, 0) = 0$.

By (7.6) and (7.7) we have, for all $g_1, g_2 \in C^k(I, \mathbb{R})$ and $x \in I$,

$$
\begin{aligned}
G\big(x, (g_1 + g_2)(x), \dots, (g_1 + g_2)^{(k)}(x)\big) &= G\big(x, g_1(x), \dots, g_1^{(k)}(x)\big) \\
+ G\big(x, g_2(x), \dots, g_2^{(k)}(x)\big) + H\big(x, g_1(x), \dots, g_1^{(k)}(x)\big) &\cdot H\big(x, g_2(x), \dots, g_2^{(k)}(x)\big).
\end{aligned}
$$

For any $\alpha = (\alpha_\ell)_{\ell=0}^k, \beta = (\beta_\ell)_{\ell=0}^k \in \mathbb{R}^{N(n,k)}$ with $\alpha_\ell, \beta_\ell \in \mathbb{R}^{M(n,\ell)}$ and $x \in I$, we may choose functions $g_1, g_2 \in C^k(I, \mathbb{R})$ such that $g_1^{(\ell)}(x) = \alpha_\ell$ and $g_2^{(\ell)}(x) = \beta_\ell$ for all $\ell \in \{0, \dots, k\}$, recalling that we represent $g_1^{(\ell)}(x)$ and $g_2^{(\ell)}(x)$ by $M(n, \ell)$ independent ℓ-th order iterated partial derivatives. Therefore, (7.6) is equivalent to the functional equation

$$G(x, \alpha + \beta) = G(x, \alpha) + G(x, \beta) + H(x, \alpha)H(x, \beta), \quad x \in I, \ \alpha, \beta \in \mathbb{R}^{N(n,k)} \tag{7.8}$$

for two unknown functions G and H. By Proposition 2.9, for each $x \in I$ there are additive functions $C(x), D(x) : \mathbb{R}^{N(n,k)} \to \mathbb{R}$ and constants $\gamma(x) \in \mathbb{R}$ such that any solution (G, H) of (7.8) has one of the following three forms: either

$$G(x, \alpha) = \tfrac{1}{2}C(x)(\alpha)^2 + D(x)(\alpha),$$
$$H(x, \alpha) = C(x)(\alpha), \tag{7.9}$$

or

$$G(x, \alpha) = \gamma(x)^2 \left[\exp(C(x)(\alpha)) - 1 \right] + D(x)(\alpha),$$
$$H(x, \alpha) = \gamma(x) \left[\exp(C(x)(\alpha)) - 1 \right], \tag{7.10}$$

or

$$G(x, \alpha) = -\gamma(x) + D(x)(\alpha),$$
$$H(x, \alpha) = \gamma(x).$$

In the third case, $Bg(x) = \gamma(x)$ for all $g \in C^k(I, \mathbb{R})$, and hence $Af = \gamma f$ would be a multiple of the identity, first on positive functions, but then also on all functions, by an argument given in part (e) below. This map A would not be non-degenerate. Therefore, we only have to investigate the specific solutions (7.9) and (7.10) of (7.8).

By additivity, the functions $C(x), D(x)$ split as a sum

$$C(x)(\alpha) = \sum_{\ell=0}^{k} c_\ell(x)(\alpha_\ell), \quad D(x)(\alpha) = \sum_{\ell=0}^{k} d_\ell(x)(\alpha_\ell), \quad \alpha = (\alpha_\ell)_{\ell=0}^k,$$

where $c_\ell(x), d_\ell(x) : \mathbb{R}^{M(n,\ell)} \to \mathbb{R}$ are additive functions, too. Hence, for all $g \in C^k(I, \mathbb{R})$,

$$C(x)\big(g(x), \ldots, g^{(k)}(x)\big) = \sum_{\ell=0}^{k} c_\ell(x)(g^{(\ell)}(x)),$$

$$D(x)\big(g(x), \ldots, g^{(k)}(x)\big) = \sum_{\ell=0}^{k} d_\ell(x)(g^{(\ell)}(x)).$$

Using this, (7.7) and (7.9) or (7.10), we have the following two possibilities for the operator B: either

$$Bg(x) = H\big(x, g(x), \ldots, g^{(k)}(x)\big) = C(x)\big(g(x), \ldots, g^{(k)}(x)\big)$$
$$= \sum_{\ell=0}^{k} c_\ell(x)(g^{(\ell)}(x)), \quad g \in C^k(I, \mathbb{R}) \tag{7.11}$$

or

$$Bg(x) = H\big(x, g(x), \ldots, g^{(k)}(x)\big)$$

$$= \gamma(x)\left[\exp\left(\sum_{\ell=0}^{k} c_\ell(x)(g^{(\ell)}(x))\right) - 1\right], \quad g \in C^k(I, \mathbb{R}). \tag{7.12}$$

We also know that $Bg(x) = H(x, g(x), \ldots, g^{(k)}(x))$ is continuous for all $g \in C^k(I)$. As indicated by the example after the remarks on Theorem 7.2, the solution formulas may possibly change continuously in a "phase transition". Therefore, we put

$$I_1 := \big\{x \in I \mid (7.11) \text{ provides the formula for } Bg(x) \text{ for all } g \in C^k(I, \mathbb{R})\big\},$$

$$I_2 := \big\{x \in I \mid (7.12) \text{ provides the formula for } Bg(x) \text{ for all } g \in C^k(I, \mathbb{R})\big\}.$$

Obviously, $I_1 \cap I_2 = \emptyset$, $I_1 \cup I_2 = I$. If $I_2 = \emptyset$ or $I_1 = \emptyset$, the solution for B, and later also for T and A, is given by a single formula, (7.11) or (7.12). We claim that I_2 is open. If this would be false, there would be $x_0 \in I_2$ and a sequence $x_m \in I_1$ with $\lim_{m \to \infty} x_m = x_0$. We know that for all $g \in C^k(I, \mathbb{R})$

$$Bg(x_0) = \gamma(x_0)\left[\exp\left(\sum_{\ell=0}^{k} c_\ell(x_0)(g^{(\ell)}(x_0))\right) - 1\right],$$

$$\gamma(x_0) \neq 0, \ (c_\ell(x_0))_{\ell=0}^{k} \neq 0.$$

We will show that a contradiction to the assumption $x_m \to x_0$ follows from the fact that a rapidly growing exponential function cannot be well approximated by additive functions as in (7.11).

Indeed, to show this, let us assume that $c_0(x_0) : \mathbb{R} \to \mathbb{R}$ is non-zero, to keep the argument and the notation simple. Then there is a fixed $\alpha \in \mathbb{R}$ such that $c_0(x_0)(\alpha) > 0$ (just multiplication of $c_0(x_0)$ by α). By additivity, for any $r \in \mathbb{N}$, $c_0(x_0)(r\alpha) = r c_0(x_0)(\alpha)$ Hence, for the constant functions g_r, $g_r(x) := r\alpha$, we have

$$Bg_r(x_0) = \gamma(x_0)\big[\exp(r\, c_0(x_0)(\alpha)) - 1\big],$$

using $c_\ell(x_0)(0) = 0$ for all $l \in \mathbb{N}$. By assumption, all $Bg_r(x_m)$ are given by (7.11). Hence there are additive functions $\widetilde{c}_0(x_m) : \mathbb{R} \to \mathbb{R}$ such that, for any $r \in \mathbb{N}$,

$$Bg_r(x_m) = \widetilde{c}_0(x_m)(r\alpha) = r\,\widetilde{c}_0(x_m)(\alpha)$$

approximates $Bg_r(x_0)$ well as $m \to \infty$. Hence there is $m(r) \in \mathbb{N}$ such that

$$\widetilde{c}_0(x_{m(r)})(\alpha) \geq \frac{\gamma(x_0)}{2}\frac{1}{r}\big[\exp(r\, c_0(x_0)(\alpha)) - 1\big],$$

and therefore $\sup_{m \in \mathbb{N}} \widetilde{c}_0(x_m)(\alpha) = \infty$. We conclude that

$$\sup_{m \in \mathbb{N}} Bg_1(x_m) = \sup_{m \in \mathbb{N}} \widetilde{c}_0(x_m)(\alpha) = \infty,$$

contradicting the continuity of Bg_1, which requires that $\lim_{m \to \infty} Bg_1(x_m) = Bg_1(x_0)$ exists in $\mathbb{R}$. Hence, I_2 is open.

If I_2 is dense in I, but $I_2 \neq I$, we will later get the formula for Bg (and Tf, Af) by continuous extension from I_2 to $\partial I_2 = I \setminus I_2$, using that all functions Bg are continuous. Assume that I_2 is not dense. Then $I_1 = I \setminus I_2$ has non-empty interior $\overset{\circ}{I}_1$, and it suffices to prove the formulas for Bg, Af and Tf – as stated in Theorem 7.2 – for $x \in I_2$ and $x \in \overset{\circ}{I}_1$. Both sets are open, and we will denote this open set by $J \in \{\overset{\circ}{I}_1, I_2\}$, $J \subset I \subset \mathbb{R}^n$.

(b) We now analyze the structure and regularity of the operator B given by (7.11) and (7.12) on J.

By the remarks before, we may assume that (7.11) and (7.12) hold for all $g \in C^k(J, \mathbb{R})$ on the open sets $J = \overset{\circ}{I}_1$ and $J = I_2$, respectively. We also know that

$$Bg(x) = H\big(x, g(x), \ldots, g^{(k)}(x)\big) \quad \text{and} \quad Sg(x) = G\big(x, g(x), \ldots, g^{(k)}(x)\big)$$

are continuous functions for any $g \in C^k(J, \mathbb{R})$.

In the case of $J = \overset{\circ}{I}_1$, (7.9) and (7.11), Theorem 2.6, with $(k-1)$ replaced by k, yields directly that $C(x) : \mathbb{R}^{N(n,k)} \to \mathbb{R}$ is linear and depends continuously on $x \in J$. Hence, the $c_\ell(x) : \mathbb{R}^{M(n,l)} \to \mathbb{R}$ are linear, continuous and

$$Bg(x) = C(x)\big(g(x), \ldots, g^{(k)}(x)\big) = \sum_{\ell=0}^{k} \langle c_\ell(x), g^{(\ell)}(x) \rangle.$$

Using this, (7.9) and the continuity of $Sg(x) = G(x, g(x), \ldots, g^{(k)}(x))$ in $x \in J$ for all $g \in C^k(J, \mathbb{R})$, $D(x)(g(x), \ldots, g^{(k)}(x))$ is continuous for all $g \in C^k(J, \mathbb{R})$, too. Hence, again by Theorem 2.6, the functions $d_\ell(x) : \mathbb{R}^{M(n,\ell)} \to \mathbb{R}$ representing $D(x)$ are linear and depend continuously on $x \in J$, too. Therefore, by (7.7) and (7.9)

$$Sg(x) = \frac{1}{2} \left(\sum_{\ell=0}^{k} \langle c_\ell(x), g^{(\ell)}(x) \rangle \right)^2 + \sum_{\ell=0}^{k} \langle d_\ell(x), g^{(\ell)}(x) \rangle,$$

$$Bg(x) = \sum_{\ell=0}^{k} \langle c_\ell(x), g^{(\ell)}(x) \rangle. \tag{7.13}$$

In the case of $J = I_2$, $C(x)$ is non-zero for any $x \in J$ since otherwise A would be degenerate. Therefore, for any x, $C(x)$ attains infinitely many different values for different functions g_m. Forming quotients, we first find that the function $\gamma : J \to \mathbb{R}$ in (7.12) is continuous and has no zero in J, which in turn implies that once more $\sum_{\ell=0}^{k} c_\ell(x)(g^{(\ell)}(x))$ depends continuously on $x \in J$ for all $g \in C^k(J, \mathbb{R})$.

Again Theorem 2.6 yields that the $c_\ell(x) : \mathbb{R}^{M(n,\ell)} \to \mathbb{R}$ are linear and depend continuously on $x \in J$, so that by (7.12),

$$Bg(x) = \gamma(x)\left[\exp\left(\sum_{\ell=0}^{k}\langle c_\ell(x), g^{(\ell)}(x)\rangle\right) - 1\right], \quad x \in J. \tag{7.14}$$

(c) We now analyze in detail the form of B, S, A and T in the first case $J = \overset{\circ}{I}_1$, where S and B are given by (7.13).

Then by (7.13) and the definition of B, for any $f \in C^k(J, \mathbb{R})$ with $f > 0$, we get, putting $g := \ln f$, $f = \exp(g)$, that

$$Af(x) = f(x)Bg(x) = f(x)\sum_{\ell=0}^{k}\langle c_\ell(x), (\ln f)^{(\ell)}(x)\rangle, \quad x \in J.$$

The ℓ-th derivative of $\ln f$ has a singularity of order $\mathcal{O}\left(\frac{1}{f^\ell}\right)$ as f tends to zero, if $f' \neq 0$. More precisely, for $\ell \geq 2$,

$$(\ln f)^{(\ell)} = \left(\frac{f'}{f}\right)^{(\ell-1)} = (-1)^{\ell-1}(\ell-1)!\left(\frac{f'}{f}\right)^\ell + P_\ell(f,\dots,f^{(\ell)}),$$

where P_ℓ is a sum of quotients with powers of f of order $\leq \ell-1$ in the denominator and product terms of derivatives of f of order $\leq \ell$ in the numerator. Therefore, the order of singularity of $f(\ln f)^{(\ell)}$ is $\mathcal{O}\left(\frac{1}{f^{\ell-1}}\right)$ as f tends to zero, if $f' \neq 0$. Since Af is continuous and hence also bounded in neighborhoods of points where f is zero, the above formula for Af requires that $c_k(x) = 0, \dots, c_2(x) = 0$ if $k \geq 2$. This argument is the same as in the proof of Theorem 3.5. Hence, for all $f \in C^k(J, \mathbb{R})$ with $f > 0$,

$$Af(x) = f(x)B(\ln f)(x) = c_0(x)f(x)\ln f(x) + \langle c_1(x), f'(x)\rangle, \quad x \in J.$$

Using this, (7.13) and the definition of S, we get for Tf, $f > 0$,

$$Tf(x) = f(x)S(\ln f)(x) = \frac{1}{2}\frac{\left(c_0(x)f(x)\ln f(x) + \langle c_1(x), f'(x)\rangle\right)^2}{f(x)}$$

$$+ f(x)\sum_{\ell=0}^{k}\langle d_\ell(x), (\ln f)^{(\ell)}(x)\rangle, \quad x \in J. \tag{7.15}$$

Since also Tf is bounded in neighborhoods of points where f is zero, we find that, if $c_1 \neq 0$, the order $\mathcal{O}(1/f)$ of singularity in Tf in the first term on the right of (7.15) has to be canceled by an opposite singularity in the second term on the right. Since $f(\ln f)^{(\ell)}$ has a singularity of order $\mathcal{O}\left(\frac{1}{f^{\ell-1}}\right)$, only d_0, d_1 and d_2 could possibly be non-zero. This yields two possibilities for solutions (T, A), as we will show now. As a multilinear form, $(\ln f)''(x)$ is given by

$$(u, v) \in \mathbb{R}^{2n} \longmapsto \frac{\langle f''(x)u, v \rangle}{f(x)} - \frac{\langle f'(x), u \rangle \langle f'(x), v \rangle}{f(x)^2},$$

returning to regular derivative notation. Hence, symmetrizing our second derivative notation – and modifying $d_2(x)$ accordingly – we need in (7.15) that $d_2(x) = \frac{1}{2}c_1(x) \otimes c_1(x)$, if $d_2(x) \neq 0$. In this case (7.15) gives

$$Tf(x) = \left(\frac{1}{2}c_0(x)^2 f(x)(\ln f(x))^2 + c_0(x) \ln f(x)\langle c_1(x), f'(x) \rangle \right.$$
$$\left. + \frac{1}{2}\frac{\langle c_1(x), f'(x) \rangle^2}{f(x)} \right) + \left(\frac{1}{2}\langle f''(x)c_1(x), c_1(x) \rangle - \frac{1}{2}\frac{\langle c_1(x), f'(x) \rangle^2}{f(x)} \right)$$
$$+ \langle d_1(x), f'(x) \rangle + d_0(x)f(x)\ln f(x).$$

This requires that $c_0 = 0$ since otherwise the second term on the right – involving the factor $\ln f(x)$ – would possibly be unbounded. Thus we get, with $d_2 = \frac{1}{2}c_1 \otimes c_1$ and $c_0 = 0$,

$$Tf(x) = \tfrac{1}{2}\langle f''(x)c_1(x), c_1(x) \rangle + \langle f'(x), d_1(x) \rangle + d_0(x)f(x)\ln f(x),$$
$$Af(x) = \langle f'(x), c_1(x) \rangle, \quad x \in J, \ f > 0.$$

This is the first solution in Theorem 7.2. If $d_2 = 0$, no singularity is allowed in the first term on the right of (7.15), requiring $c_1 = 0$. We then obtain the second solution in Theorem 7.2,

$$Tf(x) = \tfrac{1}{2}c_0(x)^2 f(x)(\ln f(x))^2 + \langle f'(x), d_1(x) \rangle + d_0(x)f(x)\ln f(x),$$
$$Af(x) = c_0(x)f(x)\ln f(x), \quad x \in J, \ f > 0.$$

(d) Concerning the solution B of (7.6) in the second case $J = I_2$, when B is given by (7.14), we have for any $f \in C^k(J, \mathbb{R})$, $f > 0$, putting $g = \ln f$,

$$Af(x) = f(x)B(\ln f)(x) = \gamma(x)f(x) \left[\exp \left(\sum_{\ell=0}^{k} \langle c_\ell(x), (\ln f)^{(\ell)}(x) \rangle \right) - 1 \right],$$

$x \in J$. The boundedness of Af in the neighborhood of zeros of functions f requires here that $c_k(x) = 0, \ldots, c_1(x) = 0$. Only $c_0(x)$ may and should be non-zero. This yields

$$Af(x) = \gamma(x)f(x) \left[\exp(c_0(x)\ln f(x)) - 1 \right] = \gamma(x)f(x) \left[f(x)^{c_0(x)} - 1 \right].$$

For Tf we find similarly, using (7.7), (7.10) and the definition of S,

$$Tf(x) = f(x)S(\ln f)(x)$$

$$= \gamma(x)^2 f(x) \left[f(x)^{c_0(x)} - 1 \right] + f(x) \sum_{\ell=0}^{k} \langle d_\ell(x), (\ln f)^{(\ell)}(x) \rangle.$$

Here only d_0 and d_1 may possibly be non-zero, yielding

$$Tf(x) = \gamma(x)^2 f(x)\big[f(x)^{c_0(x)} - 1\big] + \langle f'(x), d_1(x)\rangle + d_0(x)f(x)\ln f(x), \quad x \in J,$$

which is the third solution in Theorem 7.1. The boundedness of Af and Tf for functions with $f \searrow 0$ requires moreover that $c_0(x) \geq -1$.

In the second and third solution $Af(x)$ depends only on x and $f(x)$, but not on $f'(x)$. In the first solution, $Af(x)$ depends on $f'(x)$ since $c_1(x) \neq 0$. However, there is no continuous approximation of derivative values $f'(x)$ by only functions of x and $f(x)$ for general $f \in C^k(J, \mathbb{R})$, $f > 0$. Therefore, the first solution cannot be continuously approximated by one of the other two solutions. Hence, the first solution has the full domain $I = \overset{\circ}{I}_1$ (or $\emptyset$), since I is connected, whereas a non-trivial combination of the last two solutions on domains I_1, I_2 with $I = I_1 \cup I_2$ is possible, as the example following Theorem 7.2 showed. The formulas for the second solution (T, A) in Theorem 7.2 can be extended from $\overset{\circ}{I}_1$ to the relative boundary $I_1 \setminus \overset{\circ}{I}_1$ in I by continuity, since both functions on the left-hand side and on the right-hand side of the formulas for (T, A) can be extended by continuity, e.g.,

$$Af(x) = c_0(x)f(x)\ln f(x).$$

First, using the continuity of Af for constant functions f, we get that c_0 can be continuously extended to $I_1 \setminus \overset{\circ}{I}_1$, and then this formula holds for all $x \in I_1$, $f \in C^k(I, \mathbb{R})$, $f > 0$. The case of T is similar, first extending d_0 and then d_1 from $\overset{\circ}{I}_1$ to $I_1 \setminus \overset{\circ}{I}_1$ by applying the formula for Tf first to constants and then to linear functions.

(e) We now study functions which also attain negative values or zeros. For constant functions $f(x) = \alpha_0$, $g(x) = \beta_0$, $\alpha_0, \beta_0 \in \mathbb{R}$, equation (7.3) means in terms of the representing functions F and E with $Tf(x) = F(x, \alpha_0, 0, \ldots, 0)$ and $Af(x) = E(x, \alpha_0, 0, \ldots, 0)$ that

$$F(x, \alpha_0\beta_0, 0, \ldots, 0) = F(x, \alpha_0, 0, \ldots, 0)\beta_0 + F(x, \beta_0, 0, \ldots, 0)\alpha_0$$
$$+ E(x, \alpha_0, 0, \ldots, 0)E(x, \beta_0, 0, \ldots, 0).$$

By Proposition 2.10 there are additive functions $\widetilde{c}_0(x), \widetilde{d}_0(x) : \mathbb{R} \to \mathbb{R}$ and there is $\widehat{\gamma}(x) \in \mathbb{R}$ such that the solutions of the equation for (F, E) have one of the following three forms:
either

$$F(x, \alpha_0, 0, \ldots, 0) = \alpha_0\left[\tfrac{1}{2}\widetilde{c}_0(x)(\ln|\alpha_0|)^2 + \widetilde{d}_0(x)(\ln|\alpha_0|)\right],$$
$$E(x, \alpha_0, 0, \ldots, 0) = \alpha_0\widetilde{c}_0(x)(\ln|\alpha_0|),$$

or

$$F(x,\alpha_0,0,\ldots,0) = \alpha_0\big(\tilde{\gamma}(x)^2[\{\operatorname{sgn}\alpha_0\}\exp(\tilde{c}_0(x)(\ln|\alpha_0|))-1]+\tilde{d}_0(x)(\ln|\alpha_0|)\big),$$
$$E(x,\alpha_0,0,\ldots,0) = \alpha_0\tilde{\gamma}(x)\big[\{\operatorname{sgn}\alpha_0\}\exp(\tilde{c}_0(x)(\ln|\alpha_0|))-1\big],$$

or

$$F(x,\alpha_0,0,\ldots,0) = \alpha_0\big[\tilde{c}_0(x)(\ln|\alpha_0|))-\tilde{\gamma}(x)^2\big],$$
$$E(x,\alpha_0,0,\ldots,0) = \tilde{\gamma}(x)\alpha_0.$$

We now investigate the formulas for general functions in $C^k(I,\mathbb{R})$. First look at $f \in C^k(I,\mathbb{R})$ in points $x_0 \in I$ where $f(x_0) = 0$. On open subsets where f is identically zero, both $Tf(x)$ and $Af(x)$ are zero.

If x_0 is a limit point of points x_n where $f(x_n) > 0$, $Tf(x_0)$ and $Af(x_0)$ are expressed by the same type formulas as in the points x_n, since for any $f \in C^k(I,\mathbb{R})$, $x \in I$ with $f(x) \neq 0$, both sides in

$$Tf(x) = F(x, f(x), \ldots, f^{(k)}(x))$$

are continuous functions of $x \in I$, and similarly for $Af(x)$. The same applies to zeros of f when the values $f(x_n)$ are negative, once we have established the formulas for negative f, which we proceed to do now.

Now consider f and $x \in I$ with $f(x) < 0$. Since T and A are localized, we may assume that $f < 0$ on the full set I, even though the original function may be positive elsewhere.

Comparing the above formulas for $F(x,\alpha_0,0,\ldots,0)$ and $E(x,\alpha_0,0,\ldots,0)$ to the already established formulas for $Tf(x)$ and $Af(x)$ when $f(x) = \alpha_0 > 0$, the first two solutions for $(F(x,\alpha_0,0,\ldots,0), E(x,\alpha_0,0,\ldots,0))$ lead – as the only possibility – to the general solutions

$$Tf(x) = \tfrac{1}{2}c_0(x)^2 f(x)(\ln|f(x)|)^2 + d_0(x)f(x)\ln|f(x)| + \langle d_1(x), f'(x)\rangle,$$
$$Af(x) = c_0(x)f(x)\ln|f(x)|,$$

and

$$Tf(x) = \gamma(x)^2 f(x)\big[\{\operatorname{sgn} f(x)\}|f(x)|^{c_0(x)} - 1\big]$$
$$\qquad + d_0(x)f(x)\ln|f(x)| + \langle d_1(x), f'(x)\rangle,$$
$$Af(x) = \gamma(x)f(x)\big[\{\operatorname{sgn} f(x)\}|f(x)|^{c_0(x)} - 1\big],$$

with $\tilde{c}_0 = c_0$, $\tilde{c}_1 = c_1$, $\tilde{\gamma} = \gamma$. Calculation shows that, conversely, these formulas define operators (T, A) which satisfy the operator equation (7.3). In the last case the term $\{\operatorname{sgn} f(x)\}$ may appear both in T and A or not at all.

The last possible solution for (F, E), $Ah(x) = \gamma(x)\alpha$ for constant functions $h(x) = \alpha$, $\alpha \in \mathbb{R}$, corresponds to the solution $Af = \gamma(x)f$ of (7.7) for positive functions $f \in C^k(I, \mathbb{R})$, $f > 0$, if $\gamma \not\equiv 0$. It therefore extends to the degenerate solution $Ag = \gamma g$ for all $g \in C^k(I, \mathbb{R})$, to be excluded. However, if $\gamma \equiv 0$, this means that A annihilates all constants functions $h(x) = \alpha$, also for $\alpha < 0$.

For $\alpha > 0$, among the three non-degenerate solutions for positive functions, only the first solution for A has the property that it annihilates the constant functions, since $Ah = \langle h', c \rangle = 0$. Then $A(-\mathbf{1}) = 0$, and by the operator equation (7.3) with $T(\mathbf{1}) = 0$, we also get that $0 = T((-\mathbf{1})^2) = -2T(-\mathbf{1}) + A(-\mathbf{1})^2$, $T(-\mathbf{1}) = 0$. Again by (7.3), we have for all $f \in C^k(I, \mathbb{R})$, choosing $g = -\mathbf{1}$ that $T(-f) = -Tf$. Therefore, T is odd, which leads to the formula for T for general functions $f \in C^k(I, \mathbb{R})$ in the case of the first solution.

The second and third solutions – without $\{\operatorname{sgn} f(x)\}$-term – are also odd. The operator T with the $\{\operatorname{sgn} f(x)\}$-term is the only solution which is not odd.

This proves Theorem 7.2. $\square$

7.2   Characterizations of the Laplacian

We now turn to characterizations of the Laplacian $\Delta = \sum_{i=1}^{n} \frac{\partial^2}{\partial x_i^2}$ on $C^2(I, \mathbb{R})$ for open sets $I \subset \mathbb{R}^n$ by the functional equation (7.2), orthogonal invariance and annihilation of affine functions, i.e., functions of the form $f(x) = \langle x, y_0 \rangle + x_0$, $x \in I$, for some fixed $x_0 \in \mathbb{R}$ and $y_0 \in \mathbb{R}^n$.

Definition. Let $n \in \mathbb{N}$, $k \in \mathbb{N}_0$ and $I := \{x \in \mathbb{R}^n \mid \|x\| < r\}$ be an open disc with $r > 0$ or $I = \mathbb{R}^n$, $r = \infty$. An operator $T : C^k(I, \mathbb{R}) \to C(I, \mathbb{R})$ is $\mathcal{O}(n)$-*invariant* if for all $f \in C^k(I, \mathbb{R})$ and all orthogonal maps $u \in \mathcal{O}(n)$, we have that $T(f \circ u) = (Tf) \circ u$.

Clearly, the Laplacian $\Delta : C^2(I, \mathbb{R}) \to C(I, \mathbb{R})$ is $\mathcal{O}(n)$-invariant.

Theorem 7.5 (Characterization of the Laplacian). *Let* $n \in \mathbb{N}$, $k \in \mathbb{N}_0$ *and* $I := \{x \in \mathbb{R}^n \mid \|x\| < r\}$ *be an open disc with* $r > 0$ *or* $I = \mathbb{R}^n$, $r = \infty$. *Suppose that* $T : C^k(I, \mathbb{R}) \to C(I, \mathbb{R})$ *and* $A : C^k(I, \mathbb{R}) \to C(I, \mathbb{R}^n)$ *are operators such that the second-order Leibniz rule equation*

$$T(f \cdot g)(x) = Tf(x) \cdot g(x) + f(x) \cdot Tg(x) + \langle Af(x), Ag(x) \rangle \qquad (7.16)$$

holds for all $f, g \in C^k(I, \mathbb{R})$ *and* $x \in I$. *Assume also that* A *is non-degenerate and that* T *is* $\mathcal{O}(n)$-*invariant and annihilates the constant functions. Then there are no solutions of* (7.16) *if* $k = 0$ *or* $k = 1$. *If* $k \geq 2$, *there are continuous functions* $c, d \in C([0, r), \mathbb{R})$, $c > 0$, *and* $U : I \to \mathcal{O}(n)$ *such that for all* $f \in C^k(I, \mathbb{R})$ *and* $x \in I$,

$$Tf(x) = \tfrac{1}{2}c(\|x\|)^2 \Delta f(x) + d(\|x\|)\langle f'(x), x \rangle, \quad Af(x) = c(\|x\|)\, U(x)f'(x). \quad (7.17)$$

If T annihilates all affine functions, $d = 0$, and then

$$Tf(x) = \tfrac{1}{2}c(\|x\|)^2\Delta f(x), \quad Af(x) = c(\|x\|)\, U(x)f'(x).$$

Conversely, these operators satisfy (7.16), and T is $\mathcal{O}(n)$-invariant and annihilates the constant or affine functions, respectively.

Remarks. (i) If $d = 0$, up to some radial function, T is the Laplacian and A is essentially the first derivative; choosing the constant function $c = \sqrt{2}$, T is precisely the Laplacian. If A is orthogonally invariant, too, U is given by a radial function $V : [0, r) \to \mathcal{O}(n)$, i.e., $U(x) = V(\|x\|)$.

The natural domain for T is the space $C^2(I, \mathbb{R})$ and for A is $C^1(I, \mathbb{R})$: If $k > 2$, the formula for T can be extended to $C^2(I, \mathbb{R})$, and if $k \geq 2$, the formula for A may be extended to $C^1(I, \mathbb{R})$.

(ii) There are many orthogonally invariant solutions T of (7.16), besides those given by (7.17), which do not annihilate the constant functions: One may take an arbitrary sum T of n solutions of equation (7.3) given in Theorem 7.2, with the only condition that the parameter functions a, b, c, d are radial functions, i.e., depend only on $\|x\|$.

If T is not orthogonally invariant but annihilates the constant functions, we may also classify the solutions of (7.16).

Theorem 7.6. Let $n \in \mathbb{N}$, $k \in \mathbb{N}_0$ and $I \subset \mathbb{R}^n$ be open. Let $T : C^k(I, \mathbb{R}) \to C(I, \mathbb{R})$ and $A : C^k(I, \mathbb{R}) \to C(I, \mathbb{R}^n)$ be operators satisfying the second-order Leibniz rule equation (7.16). Suppose that T annihilates the constant functions and that A is non-degenerate.

If $k = 0$ or $k = 1$, there are no solutions of (7.16).

If $k \geq 2$, there are continuous functions $c_i \in C(I, \mathbb{R}^n)$, $i \in \{1, \ldots, n\}$, and $d \in C(I, \mathbb{R}^n)$ such that for all $f \in C^k(I, \mathbb{R})$ and $x \in I$,

$$Tf(x) = \frac{1}{2}\sum_{i=1}^{n}\langle f''(x)c_i(x), c_i(x)\rangle + \langle f'(x), d(x)\rangle,$$

$$\|Af(x)\|^2 = \sum_{i=1}^{n}|\langle f'(x), c_i(x)\rangle|^2. \tag{7.18}$$

If T annihilates the affine functions, then $d = 0$.

Conversely, these operators satisfy (7.16), and T annihilates the constant or affine functions, respectively.

Remarks. (1) This means that T is a sum of multiples of changing second-order and first-order directional derivatives, and that A satisfies

$$\langle Af(x), Ag(x)\rangle = \sum_{i=1}^{n}\langle f'(x), c_i(x)\rangle\langle g'(x), c_i(x)\rangle,$$

$f, g \in C^k(I, \mathbb{R})$, $x \in I$. If $I = \mathbb{R}^n$ and T or A are *isotropic*, i.e., commute with all shifts S_y, $y \in \mathbb{R}^n$, $AS_y = S_y A$, where $S_y f(x) := f(x + y)$, $x \in \mathbb{R}^n$, the vectors c_i are constant. Then the directions of the directional derivatives do not depend on $x \in \mathbb{R}^n$. The same holds for d, if T is isotropic.

(2) Note that in Theorems 7.5 and 7.6 we do not impose any continuity conditions on T or A. Neither is linearity assumed; most of the solutions are non-linear, in fact.

There is a common part to the *Proof* of Theorems 7.5 and 7.6:
Suppose $T : C^k(I, \mathbb{R}) \to C(I, \mathbb{R})$ and $A : C^k(I, \mathbb{R}) \to C(I, \mathbb{R}^n)$ satisfy (7.16),

$$T(f \cdot g) = Tf \cdot g + f \cdot Tg + \langle Af, Ag \rangle, \quad f, g \in C^k(I, \mathbb{R}),$$

and that A is non-degenerate. For $i \in \{1, \ldots, n\}$ define the component operators $A_i : C^k(I, \mathbb{R}) \to C(I, \mathbb{R})$ by $Af(x) = (A_i f(x))_{i=1}^n$, in terms of the standard unit vector basis of $\mathbb{R}^n$, for all $f \in C^k(I, \mathbb{R})$, $x \in I$. Consider the equation for an operator $T_1 : C^k(I, \mathbb{R}) \to C(I, \mathbb{R})$,

$$T_1(f \cdot g) = T_1 f \cdot g + f \cdot T_1 g + A_1 f \cdot A_1 g,$$

whose general solution we know by Theorem 7.2 since A_1 is non-degenerate, too. For $f \in C^k(I, \mathbb{R})$, let $\widetilde{T} f := Tf - T_1 f$, $\widetilde{A} f := (0, A_2 f, \ldots, A_n f)$. Then

$$\widetilde{T}(f \cdot g) = \widetilde{T} f \cdot g + f \cdot \widetilde{T} g + \langle \widetilde{A} f, \widetilde{A} g \rangle, \quad f, g \in C^k(I, \mathbb{R}).$$

Continuing inductively, we see that (7.16) decomposes into n scalar equations

$$T_i(f \cdot g) = T_i f \cdot g + f \cdot T_i g + A_i f \cdot A_i g, \quad f, g \in C^k(I, \mathbb{R}), \ i \in \{1, \ldots, n\},$$

of the type considered in Theorem 7.2, $T_i, A_i : C^k(I, \mathbb{R}) \to C(I, \mathbb{R})$, such that

$$Tf(x) = \sum_{i=1}^n T_i f(x), \quad \langle Af(x), Ag(x) \rangle = \sum_{i=1}^n A_i f(x) A_i g(x), \tag{7.19}$$

$f, g \in C^k(I, \mathbb{R})$, $x \in I$, where the operators (T_i, A_i) satisfy (7.3) and the A_i are non-degenerate.

Therefore T uniquely determines the scalar products $\langle Af(x), Ag(x) \rangle$ and, in particular, $\|Af(x)\|$.

Hence, to prove Theorems 7.5 and 7.6 we have to add n solutions of the form established in Theorem 7.2 and analyze them under the specific assumptions of these theorems.

Proof of Theorem 7.6. By (7.19), T is a sum of n solutions T_i of (7.3) given in Theorem 7.2. None of the solutions T_i occurring in this sum can be of the second or third type, i.e., of the form

$$T_i f = \tfrac{1}{2} d_i^2 \ f \ (\ln |f|)^2 + a_i \ f \ (\ln |f|) + \langle f', b_i \rangle$$

or

$$T_i f = d_i^2 \ f \ (\{\operatorname{sgn} f\}|f|^{p_i} - 1) + a_i \ f \ (\ln|f|) + \langle f', b_i \rangle,$$

since choosing sufficiently many different constant functions $f_j = \alpha_j$ and using that T annihilates them, $Tf_j = 0$, we show that the resulting coefficient functions of the linearly independent terms $f \ (\ln|f|)^2$, $f \ (\ln|f|)$ or $f \ (\{\operatorname{sgn} f\}|f|^{p_i} - 1)$ possibly occurring in the solution formula for T all have to be zero, pointwise for any $x \in I$. Only the $T_i f = \langle f', b_i \rangle$ terms remain, but they may be considered as part of the first solution type. Since we have $f' = f'' = 0$ for constant functions f, the coefficient functions of terms involving first or second derivatives f' or f'' are not restricted, i.e., there is no limitation to add solutions of the first type.

Therefore we can and may only add solutions of the first type in Theorem 7.2. This requires $k \geq 2$ to get non-trivial solutions for T. If $k \geq 2$, there are continuous functions $b_i, c_i \in C(I, \mathbb{R}^n)$ for $i \in \{1, \ldots, n\}$ such that, with $d(x) = \sum_{i=1}^n b_i(x)$,

$$Tf(x) = \frac{1}{2} \sum_{i=1}^n \langle f''(x)c_i(x), c_i(x) \rangle + \langle f'(x), d(x) \rangle$$

and

$$\langle Af(x), Ag(x) \rangle = \sum_{i=1}^n \langle f'(x), c_i(x) \rangle \langle g'(x), c_i(x) \rangle,$$

$f, g \in C^k(I, \mathbb{R})$, $x \in I$. Since A is non-degenerate, $c = (c_i)_{i=1}^n \neq 0$. This proves Theorem 7.6. □

Proof of Theorem 7.5. (i) Theorem 7.5 is a special case of Theorem 7.6 which we just proved. We have to determine which of the solutions T of Theorem 7.6 are orthogonally invariant. Note that I is $\mathcal{O}(n)$-invariant, being a disc or $\mathbb{R}^n$. Since T is $\mathcal{O}(n)$-invariant, $T(f \circ u) = (Tf) \circ u$ for all $f \in C^k(I, \mathbb{R})$ and $u \in \mathcal{O}(n)$. By the chain rule $(f \circ u)'(x) = f'(u(x))u$, $(f \circ u)''(x) = f''(u(x))(u, u)$. Hence, using (7.18)

$$T(f \circ u)(x) = \frac{1}{2} \sum_{i=1}^n \langle (f \circ u)''(x)c_i(x), c_i(x) \rangle + \langle (f \circ u)'(x), d(x) \rangle$$

$$= \frac{1}{2} \sum_{i=1}^n \langle f''(u(x))u(c_i(x)), u(c_i(x)) \rangle + \langle f'(u(x)), u(d(x)) \rangle,$$

$$Tf(u(x)) = \frac{1}{2} \sum_{i=1}^n \langle f''(u(x))c_i(u(x)), c_i(u(x)) \rangle + \langle f'(u(x)), d(u(x)) \rangle.$$

The assumption of orthogonal invariance of T therefore means that

$$\frac{1}{2} \sum_{i=1}^n \left[\langle f''(u(x))u(c_i(x)), u(c_i(x)) \rangle - \langle f''(u(x))c_i(u(x)), c_i(u(x)) \rangle \right]$$

$$+ \langle f'(u(x)), u(d(x)) - d(u(x)) \rangle = 0 \quad (7.20)$$

holds for all $f \in C^k(I, \mathbb{R})$, $u \in \mathcal{O}(n)$ and $x \in I$. We claim that this implies $u(d(x)) = d(u(x))$ and

$$\sum_{i=1}^{n} \langle B\, u(c_i(x)), u(c_i(x)) \rangle = \sum_{i=1}^{n} \langle B\, c_i(u(x)), c_i(u(x)) \rangle \qquad (7.21)$$

for all $u \in \mathcal{O}(n)$, $x \in I$ and all matrices $B \in L(\mathbb{R}^n, \mathbb{R}^n)$. To verify this, let $b \in \mathbb{R}^n$ be an arbitrary vector and apply (7.20) to $f = \langle b, \cdot \rangle$, hence $f'(z) = \langle b, \cdot \rangle$ and $f''(z) = 0$ for all $z \in \mathbb{R}^n$, so that $\langle b, u(d(x)) - d(u(x)) \rangle = 0$ holds for all b which implies that $u(d(x)) = d(u(x))$ for all $u \in \mathcal{O}(n)$, $x \in I$. Therefore the first derivative term in (7.20) is always zero.

Let $B = B^t \in L(\mathbb{R}^n, \mathbb{R}^n)$ be an arbitrary symmetric matrix. Apply (7.20) to $f(x) = \frac{1}{2}\langle Bx, x \rangle$. Since then $f''(z) = \langle B\cdot, \cdot \rangle$ for all $z \in \mathbb{R}^n$, (7.20) implies (7.21) for all symmetric matrices B, $u \in \mathcal{O}(n)$ and $x \in I$. For all antisymmetric matrices $\widetilde{B} = -\widetilde{B}^t \in L(\mathbb{R}^n, \mathbb{R}^n)$ and all $v \in \mathbb{R}^n$ we have $\langle \widetilde{B}v, v \rangle = 0$. Decomposing an arbitrary matrix into a sum of a symmetric and an antisymmetric part, we conclude that (7.21) holds for *all* matrices $B \in L(\mathbb{R}^n, \mathbb{R}^n)$.

(ii) Fixing $x \in I$, for all $u \in \mathcal{O}_{n-1} := \{v \in \mathcal{O}(n) \mid v(x) = x\} \simeq \mathcal{O}(n-1)$, we have $d(x) = u(d(x))$. Hence, $d(x)$ must be in the direction of x, $d(x) = \Lambda(x) \cdot x$ with $\Lambda(x) \in \mathbb{R}$. To be orthogonally invariant, we need that Λ is a radial function of x, i.e., $\Lambda(x) = \lambda(\|x\|)$ for some continuous function $\lambda \in C([0, r), \mathbb{R})$. Hence $\langle f'(x), d(x) \rangle = \lambda(\|x\|)\langle f'(x), x \rangle$.

Let $c_i(x) = (c_{ip}(x))_{p=1}^{n} \in \mathbb{R}^n$ and $C(x) = (c_{ip}(x))_{i,p=1}^{n} \in L(\mathbb{R}^n, \mathbb{R}^n)$. Then (7.21) is equivalent to

$$\mathrm{trace}\big(B\, u\, C(x)^t C(x) u^t\big) = \mathrm{trace}\big(B\, C(u(x))^t C(u(x))\big),$$

for all $B \in L(\mathbb{R}^n, \mathbb{R}^n)$ and hence

$$u\, C(x)^t C(x) u = C(u(x))^t C(u(x)),$$

for all $u \in \mathcal{O}(n)$. Fixing $x \in I$ again, any $u \in \mathcal{O}_{n-1} := \{v \in \mathcal{O}(n) \mid v(x) = x\}$ $\simeq \mathcal{O}(n-1)$ maps $H := x^{\perp}$ into itself and $\mathcal{O}_{n-1}$ acts transitively on H. For $u \in \mathcal{O}_{n-1}$, $u\, C(x)^t C(x) u = C(x)^t C(x)$ and hence $C(x)^t C(x)|_H$ is a positive multiple of the identity on H and also maps x into a multiple of x. Therefore, there are $\Sigma(x), \Gamma(x) \in \mathbb{R}$ such that

$$C(x)^t C(x) = \Sigma(x)\, \mathrm{Id} + \Gamma(x) P_x,$$

where $P_x : \mathbb{R}^n \to \mathbb{R}^n$ is the projection onto x, $P_x = \langle \cdot, \frac{x}{\|x\|} \rangle \frac{x}{\|x\|}$, $x \neq 0$. We have with $\Delta f(x) = \text{trace}(f''(x))$ that

$$
\begin{aligned}
Tf(x) &= \frac{1}{2}\sum_{i=1}^{n}\langle f''(x)c_i(x), c_i(x)\rangle + \langle f'(x), d(x)\rangle \\
&= \frac{1}{2}\text{trace}\big(f''(x)C(x)^t C(x)\big) + \lambda(\|x\|)\langle f'(x), x\rangle \\
&= \frac{\Sigma(x)}{2}\text{trace}(f''(x)) + \frac{\Gamma(x)}{2}\text{trace}(f''(x)P_x) + \lambda(\|x\|)\langle f'(x), x\rangle \\
&= \frac{\Sigma(x)}{2}\Delta f(x) + \frac{\Gamma(x)}{2}\frac{\langle f''(x)x, x\rangle}{\|x\|^2} + \lambda(\|x\|)\langle f'(x), x\rangle.
\end{aligned}
$$

Since $f''(x)$ is zero on linear functions, $\langle f''(x)x, x\rangle = 0$. Further, $\Sigma(x)$ needs to be $\mathcal{O}(n)$-invariant as well. Hence there is a continuous function $\sigma \in C([0, r), \mathbb{R})$ such that $\Sigma(x) = \sigma(\|x\|)$. We find that

$$
Tf(x) = \frac{\sigma(\|x\|)}{2} \cdot \Delta f(x) + \lambda(\|x\|) \cdot \langle f'(x), x\rangle, \qquad f \in C^k(I, \mathbb{R}),\ x \in I.
$$

By (7.16), for all $f, g \in C^k(I, \mathbb{R})$, $x \in I$,

$$
\begin{aligned}
\langle Af(x), Ag(x)\rangle &= T(f \cdot g)(x) - Tf(x) \cdot g(x) - f(x) \cdot Tg(x) \\
&= \sigma(\|x\|)\langle f'(x), g'(x)\rangle. \tag{7.22}
\end{aligned}
$$

(iii) For $f = g$ this implies $\sigma \geq 0$. In fact, since A is non-degenerate, $\sigma(\|x\|) > 0$ for any $x \in I$. Let $\mu(t) := \sqrt{\sigma(t)}$ for $t \in [0, r)$. We will show that $Af(x)$ is, up to some orthogonal matrix $U(x)$, equal to $\mu(\|x\|)\, f'(x)$. To construct the orthogonal matrix, take any $c \in \mathbb{R}^n$, consider the linear function $f_c = \langle \cdot, c\rangle$ and define a map $U(x) : \mathbb{R}^n \to \mathbb{R}^n$ for any $x \in I$ by $U(x)(c) := \frac{1}{\mu(\|x\|)}A(f_c)(x)$. Since $f_c'(x) = c$ for any $x \in I$, we find for all $c, d \in \mathbb{R}^n$, $g \in C^k(I, \mathbb{R})$ and $x \in I$, using (7.22),

$$
\begin{aligned}
\langle U(x)(c) + U(x)(d), Ag(x)\rangle &= \frac{1}{\mu(\|x\|)}\langle A(f_c)(x) + A(f_d)(x), Ag(x)\rangle \\
&= \mu(\|x\|)\langle c + d, Ag(x)\rangle = \frac{1}{\mu(\|x\|)}\langle A(f_{c+d})(x), Ag(x)\rangle \\
&= \langle U(x)(c + d), Ag(x)\rangle. \tag{7.23}
\end{aligned}
$$

Let $e_i : I \to \mathbb{R}$ denote the i-th coordinate function, for $i \in \{1, \ldots, n\}$. By (7.22), the vectors $A(e_i)(x)$ are linearly independent and therefore span $\mathbb{R}^n$. Choosing $g = e_i$, equation (7.23) implies that $U(x)(c) + U(x)(d) = U(x)(c + d)$. Similarly, $U(x)(\lambda c) = \lambda U(x)(c)$. Hence $U(x)$ is linear. Since by (7.22)

$$
\|U(x)(c)\|^2 = \frac{1}{\sigma(\|x\|)}\|A(f_c)(x)\|^2 = \|f_c'(x)\|^2 = \|c\|^2,
$$

$U(x) \in \mathcal{O}(n)$ is an orthogonal matrix, and (7.22) yields for any $g \in C^k(I, \mathbb{R})$, $x \in I$ and $c \in \mathbb{R}^n$ that

$$\langle c, g'(x) \rangle = \langle f'_c(x), g'(x) \rangle = \frac{1}{\sigma(\|x\|)} \langle A(f_c)(x), Ag(x) \rangle$$

$$= \frac{1}{\mu(\|x\|)} \langle U(x)(c), Ag(x) \rangle = \frac{1}{\mu(\|x\|)} \langle c, U(x)^t Ag(x) \rangle.$$

Therefore $U(x)^t Ag(x) = \mu(\|x\|)g'(x)$, $Ag(x) = \mu(\|x\|)U(x)g'(x)$. Clearly $U : I \rightarrow \mathcal{O}(n)$, mapping x to $U(x)$, is continuous, since Af is continuous for all $f \in C^k(I, \mathbb{R})$. Hence

$$Tf(x) = \tfrac{1}{2}\mu(\|x\|)^2 \Delta f(x) + \lambda(\|x\|) \langle f'(x), x \rangle, \quad Af(x) = \mu(\|x\|) \, U(x)f'(x),$$

which is the solution (7.17) of Theorem 7.5. If T is additionally assumed to annihilate the affine functions, this requires $\lambda = 0$, and then

$$Tf(x) = \tfrac{1}{2}\mu(\|x\|)^2 \Delta f(x), \quad Af(x) = \mu(\|x\|) \, U(x)f'(x),$$

proving Theorem 7.5. $\qquad\qquad\qquad\qquad\qquad\qquad\qquad\qquad\qquad\qquad\square$

7.3 Stability of the Leibniz rule

In Theorem 5.1 we showed that the Leibniz rule equation is stable under changing each occurrence of T to different operators. There is a similar relaxation result for the second-order Leibniz rule which generalizes Theorem 7.2. It shows that even if we significantly relax the second-order Leibniz rule equation, the solutions will not change by much.

Theorem 7.7 (Relaxation of the Leibniz rule of second order). *Let $n \in \mathbb{N}$, $k \in \mathbb{N}_0$ and $I \subset \mathbb{R}^n$ be open. Suppose that $V, T_1, T_2, A : C^k(I, \mathbb{R}) \rightarrow C(I, \mathbb{R})$ are operators satisfying the equation*

$$V(f \cdot g) = T_1 f \cdot g + f \cdot T_2 g + Af \cdot Ag, \tag{7.24}$$

for all $f, g \in C^k(I, \mathbb{R})$, and that A is non-degenerate. Then there is a continuous function $\gamma \in C(I, \mathbb{R})$ such that $T_1 f - T_2 f = \gamma f$ for all $f \in C^k(I, \mathbb{R})$.

Put $T := \tfrac{1}{2}(T_1 + T_2)$. Then there are functions $e_1, e_2 \in C(I, \mathbb{R})$, $a, d, p : I \rightarrow \mathbb{R}$, $p > -1$ and $b, c : I \rightarrow \mathbb{R}^n$, which are continuous except in isolated points of I where different solutions join, such that with the homogeneous solution

$$Rf = a \, f \ln|f| + \langle b, f' \rangle, \quad f \in C^k(I, \mathbb{R}),$$

the operators V, T and A have the form

$$Vf = Uf + (e_1^2 + 2e_2)f, \quad Tf = Uf - e_1 Bf + e_2 f, \quad Af = Bf + e_1 f,$$

where (U, B) satisfy (7.3), i.e., are of one of the following three forms, possibly combined:
either

$$Uf = \tfrac{1}{2}\langle f''c, c\rangle + Rf, \quad Bf = \langle f', c\rangle, \quad \text{with } k \geq 2,$$

or

$$Uf = \tfrac{1}{2}d^2 f(\ln|f|)^2 + Rf, \quad Bf = df \ln|f|,$$

or

$$Uf = d^2 f(\{\operatorname{sgn} f\}|f|^p - 1) + Rf, \quad Bf = df(\{\operatorname{sgn} f\}|f|^p - 1).$$

In the last case $\{\operatorname{sgn} f\}$ appears simultaneously in U and B or not at all. Conversely, the operators (V, T, A) satisfy (7.24) with $T = T_1 = T_2$.

Remarks. (a) Again, we do not impose continuity conditions on any of the operators (V, T_1, T_2, A). If Af depends non-trivially on the derivative of f, only the first form of the solution is possible. Then, in dimension 1 ($n = 1$), the operators V and $T = \tfrac{1}{2}(T_1 + T_2)$ are general second-order differential operators, up to a term involving $f \ln|f|$. If V, T_1, T_2 and A are given, this type of solution contains just one form when V is of Sturm-Liouville type, $Vf = (pf')' + qf$, with $p = \tfrac{1}{2}c^2$, $p' = cc' = b$ and $q = e_1^2 + 2e_2 = (A1)^2 + 2(T1)$. Thus in a certain way, the relaxed Leibniz rule of the second order is just an algebraic understanding of general second-order differential operators, and of Sturm-Liouville operators, in particular, up to a term with $f \ln|f|$. Note that V, T are naturally defined on $C^2(I, \mathbb{R})$ and A on $C^1(I, \mathbb{R})$.

(b) In the case of the second and third solutions, V and T are naturally defined on $C^1(I, \mathbb{R})$, or if $b \equiv 0$, even on $C(I, \mathbb{R})$ whereas A is naturally defined on $C(I, \mathbb{R})$.

(c) To illustrate Theorem 7.7, suppose that $n = 1$ and $V = D^2$ is just the second derivative. Then $c = \sqrt{2}$, $b = a = 0$ and $c_2 = -\tfrac{1}{2}e_1^2$. Then $Af = \sqrt{2}f' + e_1 f$ and $Tf = f'' - \sqrt{2}e_1 f' - \tfrac{1}{2}e_1^2 f$.

Proof. Exchanging $f, g \in C^k(I, \mathbb{R})$ in (7.24) and taking differences, we find $(T_1 f - T_2 f) \cdot g = (T_1 g - T_2 g) \cdot f$. For $g := 1$ and $\gamma := T_1 1 - T_2 1$, we get that $T_1 f - T_2 f = \gamma f$. Let $T := \tfrac{1}{2}(T_1 + T_2)$. Then (7.24) holds with T_1, T_2 being both replaced by T

$$V(f \cdot g) = Tf \cdot g + f \cdot Tg + Af \cdot Ag, \quad f, g \in C^k(I, \mathbb{R}). \tag{7.25}$$

For $g = 1$ this means with $e_1 := A1$ and $e_2 := T1$ that

$$Vf = Tf + e_1 \cdot Af + e_2 \cdot f.$$

Inserting this back into (7.25), we find

$$T(f \cdot g) + e_1 \cdot A(f \cdot g) + e_2 \cdot f \cdot g = Tf \cdot g + f \cdot Tg + Af \cdot Ag. \tag{7.26}$$

Define new operators $U, B : C^k(I, \mathbb{R}) \to C(I, \mathbb{R})$ by

$$Uf := Tf + e_1 \cdot Af - (e_1^2 + e_2) \cdot f, \quad Bf := Af - e_1 \cdot f,$$

for $f \in C^k(I, \mathbb{R})$. Equation (7.26) means in terms of U and B for $f, g \in C^k(I, \mathbb{R})$,

$$U(f \cdot g) + (e_1^2 + 2e_2) \cdot f \cdot g = \big(Uf - e_1 \cdot (Bf + e_1 \cdot f) + (e_1^2 + e_2) \cdot f\big) \cdot g$$
$$+ f \cdot \big(Ug - e_1(Bg + e_1 \cdot g) + (e_1^2 + e_2) \cdot g\big) + (Bf + e_1 \cdot f)(Bg + e_1 \cdot g),$$

which leads to

$$U(f \cdot g) = Uf \cdot g + f \cdot Ug + Bf \cdot Bg. \tag{7.27}$$

This is equation (7.3) for (U, B) instead of (T, A). Theorem 7.2 gives the form of solutions, provided that B is non-degenerate. However, $Bf = Af - e_1 \cdot f$, and A was assumed to be non-degenerate. This implies that also B is non-degenerate. Now the form of solutions U and B of (7.27) follows directly from Theorem 7.2.

Using the definition of U and B and the formula $Vf = Tf + e_1 \cdot Af + e_2 f$, we reconstruct V, T and A from U and B via

$$Vf = Uf + (e_1^2 + 2e_2) \cdot f, \quad Tf = Uf - e_1 \cdot Bf + e_2 \cdot f, \quad Af = Bf + e_1 \cdot f.$$

It is easily checked by direct calculation that these maps (V, T, A) satisfy (7.24) with $T = T_1 = T_2$ using that (U, B) satisfy (7.3). This proves Theorem 7.7. $\square$

We may start from a different point of view, when investigating the structure of the Leibniz type equation for the Laplacian

$$\Delta(f \cdot g)(x) = \Delta f(x) \cdot g(x) + f(x) \cdot \Delta g(x) + 2\langle f'(x), g'(x)\rangle,$$
$$f, g \in C^2(I, \mathbb{R}), \ x \in I,$$

other than by the operator equation (7.16), namely: Consider $2\langle f'(x), g'(x)\rangle$ as a perturbation term of the Leibniz rule and replace it by a function B of the parameters $(x, f(x), f'(x), g(x), g'(x))$, leading to the equation

$$T(f \cdot g)(x) = Tf(x) \cdot g(x) + f(x) \cdot Tg(x) + B\big(x, f(x), f'(x), g(x), g'(x)\big).$$

This is similar to our perturbation scheme in Section 5.3. The equation is not directly comparable to (7.16): On the one hand, it is more special, since the perturbation is not by an operator term like $\langle Af(x), Ag(x)\rangle$ in (7.16) but is given by a locally defined function B.

On the other hand, it is more general since B is not assumed to be given in product form separating f and g. The analogue of Theorem 7.5 states in this case, in the spirit of Proposition 5.7:

Theorem 7.8 (Stability of the Laplacian). *Let* $n \in \mathbb{N}$ *and* $I := \{x \in \mathbb{R}^n \mid \|x\| < r\}$ *be an open disc with* $r > 0$ *or* $I = \mathbb{R}^n$ *with* $r = \infty$. *Assume that* $T : C^2(I, \mathbb{R}) \to C(I, \mathbb{R})$ *is an operator and* $B : I \times \mathbb{R} \times \mathbb{R}^n \times \mathbb{R} \times \mathbb{R}^n \to \mathbb{R}$ *a function such that*

$$T(f \cdot g)(x) = Tf(x) \cdot g(x) + f(x) \cdot Tg(x) + B(x, f(x), f'(x), g(x), g'(x)) \quad (7.28)$$

holds for all $f, g \in C^2(I, \mathbb{R})$ *and* $x \in I$. *Suppose further that* T *is* $\mathcal{O}(n)$*-invariant and annihilates all affine functions. Then there is a continuous function* $d \in C([0, r), \mathbb{R})$ *such that for all* $f \in C^2(I, \mathbb{R})$, $x \in I$,

$$Tf(x) = \tfrac{1}{2}d(\|x\|)\Delta f(x),$$
$$B(x, f(x), f'(x), g(x), g'(x)) = d(\|x\|)\langle f'(x), g'(x) \rangle.$$

Proof. The localization of T is quickly verified. If $J \subset I$ is open and $f_1, f_2 \in C^2(I, \mathbb{R})$ are such that $f_1|_J = f_2|_J$, we have for any $x \in J$ and any function $g \in C^2(I, \mathbb{R})$ with $g(x) \neq 0$ and $\text{supp}(g) \subset J$ that $f_1 \cdot g = f_2 \cdot g$ and $T(f_1 \cdot g) = T(f_2 \cdot g)$. Hence, by (7.28), $(Tf_1(x) - Tf_2(x)) \cdot g(x) = (f_2(x) - f_1(x)) \cdot Tg(x) = 0$, since $B(x, f_1(x), f_1'(x), g(x), g'(x)) = B(x, f_2(x), f_2'(x), g(x), g'(x))$. Therefore, $Tf_1(x) = Tf_2(x)$, $Tf_1|_J = Tf_2|_J$. Hence, by Proposition 3.6 there is a function $F : I \times \mathbb{R}^N \to \mathbb{R}$ with $N := 1 + n + \frac{n(n+1)}{2} = \frac{(n+2)(n+1)}{2} = \binom{n+2}{2}$ such that

$$Tf(x) = F(x, f(x), f'(x), f''(x)), \quad f \in C^2(I, \mathbb{R}), \ x \in I,$$

holds. Here $f''(x)$ is represented by the $\frac{n(n+1)}{2}$ independent partial derivatives $\left(\frac{\partial^2 f(x)}{\partial x_i \partial x_j}\right)_{1 \leq i \leq j \leq n}$.

For any $\alpha = (\alpha_0, \alpha_1, \alpha_2)$, $\beta = (\beta_0, \beta_1, \beta_2) \in \mathbb{R}^N = \mathbb{R} \times \mathbb{R}^n \times \mathbb{R}^{n(n+1)/2}$ and $x \in I$, we may choose $f, g \in C^2(I, \mathbb{R})$ such that $f(x) = \alpha_0$, $f'(x) = \alpha_1$, $f''(x) = \alpha_2$, $g(x) = \beta_0$, $g'(x) = \beta_1$, $g''(x) = \beta_2$, with the above representation of the second derivative. Therefore, the operator equation (7.28) is equivalent to the functional equation for F,

$$F(x, \alpha_0\beta_0, \alpha_0\beta_1 + \beta_0\alpha_1, \alpha_0\beta_2 + \beta_0\alpha_2 + 2\alpha_1\beta_1)$$
$$= F(x, \alpha_0, \alpha_1, \alpha_2)\beta_0 + F(x, \beta_0, \beta_1, \beta_2)\alpha_0 + B(x, \alpha_0, \alpha_1, \beta_0, \beta_1), \quad (7.29)$$

for all $\alpha, \beta \in \mathbb{R}^N$, where $2\alpha_1\beta_1$ has to be read as $(\alpha_{1,i}\beta_{1,j} + \alpha_{1,j}\beta_{1,i})_{1 \leq i \leq j \leq n}$. By assumption, $T\mathbf{1} = 0$, hence $F(x, 1, 0, 0) = 0$ for all $x \in I$, implying by (7.29) that $B(x, 1, 0, 1, 0) = 0$. Choosing $\alpha_0 = \beta_0 = 1$, $\alpha_1 = \beta_1 = 0$ in (7.29) yields

$$F(x, 1, 0, \alpha_2 + \beta_2) = F(x, 1, 0, \alpha_2) + F(x, 1, 0, \beta_2) + B(x, 1, 0, 1, 0)$$
$$= F(x, 1, 0, \alpha_2) + F(x, 1, 0, \beta_2).$$

Therefore, $F(x, 1, 0, \cdot)$ is additive. For $\beta_0 = 1$, $\beta_1 = 0$, $\alpha_2 = \beta_2 = 0$, we get from (7.29)

$$F(x, \alpha_0, \alpha_1, 0) = F(x, \alpha_0, \alpha_1, 0) + F(x, 1, 0, 0)\alpha_0 + B(x, \alpha_0, \alpha_1, 1, 0),$$

which implies that $B(x, \alpha_0, \alpha_1, 1, 0) = 0$ for all α_0, α_1. Next, putting $\beta_0 = 1$, $\beta_1 = 0$ and $\alpha_2 = 0$ in (7.29), we find that

$$F(x, \alpha_0, \alpha_1, \alpha_0\beta_2) = F(x, \alpha_0, \alpha_1, 0) + F(x, 1, 0, \beta_2)\alpha_0 + B(x, \alpha_0, \alpha_1, 1, 0).$$

Since T is zero on all affine functions $f(y) = \langle \alpha_1, y \rangle + (\alpha_0 - \langle \alpha_1, x_0 \rangle)$, where $x_0 \in \mathbb{R}^n$ is fixed, we have $0 = Tf(x_0) = F(x_0, \alpha_0, \alpha_1, 0)$. Therefore,

$$F(x, \alpha_0, \alpha_1, \alpha_0\beta_2) = F(x, 1, 0, \beta_2)\alpha_0,$$

and hence, for all $\alpha_0 \neq 0$,

$$F(x, \alpha_0, \alpha_1, \alpha_2) = F\left(x, 1, 0, \frac{\alpha_2}{\alpha_0}\right)\alpha_0.$$

Since $F(x, 1, 0, \cdot)$ is additive and

$$Tf(x) = F\big(x, f(x), f'(x), f''(x)\big) = F\left(x, 1, 0, \frac{f''(x)}{f(x)}\right)f(x)$$

is continuous for all $f \in C^2(I, \mathbb{R})$, $f(x) \neq 0$, Theorem 2.6 yields that there is a continuous function $c \in C(I, \mathbb{R}^M)$, $M = \frac{n(n+1)}{2}$, such that

$$F(x, 1, 0, \alpha_2) = \langle c(x), \alpha_2 \rangle,$$

implying $F(x, \alpha_0, \alpha_1, \alpha_2) = \langle c(x), \alpha_2 \rangle$. Hence F is independent of α_0 and α_1. Therefore,

$$Tf(x) = F\big(x, f(x), f'(x), f''(x)\big) = \langle f''(x), c(x) \rangle$$

$$= \sum_{1 \leq i \leq j \leq n} c_{ij}(x) \frac{\partial^2 f}{\partial x_i \partial x_j}(x).$$

The requirement of orthogonal invariance of $Tf(x)$ then yields, as in the proof of Theorem 7.5, that there is a function $d : [0, r) \to \mathbb{R}$ such that $Tf(x) = \frac{1}{2}d(\|x\|)\Delta f(x)$ for all $f \in C^2(I, \mathbb{R})$ and $x \in I$. Inserting this back into (7.28) yields for B,

$$B\big(x, f(x), f'(x), g(x), g'(x)\big)$$
$$= \tfrac{1}{2}d(\|x\|)\big(\Delta(f \cdot g)(x) - \Delta f(x) \cdot g(x) - f(x) \cdot \Delta g(x)\big)$$
$$= d(\|x\|)\langle f'(x), g'(x) \rangle.$$

This finishes the proof of Theorem 7.8.   □

Remark. If conversely, in the setting of Theorem 7.8, the function B is given by

$$B\big(x, f(x), f'(x), g(x), g'(x)\big) = d(\|x\|)\langle f'(x), g'(x) \rangle,$$

and T satisfies equation (7.28), T is just a multiple of the Laplacian,

$$Tf(x) = \tfrac{1}{2}d(\|x\|)\Delta f(x).$$

7.4 Notes and References

Theorem 7.2 is an extension of Theorem 1 in [KM4] and Theorem 2 in [KM5], where the case $I = \mathbb{R}^n$ and $k = 2$ is studied and solved.

The special case $I = \mathbb{R}^n$, $k = 2$ of Theorems 7.5 and 7.6 was proved in [KM5, Theorem 1].

Theorem 7.7 was proved for $k = 2$, $n = 1$ in [KM7, Theorem 7].

Theorem 7.8 is found in [KM8], Theorem 2.

Chapter 8

Non-localization Results

In the case of the Leibniz-type equations (7.3) and (3.7) there were easy examples that the intertwined operators T and A need not be localized, if the map A is not non-degenerate. In this chapter we study what can be said about the solutions of these equations if A is degenerate. In this situation, A is in a resonance state with respect to other operators present.

8.1 The second-order Leibniz rule equation

Let $I \subset \mathbb{R}$ be open. We now return to study the solutions of the second-order Leibniz rule operator equation

$$T(f \cdot g) = Tf \cdot g + f \cdot Tg + Af \cdot Ag, \quad f, g \in C^k(I, \mathbb{R}), \tag{8.1}$$

for operators $T, A : C^k(I, \mathbb{R}) \to C(I, \mathbb{R})$, but now without the assumption of non-degeneration of A. In this case the operators might not be localized, as the following simple example mentioned in the last chapter shows:

$$Tf(x) = f(x+1) - f(x), \ Af(x) = f(x) - f(x+1), \quad f \in C(\mathbb{R}, \mathbb{R}), \ x \in \mathbb{R}.$$

In this section we study the consequences of the non-localization of T and A for the solutions of equation (8.1) in the case of open subsets I of $\mathbb{R}$. First of all, the example mentioned may be extended in the following way.

Definition. Let $I \subset \mathbb{R}$ be open and $k \in \mathbb{N}_0$. An operator $S : C^k(I, \mathbb{R}) \to C(I, \mathbb{R})$ is *multiplicative* if $S(f \cdot g) = Sf \cdot Sg$ holds for any $f, g \in C^k(I, \mathbb{R})$.

Example. Suppose $S : C^k(I, \mathbb{R}) \to C(I, \mathbb{R})$ is multiplicative. Define $T, A : C^k(I, \mathbb{R}) \to C(I, \mathbb{R})$ by

$$Tf := Sf - f, \quad Af := f - Sf, \quad f \in C^k(I, \mathbb{R}).$$

© Springer Nature Switzerland AG 2018
H. König, V. Milman, *Operator Relations Characterizing Derivatives*,
https://doi.org/10.1007/978-3-030-00241-1_8

Then T and A satisfy (8.1) since

$$T(f \cdot g) = Sf \cdot Sg - f \cdot g = (Sf - f) \cdot g + f \cdot (Sg - g) + (f - Sf) \cdot (g - Sg)$$
$$= Tf \cdot g + f \cdot Tg + Af \cdot Ag.$$

Simple examples of multiplicative maps S can be given in the form

$$Sf(x) := \prod_{j=1}^{m} \left| f(\phi_j(x)) \right|^{p_j} \{ \operatorname{sgn} f(\phi_j(x)) \},$$

where $m \in \mathbb{N}$, $p_j > 0$ and $\phi_j : I \to I$ are continuous functions, $j \in \{1, \ldots, m\}$. If $\phi_j(x) \neq x$ for some j and some x, then S, T and A are not localized. Here the term $\{ \operatorname{sgn} f(\phi_j(x)) \}$ may appear or not, independently for each j. More generally, let (Ω, μ) be a measure space and $\phi_\omega : I \to I$ be continuous functions for all $\omega \in \Omega$ such that $\ln |f(\phi_\omega(x))|$ is μ-integrable in $\omega \in \Omega$ for all $x \in I$ and $f \in C^k(I, \mathbb{R})$. Then

$$Sf(x) := \exp \left(\int_\Omega \ln \left| f(\phi_\omega(x)) \right| d\mu(\omega) \right)$$

defines a multiplicative map $S : C^k(I, \mathbb{R}) \to C(I, \mathbb{R})$.

The paper [LS] by Lešnjak, and Šemrl describes the *continuous* multiplicative maps $S : C(X, \mathbb{R}) \to C(Y, \mathbb{R})$ for *compact* Hausdorff spaces X and Y in terms of operators of this form. However, this description only concerns the case $k = 0$ and compact spaces. *Bijective* multiplicative maps $S : C(I, \mathbb{R}) \to C(I, \mathbb{R})$ were characterized by Milgram [M], having the form $Sf(x) = |f(\varphi(x))|^{p(x)} \operatorname{sgn} f(\varphi(x))$ for some homeomorphism φ of I and some continuous function p on I. Bijective multiplicative maps $S : C^k(I, \mathbb{R}) \to C^k(I, \mathbb{R})$ for $k \in \mathbb{N}$ have the form $Sf(x) = f(\varphi(x))$ for some C^k-diffeomorphism φ of I, cf. Mrčun, Šemrl [MS] and Alesker, Artstein-Avidan, Faifman, Milman [AAFM], [AFM]. However, to the best of our knowledge, non-bijective multiplicative operators $S : C^k(I, \mathbb{R}) \to C(I, \mathbb{R})$, $k \in \mathbb{N}$, have not been classified. Hence there are very many non-localized solutions (T, A) of (8.1).

Non-localized solutions of (8.1) such as

$$Tf(x) = f(\varphi(x)) - f(x), \quad Af(x) = f(x) - f(\varphi(x)),$$

where $\varphi : I \to I$ is continuous and $\varphi(x) \neq x$ for some x, yield degenerate operators A in the sense of Chapter 7: They have the property that for such x there exists an open interval $J \subset I$ with $x \in J$ such that all functions $g \in C^k(I, \mathbb{R})$ with support in J are annihilated by S, defined by $Sf(x) := f(\varphi(x))$, and hence $Tg = -g$, $Ag = g$ near x.

Motivated by this phenomenon, we introduce the following set P of points $x \in I$ where localization of T and A might fail:

Definition. Let $k \in \mathbb{N}_0$ and $I \subset \mathbb{R}$ be open. Let $A : C^k(I, \mathbb{R}) \to C(I, \mathbb{R})$ be an operator. Define

$$P := \{x \in I \mid \exists J \subset I \text{ open with } x \in J \quad \exists \lambda \in C(J, \mathbb{R}) \quad \forall g \in C^k(I, \mathbb{R}),$$
$$\text{supp}\, g \subset J \quad Ag|_J = \lambda g|_J\}.$$

By definition, $P \subset I$ is an open set. Note that $Ag = \lambda g$ automatically implies that λ is continuous since $\text{Im}(A) \subset C(I, \mathbb{R})$. We also introduce a *localization set* L:

Definition. Let $k \in \mathbb{N}_0$ and $I \subset \mathbb{R}$ be open. Let $T, A : C^k(I, \mathbb{R}) \to C(I, \mathbb{R})$ be operators satisfying the Leibniz rule type equation (8.1). Define

$$L := \left\{ x \in I \;\middle|\; \exists F(x, \cdot), E(x, \cdot) : \mathbb{R}^{k+1} \to \mathbb{R} \right.$$
$$\left. \forall f \in C^k(I, \mathbb{R}) \begin{bmatrix} Tf(x) = F\big(x, f(x), \ldots, f^{(k)}(x)\big) \\ Af(x) = E\big(x, f(x), \ldots, f^{(k)}(x)\big) \end{bmatrix} \right\}.$$

If T and A are not localized in x, x belongs to $\overline{P}$:

Proposition 8.1. *We have that* $\overline{P} \cup L = I$. *However,* $L \cap P \neq \emptyset$ *is possible.*

Proof. Assume that $x_0 \in I \setminus \overline{P}$. We claim that $x_0 \in L$. Since $I \setminus \overline{P}$ is open, there is an open interval $\widetilde{J} \subset I \setminus \overline{P}$ with $x_0 \in \widetilde{J}$. Let $J \subset \widetilde{J}$ be an arbitrary open subinterval of $\widetilde{J}$ and suppose that $f_1, f_2 \in C^k(I, \mathbb{R})$ satisfy $f_1|_J = f_2|_J$. We claim that $Tf_1|_J = Tf_2|_J$ and $Af_1|_J = Af_2|_J$.

Take any $y \in J$. Since $y \notin P$, for any open set $J_1 \subset J$ with $y \in J_1$ we may choose $g_1, g_2 \in C^k(I, \mathbb{R})$ with $\text{supp}(g_1), \text{supp}(g_2) \subset J_1$ such that (g_1, Ag_1) and (g_2, Ag_2) are not proportional on J_1, i.e., such that there is $z_1 \in J_1$ such that $(g_1(z_1), Ag_1(z_1)), (g_2(z_1), Ag_2(z_1)) \in \mathbb{R}^2$ are linearly independent. We iterate this procedure: Choose a decreasing set of open intervals $J_{\ell+1} \subset J_\ell \subset \cdots \subset J_1 \subset J$ with $y \in J_\ell$ and lengths $|J_\ell| \to 0$ as $\ell \to \infty$. Find functions $g_1^\ell, g_2^\ell \in C^k(I, \mathbb{R})$ with $\text{supp}(g_1^\ell), \text{supp}(g_2^\ell) \subset J_\ell$ and $z_\ell \in J_\ell$ such that $(g_1^\ell(z_\ell), Ag_1^\ell(z_\ell)), (g_2^\ell(z_\ell), Ag_2^\ell(z_\ell)) \in \mathbb{R}^2$ are linearly independent. Since $f_1 \cdot g_j^\ell = f_2 \cdot g_j^\ell$ for all $\ell \in \mathbb{N}$ and $j \in \{1, 2\}$, using equation (8.1) for these functions and taking differences yields for $j = 1, 2$ and $\ell \in \mathbb{N}$

$$\big(Tf_1(z_\ell) - Tf_2(z_\ell)\big) \cdot g_j^\ell(z_\ell) + \big(Af_1(z_\ell) - Af_2(z_\ell)\big) \cdot Ag_j^\ell(z_\ell) = 0.$$

The linear independence of $(g_j^\ell(z_\ell), Ag_j^\ell(z_\ell)) \in \mathbb{R}^2$ for $j = 1, 2$ then implies $Tf_1(z_\ell) = Tf_2(z_\ell)$ and $Af_1(z_\ell) = Af_2(z_\ell)$. Since $|J_\ell| \to 0$, $\lim_{\ell \to \infty} z_\ell = y$ and the continuity of the functions Tf_j, Af_j implies that $Tf_1(y) = Tf_2(y)$ and $Af_1(y) = Af_2(y)$. Since $y \in J$ was arbitrary, $Tf_1|_J = Tf_2|_J$ and $Af_1|_J = Af_2|_J$.

Now Proposition 3.3, applied on the open interval $\widetilde{J}$, implies that T and A are localized on $\widetilde{J}$, i.e., that there are functions $F, E : \widetilde{J} \times \mathbb{R}^{k+1} \to \mathbb{R}$ such that $Tf(x) = F(x, f(x), \ldots, f^{(k)}(x))$ and $Af(x) = E(x, f(x), \ldots, f^{(k)}(x))$ for all $f \in C^k(I, \mathbb{R})$ and $x \in \widetilde{J} \subset I \setminus \overline{P}$. Hence $x_0 \in \widetilde{J} \subset L$.

The following example shows that $P \cap L \neq \emptyset$ is possible. Let $\lambda \in C(I, \mathbb{R})$ and define $T, A : C^k(I, \mathbb{R}) \to C(I, \mathbb{R})$ by $Tf := -\lambda^2 f$, $Af = \lambda f$ for $f \in C^k(I, \mathbb{R})$. Then (8.1) is satisfied for (T, A), but every point $x \in I$ is in L and in P.

Therefore, any point $x \in I$ is either in L or in P or in $\partial P = \overline{P} \setminus P$. $\square$

The following main result of this chapter explains the structure of the solution of (8.1) in these three possible cases.

Theorem 8.2 (Second-order Leibniz rule with resonance). *Suppose that $k \in \mathbb{N}_0$, $I \subset \mathbb{R}$ is an open interval and $T, A : C^k(I, \mathbb{R}) \to C(I, \mathbb{R})$ are operators satisfying the second-order Leibniz rule equation*

$$T(f \cdot g) = Tf \cdot g + f \cdot Tg + Af \cdot Ag, \quad f, g \in C^k(I, \mathbb{R}). \tag{8.1}$$

Let P be defined as above. Then there are pairwise disjoint subsets $I_1, I_2, I_3 \subset I$, some of which might be empty, I_1 and I_3 open, such that $I \setminus \overline{P} = I_1 \cup I_2 \cup I_3$, and there are functions $a, b, d : I \to \mathbb{R}$ which are continuous on $I \setminus (\partial P \cup \partial I_3)$ such that after subtracting from T the solution R of the homogeneous Leibniz rule equation given by

$$Rf(x) := a(x) f(x) \ln |f(x)| + b(x) f'(x), \quad f \in C^k(I, \mathbb{R}), \ x \in I,$$

the operators $T_1 := T - R$ and A have the following form:

(a) *On $I \setminus \overline{P}$ the operators T and A are localized and*

$$T_1 f(x) = \tfrac{1}{2} d(x)^2 f''(x), \quad Af(x) = f'(x), \quad x \in I_1 \quad (k \geq 2),$$

or

$$T_1 f(x) = \tfrac{1}{2} d(x)^2 f(x) \big(\ln |f(x)| \big)^2, \quad Af(x) = f(x) \ln |f(x)|, \quad x \in I_2,$$

or

$$T_1 f(x) = d(x) Af(x),$$
$$Af(x) = d(x) \big[\{ \operatorname{sgn} f(x) \} |f(x)|^{p(x)} - f(x) \big], \quad x \in I_3.$$

Here $p \in C(I_3, \mathbb{R})$, $p \geq 0$, and the term $\{ \operatorname{sgn} f(x) \}$ may be present or not, yielding two different solutions on I_3.

(b) *On P the operators T and A are possibly not localized, but $T + dA$ is localized and satisfies the (ordinary) Leibniz rule (3.1). Further, there is a multiplicative operator $S : C^k(I, \mathbb{R}) \to C(P, \mathbb{R})$,*

$$S(f \cdot g)(x) = Sf(x) \cdot Sg(x), \quad f, g \in C^k(I, \mathbb{R}), \ x \in P,$$

such that

$$T_1 f(x) = d(x) Af(x), \quad Af(x) = d(x) \big(Sf(x) - f(x) \big), \quad x \in P.$$

In points $x \in P$ where T and A are localized, we get the same solution as the third one in part (a), i.e., $x \in I_3$.

Conversely, any operators (T, A) described in (a) or (b) satisfy (8.1).

(c) For $x \in \partial P = \overline{P} \setminus P$, either the operators T, A and S can be continuously extended from P to x with the same formulas as in (b), or they cannot be continuously extended, and then the operator A fulfils the Leibniz rule in x

$$A(f \cdot g)(x) = Af(x) \cdot g(x) + f(x) \cdot Ag(x), \quad f, g \in C^k(I, \mathbb{R}).$$

Under the assumption of non-degeneration of A, Theorem 7.2 gave the possible solutions of (8.1): then $P = \emptyset$ and only case (a) applies. As the example following Theorem 7.2 showed, local solutions on I_2 and I_3 could be combined to a globally continuous and non-degenerate solution. In the degenerate situation of Theorem 8.2 they can be also combined with the second derivative solution on I_1, though in a degenerate and non-localized way, but yielding operators with image in the continuous functions on I.

Example 1. Let $k = 2$ and define for any $f \in C^2(\mathbb{R}, \mathbb{R})$

$$Tf(x) := \begin{cases} \frac{1}{x^2}\big(f(2x) - f(x)\big) - \frac{1}{x}f'(x), & x > 0, \\ \frac{1}{2}f''(x), & x \le 0, \end{cases}$$

$$Af(x) := \begin{cases} \frac{1}{x}\big(f(2x) - f(x)\big), & x > 0, \\ f'(x), & x \le 0. \end{cases}$$

Then $\lim_{x \searrow 0} Af(x) = f'(0)$ and $\lim_{x \searrow 0} Tf(x) = \frac{1}{2}f''(0)$. Therefore T and A map $C^2(\mathbb{R}, \mathbb{R})$ into the *continuous* functions $C(\mathbb{R}, \mathbb{R})$. They satisfy (8.1) since they have the form given in (b) on $P = (0, \infty)$ with $d(x) = \frac{1}{x}$, $Rf(x) = -\frac{1}{x}f'(x)$, $S(x) = f(2x)$ for $x \in P$, $\partial P = \{0\}$ and $I_1 = (-\infty, 0)$ with $d = 1$ and $R = 0$ on I_1. Hence $d \in C(\mathbb{R} \setminus \partial P, \mathbb{R})$, but d is not continuous in 0.

Example 2. Let $k = 0$ and $\varphi : \mathbb{R}_{\ge 0} \to \mathbb{R}_{\ge 0}$ be continuous. Define for any $f \in C(\mathbb{R}, \mathbb{R})$

$$Tf(x) := \begin{cases} \frac{1}{x^2}\big(|f(x+\varphi(x))|^{x+1}\operatorname{sgn}(f(x)+\varphi(x)) - f(x)[1+x\ln|f(x)|]\big), & x > 0, \\ \frac{1}{2}f(x)\big(\ln|f(x)|\big)^2, & x \le 0, \end{cases}$$

$$Af(x) := \begin{cases} \frac{1}{x}\big(|f(x+\varphi(x))|^{x+1}\operatorname{sgn}f(x+\varphi(x)) - f(x)\big), & x > 0, \\ f(x)\ln|f(x)|, & x \le 0. \end{cases}$$

For $\varphi = 0$, this is the localized non-degenerate example given in Chapter 7 following Theorem 7.2. Now let $\varphi(x) = x^3$. Then $\lim_{x \searrow 0} Af(x) = f(0)\ln|f(0)|$ and $\lim_{x \searrow 0} Tf(x) = \frac{1}{2}f(x)(\ln|f(x)|)^2$. On $P = (0, \infty)$, T and A are *not* localized since $x + \varphi(x) \ne x$. Again, T and A have the form given in (b) with $d(x) = \frac{1}{x}$,

$Rf(x) = -\frac{1}{x}f(x)\ln|f(x)|$, $Sf(x) = |f(x + \varphi(x))|^{x+1}\operatorname{sgn}(f(x + \varphi(x)))$ for $x \in P$. We have $\partial P = \{0\}$ and $I_2 = (-\infty, 0)$ in Theorem 8.2, with $d = 1$ and $R = 0$ on I_2. Again, $d \in C(\mathbb{R} \setminus \partial P, \mathbb{R})$ is not continuous in 0. For $\varphi = 0$, T and A are localized and $P = \emptyset$, $I_2 = (-\infty, 0]$, $I_3 = (0, \infty)$, $\partial I_3 = \{0\}$, with the same function d belonging to $C(\mathbb{R} \setminus \partial I_3, \mathbb{R})$, again discontinuous at 0. This shows that both exceptional sets ∂P and ∂I_3 for the continuity of the coefficient functions are required in Theorem 8.2.

In both examples, T and A cannot be extended by the same formulas as on $P = (0, \infty)$ from P to $\partial P = \{0\}$, but, as stated in (c), A satisfies the Leibniz rule in $x = 0$, $A(f \cdot g)(0) = Af(0) \cdot g(0) + f(0) \cdot Ag(0)$, with $Af(0) = f'(0)$ in Example 1 and $Af(0) = f(0)\ln|f(0)|$, in Example 2. Clearly, the two examples might also be used to join solutions $Tf = f''$ on one interval and $Tf = f(\ln|f|)^2$ on a disjoint interval by connecting them via some intermediate interval belonging to $\overline{P}$. This will give solutions $T, A : C^2(I, \mathbb{R}) \to C(I, \mathbb{R})$ of (8.1) which are not identically zero at any point $x \in \mathbb{R}$. However, they would not satisfy the condition of non-degeneration, being not localized.

Proposition 8.3. *Under the assumptions of Theorem 8.2, if A is localized, also T is localized.*

Proof. For $J \subset I$ open, $f_1, f_2 \in C^k(I)$ with $f_1|_J = f_2|_J$, $x \in J$ and $g \in C^k(I)$ with $\operatorname{supp} g \subset J$ and $g(x) \neq 0$, we have

$$(Tf_1(x) - Tf_2(x))g(x) + (Af_1(x) - Af_2(x))Ag(x) = 0,$$

similar as in the proof of Proposition 8.1. Since A is assumed to be localized, $Af_1(x) = Af_2(x)$. Hence $Tf_1(x) = Tf_2(x)$, showing that T is localized on intervals and hence localized by Proposition 3.3. $\qquad\square$

Example. If A is just the derivative, $Af = f'$, A is localized and by part (a) of Theorem 8.2, $T_1 f = \frac{1}{2}d^2 f''$ is essentially the second derivative.

We now turn to the proof of Theorem 8.2.

Proof of Theorem 8.2. (a) Applying (8.1) to $f = g = \mathbf{1}$ yields that $T\mathbf{1} + (A\mathbf{1})^2 = 0$. Put $d := -A\mathbf{1}$. Then $d \in C(I, \mathbb{R})$ and $T\mathbf{1} = -d^2 = -d\,A\mathbf{1}$. For $g = \mathbf{1}$ we find using (8.1) that $d\,Af = -d^2 f$. If for some $x_0 \in I$, $d(x_0) \neq 0$, the same would hold by continuity on a small open interval $J \subset I$ with $x_0 \in J$. Then for all $f \in C^k(I, \mathbb{R})$, $x \in J$, we have $Af(x) = -d(x)\,f(x)$, $Af|_J = -d\,f|_J$ and $J \subset P$ with $\lambda = -d$. This implies for $f, g \in C^k(I, \mathbb{R})$ and $x \in J$, using (8.1),

$$T(f \cdot g)(x) + d(x)^2 (f \cdot g)(x) = \big(Tf(x) + d(x)^2 f(x)\big) \cdot g(x)$$
$$+ f(x) \cdot \big(Tg(x) + d(x)^2 g(x)\big).$$

Therefore, $R := T + d^2\,\mathrm{Id}$ satisfies the Leibniz rule on J

$$R(f \cdot g)(x) = Rf(x) \cdot g(x) + f(x) \cdot Rg(x), \qquad x \in J.$$

By Theorem 3.1, applied to the open interval J, we get that there are continuous functions $a, b \in C(J, \mathbb{R})$ such that

$$Rf(x) = a(x) f(x) \ln |f(x)| + b(x) f'(x), \quad f \in C^k(I, \mathbb{R}), \ x \in J,$$

with $b = 0$ if $k = 0$. Then for any $f \in C^k(I, \mathbb{R})$, $x \in J$,

$$Af(x) = -d(x) f(x),$$
$$T_1 f(x) := Tf(x) - Rf(x) = -d(x)^2 f(x) = d(x) Af(x).$$

Hence T_1 and A have the form given in part (b) of Theorem 8.2 with $S \equiv 0$ on any interval J where $A1 \neq 0$.

(b) We may now assume that we have $A1(x_0) = T1(x_0) = 0$ for $x_0 \in I$. Assume first that $x_0 \notin \overline{P}$. Then by Proposition 8.1, $x_0 \in L$ is in the localization set, and the proof of Proposition 8.1 showed that T and A are localized in a possibly small open neighborhood J of x_0: there are functions $F, E : I \times \mathbb{R}^{k+1} \to \mathbb{R}$ such that for all $x \in J$, $f \in C^k(I, \mathbb{R})$,

$$Tf(x) = F\big(x, f(x), \ldots f^{(k)}(x)\big), \quad Af(x) = E\big(x, f(x), \ldots, f^{(k)}(x)\big).$$

The proof of Theorem 7.2 now applies without change and yields that T and A have one of the forms given in part (a) of Theorem 8.2 on J, after subtracting an appropriate homogeneous solution R, $Rf = af \ln |f| + bf'$. If two of such open intervals J intersect for different starting points $x_0 \neq x_1$, the parameter functions in the solutions can be extended by continuity to the union of both intervals, keeping the type of solution on both intervals, i.e., they are subsets of the same set I_i for $i \in \{1, 2, 3\}$. Combining the coefficient functions to single functions on $I \setminus \overline{P}$, there may be only singularities at points of ∂P or ∂I_3; one has $\partial I_1 \subset \partial P$.

(c) Assume now that $A1(x_0) = T1(x_0) = 0$ and $x_0 \in P$. By definition of P, there is an open interval $J \subset I$ with $x_0 \in J$ and a function $\lambda \in C(J, \mathbb{R})$ such that $Ag(x) = \lambda(x) g(x)$ for all $x \in J$, $g \in C^k(J, \mathbb{R})$ with $\operatorname{supp} g \subset J$. Similarly, if two open intervals J_1, J_2 associated to two points $x_1, x_2 \in P$ overlap, $J_1 \cap J_2 \neq \emptyset$, the corresponding functions λ_1 and λ_2 must coincide on $J_1 \cap J_2$, since for any $g \in C^k(I, \mathbb{R})$ supported in $J_1 \cap J_2$ we have $\lambda_1(x) g(x) = Ag(x) = \lambda_2(x) g(x)$, $x \in J_1 \cap J_2$. Therefore, a continuous function $\lambda : P \to \mathbb{R}$ is defined on the full set P, even though $Ag(x) = \lambda(x) g(x)$ only holds for $x \in P$ and functions with small support around x. Define an operator $\overline{S} : C^k(I, \mathbb{R}) \to C(P, \mathbb{R})$ by

$$\overline{S} f(x) := \lambda(x) f(x) - Af(x), \quad f \in C^k(I, \mathbb{R}), \ x \in P. \tag{8.2}$$

Hence, $\overline{S} g(x) = 0$, $x \in J$ for all $g \in C^k(I, \mathbb{R})$ with $\operatorname{supp} g \subset J$. However, for functions with larger support, in general $\overline{S} f$ will not be zero.

For $f_1, f_2 \in C^k(I, \mathbb{R})$ with $f_1|_J = f_2|_J$ and $g \in C^k(I, \mathbb{R})$ with $\operatorname{supp} g \subset J$, we have $f_1 \cdot g = f_2 \cdot g$, and applying (8.1) and taking differences, we get

$$\big(Tf_1(x) - Tf_2(x)\big) \cdot g(x) + \big(Af_1(x) - Af_2(x)\big) \cdot Ag(x) = 0, \quad x \in J,$$

so that, using $Ag(x) = \lambda(x)g(x)$ for $x \in J$,

$$[(Tf_1(x) + \lambda(x)\,Af_1(x)) - (Tf_2(x) + \lambda(x)\,Af_2(x))] \cdot g(x) = 0.$$

Choosing for $x \in J$ a function g with $g(x) \neq 0$ and $\operatorname{supp} g \subset J$, we find that

$$(Tf_1 + \lambda\,Af_1)|_J = (Tf_2 + \lambda\,Af_2)|_J,$$

provided that $f_1|_J = f_2|_J$. The same is true for smaller open subsets of J. Therefore, Proposition 3.3 yields that $T + \lambda A = T + \lambda^2\,\mathrm{Id}$ is localized on J, even though T and A may not be localized there. For $f, g \in C^k(I, \mathbb{R})$ with $\operatorname{supp}(f), \operatorname{supp}(g) \subset J$, we get using (8.1) and $Af(x) = \lambda(x)f(x)$, $Ag(x) = \lambda(x)g(x)$ for $x \in J$ that

$$T(f \cdot g)(x) = Tf(x) \cdot g(x) + f(x) \cdot Tg(x) + \lambda(x)^2 f(x) \cdot g(x), \quad x \in J.$$

Adding $\lambda^2(x)f(x) \cdot g(x)$ shows that $Rf(x) := Tf(x) + \lambda(x)^2 f(x)$ satisfies the Leibniz rule, $R(f \cdot g)(x) = Rf(x) \cdot g(x) + f(x) \cdot Rg(x)$, $x \in J$, when restricted to J. By Theorem 3.1, there are continuous functions $a, b \in C(J, \mathbb{R})$ such that for all $f \in C^k(I, \mathbb{R})$, with $\operatorname{supp} f \subset J$, $x \in J$,

$$Rf(x) = a(x)\,f(x)\ln|f(x)| + b(x)\,f'(x).$$

Again, joining the functions on different intersecting intervals of this type we may define a, b continuously on P. Hence $a, b \in C(P, \mathbb{R})$.

We now introduce an operator $B : C^k(I, \mathbb{R}) \to C(P, \mathbb{R})$ on all functions, not only those having support in J, by

$$Bf(x) := Tf(x) + \lambda(x)^2 f(x) - Rf(x), \quad f \in C^k(I, \mathbb{R}),\ x \in P,$$

with $Rf(x) := a(x)\,f(x)\ln|f(x)| + b(x)\,f'(x)$, $x \in P$. By definition of R and B, $Bf(x) = 0$ for all $x \in J$ and $f \in C^k(I, \mathbb{R})$ with $\operatorname{supp} f \subset J$. If $\operatorname{supp} f \not\subset J$, $Bf(x)$ will in general not be zero for $x \in J$.

We claim, however, that $Bf(x) = \lambda(x)\,\overline{S}f(x)$ for all $f \in C^k(I, \mathbb{R})$ and $x \in J$, where $\overline{S}$ was defined in (8.2). To verify this, take $f, g \in C^k(I, \mathbb{R})$ with $\operatorname{supp} g \subset J$, but not necessarily $\operatorname{supp} f \subset J$. Then $\operatorname{supp}(f \cdot g) \subset J$, too. Inserting the formulas for T and A in terms of $\overline{S}, B$ and R into (8.1), we find for $x \in J$

$$\begin{aligned}
B(f \cdot g)(x) - \lambda(x)^2 (f \cdot g)(x) + R(f \cdot g)(x) &= T(f \cdot g)(x) \\
&= Tf(x) \cdot g(x) + f(x) \cdot Tg(x) + Af(x) \cdot Ag(x) \\
&= \big(Bf(x) - \lambda(x)^2 f(x) + Rf(x)\big) \cdot g(x) \\
&\quad + f(x) \cdot \big(Bg(x) - \lambda(x)^2 g(x) + Rg(x)\big) \\
&\quad + \big(\lambda(x)\,f(x) - \overline{S}f(x)\big) \cdot \big(\lambda(x)\,g(x) - \overline{S}g(x)\big).
\end{aligned}$$

The terms involving R on both sides cancel, since R, as defined above, satisfies the Leibniz rule. Further, $Bg(x) = 0$, $B(f \cdot g)(x) = 0$ and $\overline{S}g(x) = 0$ for $x \in J$ since $\operatorname{supp} g \subset J$ and $\operatorname{supp}(f \cdot g) \subset J$. We are left with

$$0 = \big(Bf(x) - \lambda(x) \cdot \overline{S}f(x)\big) \cdot g(x), \quad x \in J.$$

Choosing $g \in C^k(I, \mathbb{R})$ with $g(x) \neq 0$ and $\operatorname{supp} g \subset J$ implies $Bf(x) = \lambda(x)\overline{S}f(x)$ for $x \in J$, even if $\operatorname{supp} f$ is not contained in J.

Using this and the definition of B, we have for $x \in J$,

$$Tf(x) = -\lambda(x)^2 f(x) + Bf(x) + Rf(x) = -\lambda(x)^2 f(x) + \lambda(x)\overline{S}f(x) + Rf(x),$$
$$Af(x) = \lambda(x) f(x) - \overline{S}f(x).$$

Inserting these formulas into (8.1), calculation shows

$$\lambda(x)\overline{S}(f \cdot g)(x) = \overline{S}f(x) \cdot \overline{S}g(x), \quad f, g \in C^k(I, \mathbb{R}), \ x \in J.$$

If for some $x \in J$, $\lambda(x) = 0$, then $\overline{S}f(x) = 0$, $Af(x) = 0$ and $Tf(x) = Rf(x)$ for all $f \in C^k(I, \mathbb{R})$. This is the solution given in (b) of Theorem 8.2 with $d(x) = 0$.

If for $x \in J$, $\lambda(x) \neq 0$, define $Sf(x) := \frac{1}{\lambda(x)}\overline{S}f(x)$. Then

$$S(f \cdot g)(x) = Sf(x) \cdot Sg(x), \quad f, g \in C^k(I, \mathbb{R}), \ x \in J.$$

Combining this for different intersecting intervals J_1, J_2 around points $x_1, x_2 \in P$, we obtain that S is multiplicative on the full set P. If $\lambda(x) = 0$, formally put $Sf(x) = 0$. Then $S : C^k(I, \mathbb{R}) \to C(P, \mathbb{R})$ is multiplicative for $x \in P$. With $d(x) := -\lambda(x)$, this yields

$$Af(x) = d(x)\big(Sf(x) - f(x)\big),$$
$$Tf(x) = d(x)^2\big(Sf(x) - f(x)\big) + Rf(x), \quad x \in P,$$

which is the solution for T and A given in (b) of Theorem 8.2.

(d) Finally, it remains to consider the case that $A\mathbf{1}(x_0) = T\mathbf{1}(x_0) = 0$ and $x_0 \in \partial P = \overline{P} \setminus P$. Choose any sequence (x_n) in P with $x_0 = \lim_{n \to \infty} x_n$. Since Af and Tf are continuous for any $f \in C^k(I, \mathbb{R})$,

$$Af(x_n) = d(x_n)\big(Sf(x_n) - f(x_n)\big) \to Af(x_0),$$
$$Tf(x_n) = d(x_n)^2\big(Sf(x_n) - f(x_n)\big) + Rf(x_n) \to Tf(x_0).$$

If $Af(x_0) = 0$ for all $f \in C^k(I, \mathbb{R})$, $Rf(x_0) = \lim_{n \to \infty} Rf(x_n)$ exists for all $f \in C^k(I, \mathbb{R})$ and $Tf(x_0) = Rf(x_0)$.

If there is $f \in C^k(I, \mathbb{R})$ with $Af(x_0) \neq 0$, we have $d(x_n) \neq 0$ for large n, and we may assume this for all $n \in \mathbb{N}$. Using that S is multiplicative on $x_n \in P$, we get

$$\begin{aligned}
Af(x_n)^2 &= d(x_n)^2\big(Sf(x_n) - f(x_n)\big)^2 \\
&= d(x_n)^2\big(S(f^2)(x_n) + f(x_n)^2 - 2f(x_n)Sf(x_n)\big) \\
&= d(x_n)^2\big[\big(S(f^2)(x_n) - f(x_n)^2\big) - 2f(x_n)\big(Sf(x_n) - f(x_n)\big)\big] \\
&= d(x_n)\big[A(f^2)(x_n) - 2f(x_n)\,Af(x_n)\big]. \tag{8.3}
\end{aligned}$$

If $Af(x_0) \neq 0$ and $A(f^2)(x_0) - 2f(x_0)\,Af(x_0) \neq 0$, we find

$$d(x_n) \longrightarrow \frac{Af(x_0)^2}{A(f^2)(x_0) - 2f(x_0)Af(x_0)} =: d(x_0) \neq 0.$$

In this case d can be extended by continuity to x_0, and the same is true for Sf for all $f \in C^k(I, \mathbb{R})$,

$$Sf(x_n) = \frac{1}{d(x_n)}Af(x_n) + f(x_n) \longrightarrow \frac{1}{d(x_0)}Af(x_0) + f(x_0) =: Sf(x_0).$$

Similarly, R, T_1 and T can be extended by continuity into x_0. Therefore, in this situation, the solution (T, A) of (8.1) in x_0 is as in part (b) of Theorem 8.2.

If $Af(x_0) \neq 0$, but $A(f^2)(x_0) = 2f(x_0)Af(x_0)$, (8.3) implies that $\lim\limits_{n\to\infty} |d(x_n)| = \infty$, i.e., d has a singularity at x_0. This is the case in the Examples 1 and 2 following Theorem 8.2. Using the multiplicativity of S, we find for all $g, h \in C^k(I, \mathbb{R})$

$$\begin{aligned}
A(g \cdot h)(x_n) &- Ag(x_n) \cdot h(x_n) - g(x_n) \cdot Ah(x_n) \\
&= d(x_n)\big[(S(g \cdot h)(x_n) - (g \cdot h)(x_n)) - (Sg(x_n) - g(x_n))h(x_n) \\
&\qquad\qquad - g(x_n)(Sh(x_n) - h(x_n))\big] \\
&= d(x_n)\big(Sg(x_n) - g(x_n)\big)\big(Sh(x_n) - h(x_n)\big) \\
&= \frac{1}{d(x_n)}Ag(x_n)\,Ah(x_n) \to 0 \cdot Ag(x_0) \cdot Ah(x_0) = 0.
\end{aligned}$$

Therefore, $A(g \cdot h)(x_0) = Ag(x_0)h(x_0) + g(x_0)\,Ah(x_0)$ and A satisfies the Leibniz rule at x_0 for all $g, h \in C^k(I, \mathbb{R})$. This ends the proof of Theorem 8.2. $\qquad\square$

8.2   The extended Leibniz rule equation

Let $I \subset \mathbb{R}$ be an open interval and $k \in \mathbb{N}_0$. In Chapter 3, we studied the extended Leibniz rule equation

$$T(f \cdot g) = Tf \cdot Ag + Af \cdot Tg, \quad f, g \in C^k(I, \mathbb{R}), \tag{8.4}$$

for operators $T, A : C^k(I, \mathbb{R}) \to C(I, \mathbb{R})$ under the assumption of non-degeneration of (T, A), cf. Theorem 3.7. Without this assumption there are simple solutions of (8.4) which are not localized, such as

$$Tf(x) = d(x)\big(f(x) - f(x+1)\big), \quad Af(x) = \tfrac{1}{2}\big(f(x) + f(x+1)\big).$$

Note that for functions f with small support around some point x, Tf and Af are proportional near x. The term $f(x)$ in the example may be replaced by $|f(x)|^{p(x)}\{\operatorname{sgn} f(x)\}$ and the term $f(x+1)$ by $Sf(x)$, where $S : C^k(I, \mathbb{R}) \to C(I, \mathbb{R})$

is an arbitrary multiplicative map, $S(f \cdot g) = Sf \cdot Sg$, still yielding a solution of (8.4). We consider the converse question: Does this describe the general form of solutions of (8.4), in addition to the localized solutions given in Theorem 3.7?

Equation (8.4) allows T and A to be zero on large subsets of $C^k(I, \mathbb{R})$. To avoid completely degenerate cases, we will impose a weak non-degeneration property.

Definition. Let $I \subset \mathbb{R}$ be an open interval, $k \in \mathbb{N}_0$ and $T, A : C^k(I, \mathbb{R}) \to C(I, \mathbb{R})$ be operators. The pair (T, A) is *weakly non-degenerate* if and only if

(i) $\forall x \in I \quad \exists f \in C^k(I, \mathbb{R}) : \quad Tf(x) \neq 0$ and

(ii) $\forall x \in I \quad \forall J \subset I$ open, $x \in J \quad \exists g \in C^k(I, \mathbb{R})$, supp $g \subset J : \quad Ag(x) \neq 0$.

The second condition prevents examples like

$$Tf(x) = f(\varphi(x)) - f(\psi(x)), \quad Af(x) = \tfrac{1}{2}\big[f(\varphi(x)) + f(\psi(x))\big],$$

where $\varphi, \psi : I \to I$ are maps not necessarily fixing x. For fixed x_0 with $\varphi(x_0) \neq x_0 \neq \psi(x_0)$, $Af(x)$ will necessarily be zero for all non-zero functions with very small support J around x_0. Non-localized operators of this type seem to be very difficult to classify. Non-degeneration of (T, A) in Chapter 3 required *more strongly* that T and A were not homothetic on functions with small support around some point x. Here we only assume that A is not identically zero on such functions.

Similar to the set P, introduced before Proposition 8.1, we define a set Q where the localization of the solution operators of (8.4) may possibly fail,

$$Q := \big\{ x \in I \mid \exists J \subset I \text{ open with } x \in J \; \exists \lambda \in C(J, \mathbb{R}) \; \forall g \in C^k(I, \mathbb{R}),$$

$$\text{supp } g \subset J : \quad Tg|_J = \lambda \, Ag|_J \big\}.$$

By definition, Q is open and λ is automatically continuous, since Tg and Ag are continuous on I and A is not identically zero on such functions. We use the same localization set L as in Proposition 8.1. If T and A satisfy (8.4), but are not localized in x, x belongs to $\overline{Q}$:

Proposition 8.4. *Suppose that $k \in \mathbb{N}_0$ and $T, A : C^k(I, \mathbb{R}) \to C(I, \mathbb{R})$ satisfy (8.4). Then $\overline{Q} \cup L = I$. However, $\overline{Q} \cap L \neq \emptyset$ is possible.*

Proof. The proof is very similar to the one of Proposition 8.1. We show that any point $x_0 \in I \setminus \overline{Q}$ belongs to L, $x_0 \in L$. Choose $\widetilde{J} \subset I \setminus \overline{Q}$ open with $x_0 \in \widetilde{J}$. Let $J \subset \widetilde{J}$ be an open subinterval of $\widetilde{J}$ and suppose that $f_1|_J = f_2|_J$ holds for some $f_1, f_2 \in C^k(I, \mathbb{R})$. We claim that $Tf_1|_J = Tf_2|_J$ and $Af_1|_J = Af_2|_J$, which would imply by Proposition 3.3 that T and A are localized. Let $y \in J$. Since $y \notin Q$, for any open set $J_1 \subset J$ with $y \in J_1$ we may find $g_1, g_2 \in C^k(I, \mathbb{R})$ with supports in J_1 such that (Tg_1, Ag_1) and (Tg_2, Ag_2) are not proportional on J_1, i.e., there is $z_1 \in J_1$ such that $(Tg_1(z_1), Ag_1(z_1))$, $(Tg_2(z_1), Ag_2(z_1))$ are linearly independent in $\mathbb{R}^2$. Choose a nested sequence of intervals $J_{\ell+1} \subset J_\ell \subset \cdots \subset J_1$ with length $|J_\ell| \to 0$, $y \in J_\ell$. Find functions $g_1^\ell, g_2^\ell \in C^k(I, \mathbb{R})$ with supports in J_ℓ and $z_\ell \in J_\ell$ such that $(Tg_1^\ell(z_\ell), Ag_1^\ell(z_\ell))$, $(Tg_2^\ell(z_\ell), Ag_2^\ell(z_\ell))$ are linearly independent in $\mathbb{R}^2$. Then $f_1 \cdot g_j^\ell = f_2 \cdot g_j^\ell$ for all $\ell \in \mathbb{N}$, $j \in \{1, 2\}$. Using equation (8.4) for the functions (g_1^ℓ, g_2^ℓ) and taking differences, we get

$$\big(Tf_1(z_\ell) - Tf_2(z_\ell)\big)Ag_j^\ell(z_\ell) + \big(Af_1(z_\ell) - Af_2(z_\ell)\big)Tg_j^\ell(z_\ell) = 0,$$

for $j = 1, 2$, $\ell \in \mathbb{N}$. The linear independence of $(Tg_j^\ell(z_\ell), Ag_j^\ell(z_\ell))$ for $j = 1, 2$, with ℓ fixed, implies that $Tf_1(z_\ell) = Tf_2(z_\ell)$ and $Af_1(z_\ell) = Af_2(z_\ell)$. Since $y, z_\ell \in J_\ell$ and $|J_\ell| \to 0$, we have $\lim_{\ell \to \infty} z_\ell = y$, and by continuity $Tf_1(y) = Tf_2(y)$ and $Af_1(y) = Af_2(y)$. Therefore $Tf_1|_J = Tf_2|_J$, $Af_1|_J = Af_2|_J$, and T and A are localized at x_0. Hence $x_0 \in L$, and actually a small open neighborhood of x_0 is in L, as well. $\square$

We extend Theorem 3.7 to the degenerate case. To do so, we describe the general structure of the solutions of (8.4) on the three sets $I \setminus \overline{Q}$, Q and $\partial Q = \overline{Q} \setminus Q$.

Theorem 8.5 (Extended Leibniz rule with resonance). *Let $I \subset \mathbb{R}$ be an open interval, $k \in \mathbb{N}_0$ and $T, A : C^k(I, \mathbb{R}) \to C(I, \mathbb{R})$ be operators satisfying the* extended Leibniz rule *equation*

$$T(f \cdot g) = Tf \cdot Ag + Af \cdot Tg, \quad f, g \in C^k(I, \mathbb{R}). \tag{8.4}$$

Suppose that (T, A) are weakly non-degenerate and that T and A are pointwise continuous in the sense of Chapter 3. Let Q be defined as before. Then there are pairwise disjoint – possibly empty – subsets I_1, I_2, I_3 of I, where I_2, I_3 are open, with $I \setminus Q = I_1 \cup I_2 \cup I_3$, and functions $c, d, p : I \to \mathbb{R}$ which are continuous except possibly on the exceptional set $N = \partial Q \cup \partial I_2 \cup \partial I_3$ such that:

(a) *On $I \setminus \overline{Q}$ the operators T and A are localized, and for all $f \in C^k(I, \mathbb{R})$ and $x \in I_1$,*

$$Tf(x) = \left(c(x)\ln|f(x)| + d(x)\frac{f'(x)}{f(x)}\right)|f(x)|^{p(x)}\{\operatorname{sgn} f(x)\},$$

$$Af(x) = |f(x)|^{p(x)}\{\operatorname{sgn} f(x)\}, \quad p(x) > 1 \quad (k \geq 1);$$

and for $x \in I_2$,

$$Tf(x) = c(x)\sin\big(d(x)\ln|f(x)|\big)|f(x)|^{p(x)}\{\operatorname{sgn} f(x)\},$$

$$Af(x) = \cos\big(d(x)\ln|f(x)|\big)|f(x)|^{p(x)}\{\operatorname{sgn} f(x)\}, \quad p(x) > 0;$$

and for $x \in I_3$,

$$Tf(x) = \tfrac{1}{2}c(x)\big(|f(x)|^{p(x)}\{\operatorname{sgn} f(x)\} - |f(x)|^{d(x)}[\operatorname{sgn} f(x)]\big),$$

$$Af(x) = \tfrac{1}{2}\big(|f(x)|^{p(x)}\{\operatorname{sgn} f(x)\} + |f(x)|^{d(x)}[\operatorname{sgn} f(x)]\big), \quad \min(p(x), d(x)) > 0.$$

(b) *On Q, the operators T and A are possibly not localized, but $T + cA$ is localized and multiplicative. Moreover, there is another multiplicative operator $S : C^k(I, \mathbb{R}) \to C(Q, \mathbb{R})$,*

$$S(f \cdot g) = Sf \cdot Sg, \quad f, g \in C^k(I, \mathbb{R}),$$

so that for all $x \in Q$

$$Tf(x) = \tfrac{1}{2}c(x)\big(|f(x)|^{p(x)}\{\operatorname{sgn} f(x)\} - Sf(x)\big),$$
$$Af(x) = \tfrac{1}{2}\big(|f(x)|^{p(x)}\{\operatorname{sgn} f(x)\} + Sf(x)\big), \quad x \in Q.$$

In points $x \in Q$ where T and A are localized, we get the same solution as the third one in part (a), i.e., $x \in I_3$.

Conversely, the operators T and A described in (a) or (b) satisfy (8.4) on $I \setminus \partial Q$.

(c) *For $x \in \partial Q = \overline{Q} \setminus Q$,*

either the operators T, A and S can be continuously extended from Q to x with the same formulas as in (b), or they cannot be extended, in which case the operator A is multiplicative on x, $A(f \cdot g)(x) = Af(x) \cdot Ag(x)$.

More precisely, in this case there is $p(x) \geq 0$ such that

$$Af(x) = |f(x)|^{p(x)}\{\operatorname{sgn} f(x)\}, \quad f, g \in C^k(I, \mathbb{R}), \ x \in \partial Q.$$

As usual, the term $\{\operatorname{sgn} f(x)\}$ in each solution is present in T and A always or not at all. We showed by an example after Theorem 3.7 that local solutions without a derivative term, i.e., when $d = 0$ on I_1, could be combined to a globally defined, non-degenerate solution of (8.4) with image in the continuous functions on I. In the degenerate situation of Theorem 8.5 solutions involving the first derivative term can also be combined with other solutions to yield well-defined operators with image in $C(I, \mathbb{R})$. These solutions, derived from part (b), however, are not non-degenerate in the sense of Chapter 3.

Example. Let $I = \mathbb{R}$ and define $S : C^1(\mathbb{R}, \mathbb{R}) \to C(\mathbb{R}, \mathbb{R})$ by $Sf(x) := f(2x)$. Choose in case (b) of Theorem 8.5 $p(x) = 1$ and $c(x) = -2/x$ for $x > 0$. Then

$$Tf(x) = \begin{cases} -\tfrac{1}{x}\big(f(x) - f(2x)\big), & x > 0, \\ f'(x), & x \leq 0, \end{cases}$$

$$Af(x) = \begin{cases} \tfrac{1}{2}\big(f(x) + f(2x)\big), & x > 0, \\ f(x), & x \leq 0 \end{cases}$$

satisfies (8.4) on $\mathbb{R}$. On $Q = (0, \infty)$ it is a non-localized solution of the type explained in (b); on $I_1 = (-\infty, 0]$ it is a localized solution. We have $\partial Q = \partial I_1 = \{0\}$. Since $\lim_{x \searrow 0} \frac{f(2x) - f(x)}{x} = f'(0)$, the ranges of T and A consist of continuous functions. These operators (T, A) are not non-degenerate in the sense of Chapter 3.

Proposition 8.6. *Under the assumptions of Theorem 8.5, if A is localized, also T is localized.*

The proof of Proposition 8.6 is similar to the one of Proposition 8.3.

Example. To illustrate this, suppose that A is given by $Af = |f|^p \{\operatorname{sgn} f\}$ and that T satisfies (8.4). Then by part (a) of Theorem 8.5, $Tf = (c \ln |f| + d\frac{f'}{f})Af$.

In the proof of Theorem 8.5 we will use the following result on localized multiplicative operators.

Proposition 8.7. *Let $k \in \mathbb{N}_0$, $I \subset \mathbb{R}$ be an open interval and $A : C^k(I, \mathbb{R}) \to C(I, \mathbb{R})$ be a non-zero multiplicative operator, $A(f \cdot g) = Af \cdot Ag$ for all $f, g \in C^k(I, \mathbb{R})$. Suppose there is a function $B : I \times \mathbb{R}^{k+1} \to \mathbb{R}$ such that for all $f \in C^k(I, \mathbb{R})$, $x \in I$,*

$$Af(x) = B(x, f(x), \ldots, f^{(k)}(x)).$$

Then there is a continuous function $p \in C(I, \mathbb{R})$ with $p \geq 0$ such that

$$Af(x) = |f(x)|^{p(x)} \{\operatorname{sgn} f(x)\}.$$

The $\{\operatorname{sgn} f(x)\}$-term either appears for all f or never. If the term is present, $p > 0$ holds. In particular, Af does not depend on any derivatives of f.

Proof. Since $A(1) = A(1)^2 \geq 0$, $A(1)(x) \in \{0, 1\}$ for any $x \in I$. If for some $x_0 \in I$, $A(1)(x_0) = 0$, by continuity of $A(1)$ we would have $A(1) \equiv 0$ on I and A would be zero. Hence $A(1) = 1$. Therefore $A(-1)^2 = A((-1)^2) = A(1) = 1$, and by continuity of $A(-1)$, either $A(1) = 1$ on I or $A(1) = -1$ on I. Similarly, $A(0) = 0$, unless $Af = 1$ for all f. We have for all $f \in C^k(I, \mathbb{R})$ that $A(-f) = A(-1)A(f)$. Since A is represented pointwise by B, it suffices to determine Af for functions $f \in C^k(I, \mathbb{R})$ which are strictly positive on I. Then $Af = A(\sqrt{f})^2 \geq 0$. Therefore we may define an operator $C : C^k(I, \mathbb{R}) \to C(I, \mathbb{R})$ by $Ch := \ln A(\exp(h))$, $h \in C^k(I)$. Since A is multiplicative, C is additive,

$$C(h_1 + h_2) = \ln A(\exp(h_1) \exp(h_2))$$
$$= \ln A(\exp(h_1)) + \ln A(\exp(h_2)) = C(h_1) + C(h_2) \; ; \; h_1, h_2 \in C^k(I).$$

The derivatives of $\exp h$ are expressible in terms of $\exp h$ and the derivatives of h. Hence the local representation of A by B yields a local representation of C: there is a function $D : I \times \mathbb{R}^{k+1} \to \mathbb{R}$ such that for all $h \in C^k(I, \mathbb{R})$ and $x \in I$

$$Ch(x) = D(x, h(x), \ldots, h^{(k)}(x)).$$

For any $\alpha = (\alpha_j)_{j=0}^k$, $\beta = (\beta_j)_{j=0}^k$ in $\mathbb{R}^{k+1}$ and $x \in I$, choose $h_1, h_2 \in C^k(I, \mathbb{R})$ such that $h_1^{(j)}(x) = \alpha_j$ and $h_2^{(j)}(x) = \beta_j$ for all $j \in \{0, \ldots, k\}$. Then the additivity of C is equivalent to

$$D(x, \alpha + \beta) = D(x, \alpha) + D(x, \beta), \quad \alpha, \beta \in \mathbb{R}^{k+1}, \ x \in I.$$

Therefore $D(x, \cdot)$ is additive on $\mathbb{R}^{k+1}$ and $Ch(x) = D(x, h(x), \ldots, h^{(k)}(x))$ is a continuous function of $x \in I$ for any $h \in C^k(I, \mathbb{R})$. By Theorem 2.6 there are continuous functions $c_0, \ldots, c_k \in C(I, \mathbb{R})$ such that $D(x, \alpha) = \sum_{j=0}^k c_j(x)\alpha_j$ and hence $Ch(x) = \sum_{j=0}^k c_j(x) h^{(j)}(x)$. For $f \in C^k(I, \mathbb{R})$ with $f > 0$ and $h = \ln f$ we get

$$Af = A(\exp(h)) = \exp(Ch) = \exp\left(\sum_{j=0}^k c_j \, (\ln f)^{(j)}\right).$$

Since $(\ln f)^{(j)}$ has a singularity of order $(\frac{f'}{f})^j$ as $f \searrow 0$, if $f' \neq 0$, and A is given in localized form, the coefficients functions c_j of $(\ln f)^{(j)}$ have to be zero for all $j \geq 1$. The argument for this is again the same as in the proof of Theorem 3.1. Hence $Af = \exp(c_0 \ln f) = f^{c_0}$ if $f > 0$. Applying this to the constant function $f = 2$ shows that c_0 is continuous, $c_0 \in C(I, \mathbb{R})$. We also need $c_0 \geq 0$ to guarantee the continuity of Af for functions f having zeros. Let $p := c_0$. Then either $Af = |f|^p$ for all $f \in C^k(I)$ or $Af = |f|^p \, \text{sgn} \, f$ for all f, depending on whether $A(-1) = \mathbf{1}$ or $A(-1) = -\mathbf{1}$. This ends the proof of Proposition 8.7. $\qquad \square$

We can now prove Theorem 8.5.

Proof of Theorem 8.5. (i) For $f = g = \mathbf{1}$, equation (8.4) yields that $T\mathbf{1}(x)(1 - 2A\mathbf{1}(x)) = 0$ for any $x \in I$. In the non-degenerate case, $T\mathbf{1} = 0$ and $A\mathbf{1} = 1$ holds. In general, however, $T(\mathbf{1}) \neq 0$ is possible. Then $A\mathbf{1}(x) = 1/2$. Conversely, if $A\mathbf{1}(x) = 1/2$, choosing $g = \mathbf{1}$ in (8.4) we find $Tf(x) = \frac{1}{2}Tf(x) + T\mathbf{1}(x) \, Af(x)$, $Tf(x) = d(x) \, Af(x)$ with $d(x) = 2T\mathbf{1}(x)$ for all $f \in C^k(I, \mathbb{R})$. By assumption, for any $x \in I$ there is f with $Tf(x) \neq 0$. Therefore, $d(x) \neq 0$. Let $\mathcal{O} := \{x \in I \mid A\mathbf{1}(x) = 1/2\}$. Then $\mathcal{O} = \{x \in I \mid T\mathbf{1}(x) \neq 0\}$ and $\mathcal{O}$ is open with $\mathcal{O} \subset Q$. Inserting $Tf(x) = d(x) \, Af(x)$ into (8.4), we get $d(x) \, A(f \cdot g)(x) = 2d(x) \, Af(x) \cdot Ag(x)$ for any $f, g \in C^k(I, \mathbb{R})$, $x \in \mathcal{O}$, i.e., $R := 2A$ is multiplicative on $x \in \mathcal{O}$, $R(f \cdot g)(x) = Rf(x) \cdot Rg(x)$, $x \in \mathcal{O}$.

We claim that T, A and R are localized on $\mathcal{O}$. Let $J \subset \mathcal{O}$ be open and $f_1, f_2 \in C^k(I, \mathbb{R})$ satisfy $f_1|_J = f_2|_J$. Let $x \in J$. By the assumption of weak non-degeneracy, there is $g \in C^k(I, \mathbb{R})$ with $\text{supp} \, g \subset J$ and $Ag(x) \neq 0$. Since $f_1 \cdot g = f_2 \cdot g$, an application of (8.4) yields

$$(Tf_1 - Tf_2) \cdot Ag + (Af_1 - Af_2) \cdot Tg = 0. \tag{8.5}$$

Since $Tg = d \, Ag$ and $Tf_i = d \, Af_i$ for $i = 1, 2$, we find at x

$$2d(x)\big(Af_1(x) - Af_2(x)\big) \cdot Ag(x) = 0.$$

Since $Ag(x) \neq 0$ and $d(x) \neq 0$ in view of $x \in J \subset \mathcal{O}$, $Af_1(x) = Af_2(x)$, i.e., $Af_1|_J = Af_2|_J$. By Proposition 3.3, A is localized in $\mathcal{O}$, i.e., there is $B : \mathcal{O} \times \mathbb{R}^{k+1} \to \mathbb{R}$ such that for any $f \in C^k(I, \mathbb{R})$ and $x \in \mathcal{O}$

$$Af(x) = B\big(x, f(x), f'(x), \dots, f^{(k)}(x)\big).$$

Since $2A$ is multiplicative, Proposition 8.7 implies that there is a continuous function $p \in C(\mathcal{O}, \mathbb{R})$, $p \geq 0$ such that

$$Af(x) = \tfrac{1}{2}|f(x)|^{p(x)}, \quad x \in \mathcal{O} \quad \text{or} \quad Af(x) = \tfrac{1}{2}|f(x)|^{p(x)} \operatorname{sgn} f(x), \quad x \in \mathcal{O},$$

in the second case with $p > 0$. Hence on $\mathcal{O} \subset Q$, A and $T = d\,A$ have the form described in (b) of Theorem 8.5 with $S = 0$.

(ii) From now on, we may assume that $x \notin \mathcal{O}$. Then $T\mathbf{1}(x) = 0$ and $A\mathbf{1}(x) = 1$. Assume first that $x \notin \overline{Q}$. By Proposition 8.4, x is in the localization set L, and the same is true for all points in a suitable open neighborhood J of x. Theorem 3.7 now applies on J and yields that T and A are of one of the forms given in (a) of Theorem 8.5 for all $y \in J$. We note that in the proof of Theorem 3.7, $A\mathbf{1}(y) = 1$ is used. If two such open sets intersect, the solutions coincide on the intersection. They may be extended by continuity to the union and thus to $I \setminus \overline{Q}$, although the coefficient functions of the three possible solutions may possibly become singular at the exceptional set $(\partial I_2 \cup \partial I_3) \cap (I \setminus \overline{Q})$.

(iii) Now consider $x \in Q$, and again $x \notin \mathcal{O}$, i.e., $T\mathbf{1}(x) = 0$, $A\mathbf{1}(x) = 1$. By definition of Q, there is an open interval $J \subset I$ with $x \in J$ and $\lambda \in C(J, \mathbb{R})$ such that for all $g \in C^k(I, \mathbb{R})$ with $\operatorname{supp} g \subset J$, we have $Tg = \lambda\,Ag$. If two such intervals intersect, the corresponding λ-functions must coincide, just using functions g supported in the intersection. Therefore, λ may be extended to Q, yielding a continuous function $\lambda \in C(Q, \mathbb{R})$.

Define two operators $C_\pm : C^k(I, \mathbb{R}) \to C(Q, \mathbb{R})$ by

$$C_\pm f(x) := \lambda(x)\,Af(x) \pm Tf(x), \quad f \in C^k(I, \mathbb{R}),\ x \in Q.$$

Note that for g with $\operatorname{supp} g \subset J$ we have $C_- g = 0$. We will show that $C_\pm$ are homothetic to multiplicative operators on Q and that C_+ is localized. Using (8.4), calculation shows

$$
\begin{aligned}
C_\pm f \cdot C_\pm g &= (\lambda Af \pm Tf) \cdot (\lambda Ag \pm Tg) \\
&= (Tf \cdot Tg + \lambda^2 Af \cdot Ag) \pm \lambda(Tf \cdot Ag + Af \cdot Tg) \\
&= \lambda^2 Af \cdot Ag + Tf \cdot Tg \pm \lambda T(f \cdot g), \\
\lambda C_\pm(f \cdot g) &= \lambda^2 A(f \cdot g) \pm \lambda T(f \cdot g).
\end{aligned}
$$

Therefore, $C_\pm f(x)\,C_\pm g(x) = \lambda\,C_\pm(f \cdot g)(x)$ for $x \in J$ is equivalent to

$$Tf(x) \cdot Tg(x) = \lambda^2\big(A(f \cdot g)(x) - Af(x) \cdot Ag(x)\big), \quad x \in J, \qquad (8.6)$$

for *all* functions $f, g \in C^k(I, \mathbb{R})$, and not only those supported in J.

We now prove (8.6). Start with three functions $f, g, h \in C^k(I, \mathbb{R})$. Repeated application of (8.4) yields

$$T((fg)h) = T(fg)\,A(h) + A(fg)\,T(h)$$
$$= T(f)\,A(g)\,A(h) + T(g)\,A(f)\,A(h) + A(fg)\,T(h),$$
$$T(f(gh)) = T(f)\,A(gh) + A(f)\,T(gh)$$
$$= T(f)\,A(gh) + T(g)\,A(f)\,A(h) + A(f)\,A(g)\,T(h).$$

Since both are equal, we find

$$T(f)\big(A(gh) - A(g)\,A(h)\big) = T(h)\big(A(fg) - A(f)\,A(g)\big). \qquad (8.7)$$

By weak non-degeneration, for any $x \in J$ there is $h \in C^k(I, \mathbb{R})$ with $\operatorname{supp} h \subset J$ and $Ah(x) \neq 0$. Then $\operatorname{supp}(g \cdot h) \subset J$, too, and by definition of Q, $T(h)|_J = \lambda\,A(h)|_J$, $T(gh)|_J = \lambda\,A(gh)|_J$. Multiplying (8.7) by λ and inserting this, we get

$$T(f)|_J\big(T(gh)|_J - A(g)|_J\,T(h)|_J\big) = \lambda^2 A(h)|_J\big(A(fg)|_J - A(f)|_J\,A(g)|_J\big),$$

and by (8.4)

$$T(f)|_J\,T(g)|_J\,A(h)|_J = \lambda^2 A(h)|_J\big(A(fg)|_J - A(f)|_J\,A(g)|_J\big).$$

Since $Ah(x) \neq 0$,

$$Tf(x)\,Tg(x) = \lambda^2\big(A(fg)(x) - Af(x) \cdot Ag(x)\big),$$

which proves (8.6). Hence we have shown that

$$\lambda\,C_\pm(f \cdot g)|_Q = C_\pm f|_Q \cdot C_\pm g|_Q,$$

where, of course, Q is the union of smaller sets $J \subset Q$, for which this was really verified. For $x \in J$, h with $\operatorname{supp} h \subset J$, $\operatorname{supp}(fh) \subset J$ and $Ah(x) \neq 0$, we find using (8.4) and the definition of Q,

$$\lambda(x)\,A(f \cdot h)(x) = T(f \cdot h)(x) = Tf(x) \cdot Ah(x) + Af(x) \cdot Th(x)$$
$$= Tf(x) \cdot Ah(x) + \lambda(x)\,Af(x) \cdot Ah(x),$$
$$\lambda(x)\big(A(f \cdot h)(x) - Af(x) \cdot Ah(x)\big) = Tf(x)\,Ah(x).$$

If $\lambda(x) = 0$, $Tf(x) = 0$ would follow for all $f \in C^k(I, \mathbb{R})$, which contradicts the assumption of weak non-degeneration of (T, A). Therefore, $\lambda(x) \neq 0$ for all $x \in Q$ and both operators $\widetilde{C}_\pm := \frac{1}{\lambda}C_\pm$ are multiplicative on Q,

$$\widetilde{C}_\pm(f \cdot g)(x) = \widetilde{C}_\pm(f)(x) \cdot \widetilde{C}_\pm(g)(x), \quad f, g \in C^k(I, \mathbb{R}),\ x \in Q,$$

$$\widetilde{C}_\pm f = A \pm \frac{1}{\lambda}T.$$

(iv) We now show that $C_+ = T + \lambda A$ is localized on Q. Let $\widetilde{J} \subset Q$ be an open interval and $f_1, f_2 \in C^k(I, \mathbb{R})$ satisfy $f_1|_{\widetilde{J}} = f_2|_{\widetilde{J}}$. Let $x \in \widetilde{J}$. Since $x \in Q$, there is an open interval $J \subset \widetilde{J}$, $x \in J$ with $Tg|_J = \lambda Ag|_J$ for all $g \in C^k(I, \mathbb{R})$ with $\operatorname{supp} g \subset J$. By the assumption of weak non-degeneration, there is g with $\operatorname{supp} g \subset J$ and $Ag(x) \neq 0$. Then $f_1 \cdot g = f_2 \cdot g$ and (8.5) implies

$$0 = \big(Tf_1(x) - Tf_2(x)\big)Ag(x) + \big(Af_1(x) - Af_2(x)\big)Tg(x)$$
$$= \big((T + \lambda A)f_1(x) - (T + \lambda A)f_2(x)\big)Ag(x).$$

Since $Ag(x) \neq 0$, $(T + \lambda A)f_1(x) = (T + \lambda A)f_2(x)$, $(T + \lambda A)f_1|_J = (T + \lambda A)f_2|_J$. By Proposition 3.3, $C_+ := T + \lambda A$ is localized on Q. Since $\widetilde{C}_+ = \frac{1}{\lambda}C_+$ is also multiplicative, Proposition 8.7 implies that there is a continuous function $p \in C(Q, \mathbb{R})$ with $p \geq 0$ such that for all $f \in C^k(I, \mathbb{R})$

$$\widetilde{C}_+ f(x) = |f(x)|^{p(x)}\{\operatorname{sgn} f(x)\}, \quad x \in Q.$$

Let $S := \widetilde{C}_-$. As we have seen, S is multiplicative and for $f \in C^k(I, \mathbb{R})$

$$Af(x) = \tfrac{1}{2}(\widetilde{C}_+ + \widetilde{C}_-)f(x) = \tfrac{1}{2}\big(|f(x)|^{p(x)}\{\operatorname{sgn} f(x)\} + Sf(x)\big),$$
$$Tf(x) = \frac{\lambda(x)}{2}(\widetilde{C}_+ - \widetilde{C}_-)f(x) = \frac{\lambda(x)}{2}\big(|f(x)|^{p(x)}\{\operatorname{sgn} f(x)\} - Sf(x)\big), \ x \in Q.$$

This is the form of T and A given in (b) with $c = \lambda$. Calculation shows that, conversely, these operators satisfy the extended Leibniz rule (8.4) on Q.

(v) Finally, consider $x_0 \in \partial Q = \overline{Q} \setminus Q$, and again $T\mathbf{1} = 0$, $A\mathbf{1} = 1$. Choose a sequence $x_n \in Q$ with $x_n \to x_0$. Since Af and Tf are continuous for all $f \in C^k(I, \mathbb{R})$, so are $\widetilde{C}_+ f$ and $Sf = \widetilde{C}_- f$. Therefore,

$$Af(x_n) = \tfrac{1}{2}\big(\widetilde{C}_+ f(x_n) + Sf(x_n)\big) \to Af(x_0) = \tfrac{1}{2}\big(\widetilde{C}_+ f(x_0) + Sf(x_0)\big),$$
$$Tf(x_n) = \frac{\lambda(x_n)}{2}\big(\widetilde{C}_+ f(x_n) - Sf(x_n)\big) \to Tf(x_0).$$

Choose $g \in C^k(I, \mathbb{R})$ with $Tg(x_0) \neq 0$. If $\widetilde{C}_+ g(x_0) \neq Sg(x_0)$, the limit

$$\lim_{n\to\infty} \lambda(x_n) = \frac{2Tg(x_0)}{\widetilde{C}_+ g(x_0) - Sg(x_0)} \neq 0$$

exists. Put $\lambda(x_0) := \lim_{n\to\infty} \lambda(x_n)$. Then $Tf(x_0) = \frac{\lambda(x_0)}{2}(\widetilde{C}_+ f(x_0) - Sf(x_0))$ holds by continuity of Tf, $\widetilde{C}_+ f$ and Sf for all $f \in C^k(I, \mathbb{R})$, and the formulas from (b) for Af and Tf in Q extend to $x_0 \in \partial Q$.

If $\widetilde{C}_+ g(x_0) = Sg(x_0)$, $\sup_{n\in\mathbb{N}} |\lambda(x_n)| = \infty$ since $Tg(x_0) \neq 0$. In this case, the formulas from (b) do not extend to $x_0 \in \partial Q$. However, since

$$Tf(x_0) = \lim_{n\to\infty} Tf(x_n) = \lim_{n\to\infty} \frac{\lambda(x_n)}{2}\big(\widetilde{C}_+ f(x_n) - Sf(x_n)\big),$$

and $\lim_{n\to\infty} \widetilde{C}_+ f(x_n) = \widetilde{C} f(x_0)$, $\lim_{n\to\infty} S f(x_n) = S f(x_0)$ exist for all $f \in C^k(I, \mathbb{R})$, it follows that $\widetilde{C}_+ f(x_0) = S f(x_0)$ for all $f \in C^k(I, \mathbb{R})$. Therefore, $Af(x_0) = \frac{1}{2}(\widetilde{C}_+ f(x_0) + S f(x_0)) = \widetilde{C}_+ f(x_0)$. For the constant function $f_0 = 2$, $\widetilde{C}_+ f_0(x) = 2^{p(x)}$, $x \in Q$, and p has a continuous extension into $x_0 \in \partial Q$. Then $Af(x_0) = \widetilde{C}_+ f(x_0) = |f(x_0)|^{p(x_0)}\{\operatorname{sgn} f(x_0)\}$ for all $f \in C^k(I, \mathbb{R})$. In particular, A is multiplicative, i.e., $A(f \cdot g)(x_0) = Af(x_0) \cdot Ag(x_0)$. This proves the last part (c) of Theorem 8.5. $\qquad\square$

In the example following the formulation of Theorem 8.5, $\lambda(x) = c(x) = -\frac{2}{x}$ is unbounded on $x \in Q = (0, \infty)$ and $Af(0) = f(0)$ is multiplicative on $\partial Q = \{0\}$.

8.3 Notes and References

Proposition 8.1 and Theorem 8.2 were shown by König, Milman in [KM8] in the case $k = 2$ and $I = \mathbb{R}$.

One might compare Proposition 8.7 with the result of Milgram [M] that *bijective* multiplicative maps $A : C(I, \mathbb{R}) \to C(I, \mathbb{R})$ have a similar form as in Proposition 8.7, up to some homeomorphism $u : I \to I$,

$$Af(u(x)) = |f(x)|^{p(x)}\{\operatorname{sgn} f(x)\}, \quad f \in C(I, \mathbb{R}), \ x \in I.$$

In Proposition 8.7 we do not assume the bijectivity, but the localization of the operator A, and then u is not needed.

Chapter 9

The Second-Order Chain Rule

Applying the chain rule twice to functions $f, g \in C^2(\mathbb{R})$ yields

$$D^2(f \circ g) = D^2 f \circ g \cdot (Dg)^2 + Df \circ g \cdot D^2 g.$$

We use this identity as a model for a more general operator equation. Replacing D^2 by T and the first derivative expressions by A_1 and A_2, we study in this chapter the operator equation

$$T(f \circ g) = Tf \circ g \cdot A_1 g + A_2 f \circ g \cdot Tg, \quad f, g \in C^k(\mathbb{R}), \tag{9.1}$$

for general $k \in \mathbb{N}$. In the case of the second derivative, $Tf = f''$, we have that $A_1 f = f'^2$ and $A_2 f = f'$. Therefore we consider A_1, A_2 to be "of lower order" than T, and we will assume that T maps $C^k(\mathbb{R})$ into $C(\mathbb{R})$ while A_1, A_2 operate from $C^\ell(\mathbb{R})$ to $C(\mathbb{R})$, $\ell = \max(k-1, 1)$. It turns out that under reasonable assumptions on T, A_1, A_2, equation (9.1) does not admit too many types of solutions.

By Theorem 4.1, the chain rule equation $R(f \circ g) = Rf \circ g \cdot Rg$ for maps $R : C^k(\mathbb{R}) \to C(\mathbb{R})$, $k \in \mathbb{N}$, has the solutions

$$Rf = \frac{K \circ f}{K} |f'|^q \{\operatorname{sgn} f'\},$$

with $K \in C(\mathbb{R})$, $K > 0$ and $q \geq 0$, if R is not identically zero on the half-bounded C^k-functions. On the functions f with strictly positive images Rf, i.e., $f' = |f'| > 0$, we may consider $Tf := \ln Rf$ which will satisfy the equation

$$T(f \circ g) = Tf \circ g + Tg,$$

with the solution

$$Tf = q \ln |f'| + (H \circ f - H),$$

© Springer Nature Switzerland AG 2018
H. König, V. Milman, *Operator Relations Characterizing Derivatives*,
https://doi.org/10.1007/978-3-030-00241-1_9

where $H := \ln K$. This solves a special case of (9.1), namely for $A_1 = A_2 = 1$. However, the solution does not extend to $C^1(\mathbb{R})$-functions f with $f'(x) = 0$ for some $x \in \mathbb{R}$. But we may replace this by

$$Tf = \left(q\ln|f'| + H \circ f - H\right) \cdot |f'|^p, \quad A_1 f = A_2 f = |f'|^p,$$

with $p > 0$. These three operators are well defined on $C^1(\mathbb{R})$ and satisfy (9.1). It is this example which motivates setting $\ell = 1$ if $k = 1$, where T is defined on $C^k(\mathbb{R})$ and A on $C^\ell(\mathbb{R})$. Otherwise, for $k \geq 2$, we put $\ell = k - 1$.

Besides the second derivative and the $\ln|f'|$-solution there is also the Schwarzian derivative $T = S$ which satisfies (9.1) with suitable operators A_1, A_2. The Schwarzian derivative of a function $f \in C^3(\mathbb{R})$ with $f' \neq 0$ is defined by

$$Sf = \left(\frac{f''}{f'}\right)' - \frac{1}{2}\left(\frac{f''}{f'}\right)^2 = \frac{f'''}{f'} - \frac{3}{2}\left(\frac{f''}{f'}\right)^2.$$

The kernel of S on suitable function spaces consists of the fractional linear transformations $f(x) = \frac{ax+b}{cx+d}$. Although the Schwarzian derivative is mainly used in complex analysis, when studying conformal mappings, univalent functions or complex dynamics, we will consider it here from the perspective of real analysis composition formulas. It satisfies

$$S(f \circ g) = Sf \circ g \cdot (g')^2 + Sg, \quad f, g \in C^3(\mathbb{R}),$$

if $f' \neq 0 \neq g'$. This, too, is of the form (9.1) with $A_1 g = g'^2$, $A_2 g = 1$. However, as in the example of $\ln|f'|$, it is not defined if $f' = 0$. We may compensate for this fact by multiplying Sf with f'^2, and then

$$Tf = f'^2 Sf = f'f''' - \tfrac{3}{2}(f'')^2, \quad A_1 f = f'^4, \quad A_2 f = f'^2$$

define maps $T : C^3(\mathbb{R}) \to C(\mathbb{R})$, $A_1, A_2 : C^2(\mathbb{R}) \to C(\mathbb{R})$ which satisfy (9.1). This raises the question whether there are solutions T of (9.1) with associated suitable operators A_1, A_2 which depend non-trivially on the fourth or higher derivatives. Under natural assumptions, it turns out that no such operators exist, as we will show, and we will find all solutions of (9.1).

Besides the "basic" solutions f'', $f'^2 Sf$ and $f'\log|f'|$ there are two additional solutions of (9.1) when $k = 1$, i.e., when T, A_1, A_2 are all defined on $C^1(\mathbb{R})$.

Equation (9.1) resembles the addition formula of the sin-function, though in a multiplicative setting, and thus allows for a solution of the form $Tf = \sin(\ln|f'|)$, $A_1 f = A_2 f = \cos(\ln|f'|)$, which again would have to be multiplied by terms $|f'|^p$, $p > 0$, to be well defined on $C^1(\mathbb{R})$. The second additional solution for $k = 1$ is based on a cancelation effect. This is similar to the cancelation of terms in the (non-localized) example $Tf(x) = -f(x) + f(x + 1)$, $Af(x) = f(x) - f(x + 1)$ in the case of the second-order Leibniz rule.

9.1 The main result

If $A_1 = A_2 = \frac{1}{2}T$, (9.1) would be the ordinary chain rule $T(f \circ g) = Tf \circ g \cdot Tg$. To exclude this reduction to a previously studied case, we will make the following assumption of non-degeneration, which prevents T and $A_1 = A$ to be proportional.

Definition. Let $k \in \mathbb{N}$, $\ell := \max(k-1, 1)$, $T : C^k(\mathbb{R}) \to C(\mathbb{R})$ and $A : C^\ell(\mathbb{R}) \to C(\mathbb{R})$ be operators. The pair (T, A) is C^k-*non-degenerate* provided that

(a) for every $x \in \mathbb{R}$ there is $g \in C^k(\mathbb{R})$ such that $Tg(x) \neq 0$;

(b) for any open interval $J \subset \mathbb{R}$ and any $x \in J$ there exist functions $g_1, g_2 \in C^k(\mathbb{R})$ with image in J and points $y_1, y_2 \in \mathbb{R}$ with $g_1(y_1) = g_2(y_2) = x$ such that the vectors $(Tg_i(y_i), Ag_i(y_i)) \in \mathbb{R}^2$ for $i = 1, 2$, are linearly independent.

We also need the following definitions, which are similar to notions which already appeared in Chapter 7.

Definition. For $\ell \in \mathbb{N}_0$, an operator $A : C^\ell(\mathbb{R}) \to C(\mathbb{R})$ is *isotropic* if it commutes with shift functions $S_y : \mathbb{R} \to \mathbb{R}$, $S_y(x) := x + y$, $x, y \in \mathbb{R}$, i.e., if $A(f \circ S_y) = (Af) \circ S_y$ for any $f \in C^\ell(\mathbb{R})$, $y \in \mathbb{R}$.

Definition. For $k \in \mathbb{N}$, an operator $A : C^{k-1}(\mathbb{R}) \to C(\mathbb{R})$ is C^{k-1}-*pointwise continuous* if for any sequence $(f_n)_{n \in \mathbb{N}}$ of $C^k(\mathbb{R})$-functions and $f \in C^{k-1}(\mathbb{R})$, such that $\lim_{n \to \infty} f_n^{(j)} = f^{(j)}$ converges uniformly on $\mathbb{R}$ for all $j \in \{0, \ldots, k-1\}$, we have pointwise convergence $\lim_{n \to \infty} Af_n(x) = Af(x)$ for every $x \in \mathbb{R}$.

Definition. For $k \in \mathbb{N}$, an operator $T : C^k(\mathbb{R}) \to C(\mathbb{R})$ *depends on the k-th derivative*, if there are $x \in \mathbb{R}$ and functions $g_1, g_2 \in C^k(\mathbb{R})$ with $g_1^{(j)}(x) = g_2^{(j)}(x)$ for all $j \in \{0, \ldots, k-1\}$ and $Tg_1(x) \neq Tg_2(x)$.

We may now state the main result of this chapter.

Theorem 9.1 (Second-order chain rule)*. Let $k \in \mathbb{N}$, $\ell := \max(k-1, 1)$. Suppose that $T : C^k(\mathbb{R}) \to C(\mathbb{R})$ and $A_1, A_2 : C^\ell(\mathbb{R}) \to C(\mathbb{R})$ are operators such that the second-order chain rule*

$$T(f \circ g) = Tf \circ g \cdot A_1 g + A_2 f \circ g \cdot Tg, \quad f, g \in C^k(\mathbb{R}), \qquad (9.1)$$

holds. Assume that the pair (T, A_1) is C^k-non-degenerate and that A_1 and A_2 are isotropic and, if $k \geq 2$, are C^{k-1}-pointwise continuous. Then T, A_1 and A_2 are localized and

(a) *if $k \geq 4$, T does not depend on the k-th derivative;*

(b) *if $k \in \{1, 2, 3\}$ and T depends on the k-th derivative, there exist constants $c, d, p \in \mathbb{R} \setminus \{0\}$, $q, r \in \mathbb{R}$, $q, r \geq 0$, $p \geq k-1$ and a continuous function $H \in C(\mathbb{R})$ such that either,*

(b1)

$$Tf = \left[d\, T_k f + ((f')^{k-1} H \circ f - H)(f')^{k-1}\right] |f'|^{p-k+1} \{\operatorname{sgn} f'\},$$
$$A_1 f = (f')^{k-1} A_2 f, \quad A_2 f = |f'|^p \{\operatorname{sgn} f'\}, \tag{9.2}$$

where $T_1 f = \ln |f'|$, $T_2 f = f''$ and $T_3 f = (f')^2 Sf = f' f''' - \frac{3}{2}(f'')^2$,

or, if $k = 1$, additionally the following solutions are possible,

(b2)

$$Tf = d\, \sin\bigl(c \ln |f'|\bigr)\, |f'|^p \{\operatorname{sgn} f'\},$$
$$A_1 f = A_2 f = \cos\bigl(c \ln |f'|\bigr)\, |f'|^p \{\operatorname{sgn} f'\}, \tag{9.3}$$

or

$$Tf = (H \circ f)\, |f'|^q \{\operatorname{sgn} f'\} - H\, |f'|^r\, [\operatorname{sgn} f'],$$
$$A_1 f = |f'|^q \{\operatorname{sgn} f'\}, \quad A_2 f = |f'|^r\, [\operatorname{sgn} f']. \tag{9.4}$$

The terms $\{\operatorname{sgn} f'\}$ or $[\operatorname{sgn} f']$ should be simultaneously present in (T, A_1, A_2) or not at all, yielding two possible solutions, in the last case even four solutions. If the function H is constant in (9.4), the form of the operators (T, A_1, A_2) satisfying (9.1) would be slightly more general, namely $(T, A_1 + \gamma T, A_2 - \gamma T)$ where $\gamma \in \mathbb{R}$ is a suitable constant.

Conversely, all operators in (b) satisfy the second-order chain rule (9.1).

Corollary 9.2. *Suppose that $k \in \{1, 2, 3\}$ and that the operators (T, A_1, A_2) satisfy the assumptions of Theorem 9.1.*

(a) *Assume also that T annihilates all affine functions on $\mathbb{R}$. Then there exist $d, p \in \mathbb{R}$ such that*

either

$$Tf = d\, Sf\, |f'|^p \{\operatorname{sgn} f'\}, \; A_1 f = (f')^2\, A_2 f, \; A_2 f = |f'|^p \{\operatorname{sgn} f'\}, \; p \geq 2,$$

or

$$Tf = d\, f''\, |f'|^{p-1} \{\operatorname{sgn} f'\}, \; A_1 f = f'\, A_2 f, \; A_2 f = |f'|^p \{\operatorname{sgn} f'\}, \; p \geq 1.$$

(b) *If, in addition to (a), T satisfies the initial conditions*

$$T(\frac{x^2}{2}) = 1, \quad T(\frac{x^3}{6}) = x,$$

then $Tf = f''$, $A_1 f = f'^2$ and $A_2 f = f'$.

(c) If, in addition to (a), T satisfies the initial conditions

$$T(\frac{x^2}{2}) = -\frac{3}{2}, \quad T(\frac{x^3}{6}) = -x^2,$$

then $Tf = f'^2 Sf = f'f''' - \frac{3}{2}f''^2$, $A_1f = f'^4$ and $A_2f = f'^2$.

Proof. (a) The first part of Corollary 9.2 follows directly from Theorem 9.1 since the solutions in formulas (9.3) and (9.4) and the one in (9.2) for $k = 1$ do not annihilate the affine functions. Therefore T depends non-trivially on f'' or f'''. Moreover, the function H has to be zero so that the term involving H is zero, to guarantee that T annihilates all affine functions.

(b) Assuming in addition, that $T(\frac{x^2}{2}) = 1$ and $T(\frac{x^3}{6}) = x$ holds, the first solution in (a) would require that $1 = T(\frac{x^2}{2}) = d(-\frac{3}{2}x^{p-2})$ and $x = T(\frac{x^3}{6}) = d(-x^2x^{p-2})$ for all $x > 0$, yielding the contradiction $2 = p = 1$. The second solution in (a) satisfies these initial conditions with $p = 1$, $d = 1$ and the $\{\text{sgn } f'\}$-term being present, i.e., $Tf = f''$, $A_1f = f'^2$ and $A_2f = f'$.

(c) Assuming in addition, that $T(\frac{x^2}{2}) = -\frac{3}{2}$ and $T(\frac{x^3}{6}) = -x^2$ holds, the second solution in (a) would require that $-\frac{3}{2} = T(\frac{x^2}{2}) = dx^{p-1}$ and $-x^2 = T(\frac{x^3}{6}) = d(x(\frac{x^2}{2})^{p-1})$ for all $x > 0$, yielding the contradiction $1 = p = \frac{3}{2}$. In this case, the first solution satisfies these initial conditions with $p = 2$, $d = 1$ and the $\{\text{sgn } f'\}$-term not being present, i.e., $Tf = f'^2 Sf = f'f''' - \frac{3}{2}f''^2$, $A_1f = f'^4$ and $A_2f = f'^2$. $\square$

Remarks. (a) The assumption that A_1, A_2 are isotropic is not needed in Theorem 9.1. However, it simplifies the proof considerably, which even in the isotropic case is technical and lengthy. Under the assumptions of Theorem 9.1, but without the isotropy condition on A_1, A_2, the general solution of (9.1) can be obtained as follows:

There are strictly positive functions $K_1, K_2 \in C(\mathbb{R})$ so that for any solution $(\widetilde{T}, \widetilde{A}_1, \widetilde{A}_2)$ of (9.1) there is an isotropic solution (T, A_1, A_2) of (9.1) such that

$$\widetilde{T}f(x) = \frac{K_2(f(x))}{K_1(x)}Tf(x),$$

$$\widetilde{A}_1f(x) = \frac{K_1(f(x))}{K_1(x)}A_1f(x), \quad \widetilde{A}_2f(x) = \frac{K_2(f(x))}{K_2(x)}A_2f(x).$$

It is quickly checked that $(\widetilde{T}, \widetilde{A}_1, \widetilde{A}_2)$ satisfy (9.1) provided that (T, A_1, A_2) does. Conversely, this gives the form of all possible non-isotropic solutions of (9.1). The assumption that (A_1, A_2) are isotropic is not used to prove (a) of Theorem (9.1). The proof simplifies under the isotropy condition since then the functions representing A_1, A_2 do not depend on the independent variable. Therefore, we stick to this simpler case in our proof.

(b) If $k \geq 2$, the assumption of C^{k-1}-pointwise continuity of A_1, A_2 most likely can be eliminated. However, it also simplifies the proof of Theorem 9.1, since then the representing functions of the operators A_1, A_2 do not depend on the k-th derivative variable.

(c) The results of Theorem 9.1 show that $C^k(\mathbb{R})$ for $k = 0, 1, 2, 3$ constitute the "natural" domains for T and that $C^k(\mathbb{R})$ for $k = 0, 1$ are the "natural" domains for A_1 and A_2. The case $k = 0$ corresponds to the degenerate case in (b1) when formally putting there $d = 0$ and $k = 1$, i.e., when $Tf = H \circ f - H$, $A_1 f = A_2 f = \mathbf{1}$. This solution was already mentioned in the introduction.

The operators A_1 and A_2 are closely related by $A_1 f = (f')^{k-1} A_2 f$ in the main cases of (b1); the motivating examples therefore showed the typical phenomenon. The operators A_1 or A_2 cannot be zero, due to the assumption of C^k-non-degeneration.

(d) The functions $f_3(x) = x + \frac{x^3}{6}$, $f_2(x) = x + \frac{x^2}{2}$ may be used to determine the constant d in the described form of T in (b1) from $d = Tf_k(0)$ for $k = 3, 2$. The function H in (b1) is completely determined by the function $T(2\,\mathrm{Id})$, similar as in the case of the chain rule equation, cf. Remark (b) following Theorem 4.1.

(e) The structure of equation (9.1), $T(f \circ g) = Tf \circ g \cdot A_1 g + A_2 f \circ g \cdot Tg$, is somewhat similar to the one of the operator equation $T(f \cdot g) = Tf \cdot A_1 g + A_2 f \cdot Tg$ studied in Theorem 3.7 as an extension of the Leibniz rule, except that the product $Tf \cdot A_1 g$ there is replaced by the "compound" product $Tf \circ g \cdot A_1 g$. There is a certain similarity in some of the solutions. However, the function variable $\alpha_0 = f(x)$ in Theorem 3.7 is replaced by the derivative variable $\alpha_1 = f'(x)$ here. The difference between these equations is that (9.1) does not have any solutions depending non-trivially on the k-th derivative $f^{(k)}$ for any $k \geq 4$, while the extended Leibniz rule has solutions which depend non-trivially on $f^{(k)}$ for all $k \in \mathbb{N}$.

(f) Solving the second-order chain rule, no "phase transition" between two of the solutions in Theorem 9.1 is possible, contrary to the case of the solutions for the extended Leibniz rule (Theorem 3.7), or the second-order Leibniz rule (Theorem 7.2), cf. the examples there. This is also true if A_1 and A_2 are not assumed to be isotropic. It is essentially a consequence of the fact that $c, d, p \in \mathbb{R} \setminus \{0\}$ in (9.2), (9.3) and (9.4) are constants, and not functions of x, which could have a singularity or decay to zero at a point of phase transition, as in the examples following Theorems 3.7 and 7.2.

(g) Concerning a related cohomological result for diffeomorphisms on the projective line, we refer to Section 9.4.

9.2   Proof of Theorem 9.1

We first show that (T, A_1, A_2) are localized.

Proposition 9.3. *Let $k \in \mathbb{N}$, $k \geq 2$. Assume that $T : C^k(\mathbb{R}) \to C(\mathbb{R})$ and $A_1, A_2 : C^{k-1}(\mathbb{R}) \to \mathbb{R}$ satisfy (9.1), that (T, A_1) is C^k-non-degenerate and that A_1 and A_2 are isotropic and C^{k-1}-pointwise continuous. Then there are functions $F : \mathbb{R}^{k+2} \to \mathbb{R}$ and $B_1, B_2 : \mathbb{R}^k \to \mathbb{R}$ such that for all $f \in C^k(\mathbb{R})$ and $x \in \mathbb{R}$*

$$Tf(x) = F\big(x, f(x), \ldots, f^{(k)}(x)\big),$$
$$A_i f(x) = B_i\big(f(x), \ldots, f^{(k-1)}(x)\big), \quad i = 1, 2.$$

For $k = 1$, without the pointwise continuity assumption, there are functions $F : \mathbb{R}^3 \to \mathbb{R}$ and $B_1, B_2 : \mathbb{R}^2 \to \mathbb{R}$ such that for all $f \in C^1(\mathbb{R})$ and $x \in \mathbb{R}$

$$Tf(x) = F\big(x, f(x), f'(x)\big), \quad A_i f(x) = B_i\big(f(x), f'(x)\big), \quad i = 1, 2.$$

Proof. (a) We first show that $T(\mathrm{Id}) = 0$ and $A_1(\mathrm{Id}) = A_2(\mathrm{Id}) = \mathbf{1}$. Choosing $f = \mathrm{Id}$ in (9.1), we find for all $g \in C^k(\mathbb{R})$, $y \in \mathbb{R}$,

$$Tg(y) = T(\mathrm{Id})(g(y))\, A_1 g(y) + A_2(\mathrm{Id})(g(y))\, Tg(y),$$
$$Tg(y)\big(1 - A_2(\mathrm{Id})(g(y))\big) = A_1 g(y)\, T(\mathrm{Id})(g(y)).$$

Since (T, A_1) is C^k-non-degenerate, for any $x \in \mathbb{R}$ we may find $g_1, g_2 \in C^k(\mathbb{R})$ and $y_1, y_2 \in \mathbb{R}$ with $g_1(y_1) = g_2(y_2) = x$ such that the two vectors $(Tg_i(y_i), A_1 g_i(y_i)) \in \mathbb{R}^2$ are linearly independent for $i = 1, 2$. The resulting two linear equations

$$Tg_i(y_i)\big(1 - A_2(\mathrm{Id})(x)\big) = A_1 g_i(y_i)\, T(\mathrm{Id})(x), \quad i = 1, 2,$$

therefore only admit the trivial solution $T(\mathrm{Id})(x) = 0$, $1 - A_2(\mathrm{Id})(x) = 0$. Hence $T(\mathrm{Id}) = 0$, $A_2(\mathrm{Id}) = \mathbf{1}$. Choosing $g = \mathrm{Id}$ in (9.1), we get for all $f \in C^k(\mathbb{R})$ and $x \in \mathbb{R}$,

$$Tf(x) = Tf(x)\, A_1(\mathrm{Id})(x) + A_2 f(x) T(\mathrm{Id})(x) = Tf(x)\, A_1(\mathrm{Id})(x).$$

By non-degeneracy, choose $f \in C^k(\mathbb{R})$ such that $Tf(x) \neq 0$. This implies that also $A_1(\mathrm{Id}) = \mathbf{1}$.

(b) Let $J \subset \mathbb{R}$ be an open interval and $f_1, f_2 \in C^k(\mathbb{R})$ be such that $f_1|_J = f_2|_J$. We claim that $Tf_1|_J = Tf_2|_J$, $A_i f_1|_J = A_i f_2|_J$, $i = 1, 2$.

Take any $x \in J$. By assumption, there are $g_1, g_2 \in C^k(\mathbb{R})$ with images in J and points $y_1, y_2 \in \mathbb{R}$ with $g_i(y_i) = x$ such that $(Tg_i(y_i), A_1 g_i(y_i)) \in \mathbb{R}$ are linearly independent for $i = 1, 2$. Since $f_1|_J = f_2|_J$, we have $f_1 \circ g_i = f_2 \circ g_i$ for $i = 1, 2$. By (9.1)

$$0 = T(f_1 \circ g_i)(y_i) - T(f_2 \circ g_i)(y_i)$$
$$= \big(Tf_1(x) - Tf_2(x)\big) A_1 g_i(y_i) + \big(A_2 f_1(x) - A_2 f_2(x)\big) Tg_i(y_i)$$

for $i = 1, 2$. This implies $Tf_1(x) = Tf_2(x)$, $A_2 f_1(x) = A_2 f_2(x)$ and hence $Tf_1|_J = Tf_2|_J$ and $A_2 f_1|_J = A_2 f_2|_J$.

For $x \in J$, let $y := f_1(x) = f_2(x)$. Choose $h \in C^k(\mathbb{R})$ with $Th(y) \neq 0$. Since also $(h \circ f_1)|_J = (h \circ f_2)|_J$, we know that $T(h \circ f_1)|_J = T(h \circ f_2)|_J$ and $Tf_1|_J = Tf_2|_J$. By (9.1)

$$
\begin{aligned}
0 &= T(h \circ f_1)(x) - T(h \circ f_2)(x) \\
&= Th(y)\big(A_1 f_1(x) - A_1 f_2(x)\big) + A_2 h(y)\big(Tf_1(x) - Tf_2(x)\big) \\
&= Th(y)\big(A_1 f_1(x) - A_2 f_2(x)\big),
\end{aligned}
$$

hence $A_1 f_1(x) = A_1 f_2(x)$, i.e., $A_1 f_1|_J = A_1 f_2|_J$, too.

(c) By part (b) and Proposition 3.3 there are functions $F, \widetilde{B}_i : \mathbb{R}^{k+2} \to \mathbb{R}$, $i = 1, 2$, such that for all $f \in C^k(I)$ and all $x \in I$

$$
Tf(x) = F\big(x, f(x), \ldots, f^{(k)}(x)\big),
$$
$$
A_i f(x) = \widetilde{B}_i\big(x, f(x), \ldots, f^{(k)}(x)\big), \quad i = 1, 2.
$$

We claim that the $\widetilde{B}_i$, $i \in \{1, 2\}$, do not depend on $f^{(k)}(x)$ for all $f \in C^k(\mathbb{R})$, $x \in \mathbb{R}$. Let $f \in C^k(I)$, $x_0 \in \mathbb{R}$ and $g(x) := \sum_{j=0}^{k-1} \frac{f^{(j)}(x_0)}{j!}(x - x_0)^j$ be the $(k-1)$-st Taylor approximation to f at x_0. We will show by C^{k-1}-smooth approximations that $Tf(x_0) = Tg(x_0)$ holds, and obviously $Tg(x_0)$ depends only on x_0, $f(x_0)$ and all derivatives up to $f^{(k-1)}(x_0)$, but not on $f^{(k)}(x_0)$.
For $n \in \mathbb{N}$, let $x_n := x_0 + \frac{1}{n}$, $y_n := x_0 - \frac{1}{n}$ and define $\phi_n : \mathbb{R} \to \mathbb{R}$ by

$$
\phi_n(x) := \begin{cases} g(x) + f^{(k)}(x_0)(x - x_0)^k \left(\frac{x_n - x}{x_n - x_0}\right)^{k+1}, & x \geq x_0, \\ g(x) + f^{(k)}(x_0)(x - x_0)^k \left(\frac{x - y_n}{x_n - y_n}\right)^{k+1}, & x < x_0, \end{cases}
$$

and functions $g_n, h_n : \mathbb{R} \to \mathbb{R}$ by

$$
g_n(x) := \begin{cases} g(x), & x \leq y_n \text{ or } x_n \leq x, \\ \phi_n(x), & y_n < x < x_n, \end{cases} \qquad h_n(x) := \begin{cases} f(x), & x < x_0, \\ g_n(x), & x \geq x_0. \end{cases}
$$

Since ϕ_n is in $C^k(\mathbb{R})$ with $\phi_n^{(j)}(x_0) = g^{(j)}(x_0) = f^{(j)}(x_0)$, $\phi_n^{(j)}(x_n) = g^{(j)}(x_n)$, $\phi_n(y_n) = g^{(j)}(y_n)$ for all $j \in \{0, \ldots, k\}$, g_n and h_n are in $C^k(\mathbb{R})$ as well, and g_n converges to g uniformly in C^{k-1}, i.e., $g_n^{(j)} \to g^{(j)}$ uniformly for all $j \in \{0, \ldots, k-1\}$. The C^{k-1}-pointwise continuity assumption for A_1 and A_2 implies that $A_i g_n(x_0) \to A_i g(x_0)$, $i \in \{1, 2\}$. Let $I_- := (-\infty, x_0)$, $I_+ := (x_0, \infty)$. Then $f|_{I_-} = h_n|_{I_-}$ and $h_n|_{I_+} = g_n|_{I_+}$. By part (b), $A_i f|_{I_-} = A_i h_n|_{I_-}$ and $A_i h_n|_{I_+} = A_i g_n|_{I_+}$, $i \in \{1, 2\}$. Since the images of A_1 and A_2 consist of continuous functions, and $\{x_0\} = \overline{I_-} \cap \overline{I_+}$, we get $A_i f(x_0) = A_i h_n(x_0) = A_i g_n(x_0)$. With $A_i g_n(x_0) \to A_i g(x_0)$ we have $A_i f(x_0) = A_i g(x_0)$ for $i \in \{1, 2\}$. However, the $(k-1)$-st Taylor polynomial g depends only on $(x_0, f(x_0), \ldots, f^{(k-1)}(x_0))$. Thus the $A_i f(x_0)$ do not depend on $f^{(k)}(x_0)$ and

$$
(A_i f)(x_0) = \widetilde{B}_i(x_0, f(x_0), \ldots, f^{(k)}(x_0)) =: \bar{B}_i(x_0, f(x_0), \ldots, f^{(k-1)}(x_0)),
$$

for $i \in \{1, 2\}$ and all $x_0 \in \mathbb{R}$.

(d) The assumption that A_1, A_2 are isotropic means that for all $x, y \in \mathbb{R}$

$$\bar{B}_i\big(x, f(y + x), \ldots, f^{(k-1)}(y + x)\big) = \bar{B}_i\big(y + x, f(y + x), \ldots, f^{(k-1)}(y + x)\big).$$

This implies that $\bar{B}_i$ is independent of the first variable x, i.e.,

$$A_i f(x) = \bar{B}_i\big(x, f(x), \ldots, f^{(k-1)}(x)\big) =: B_i\big(f(x), \ldots, f^{(k-1)}(x)\big). \qquad \square$$

We now prove part (b) of Theorem 9.1, finding all solutions of the second-order chain rule if $k \in \{1, 2, 3\}$. Afterwards, we will show part (a), that there are no solutions which depend on the k-th derivative, if $k \geq 4$.

Proof of (b) *of Theorem* 9.1. (i) We first consider the case $k = 3$, $T : C^3(\mathbb{R}) \to C(\mathbb{R})$ and $A_1, A_2 : C^2(\mathbb{R}) \to C(\mathbb{R})$ satisfying (9.1), and such that Tf depends non-trivially on f'''. The case $k = 2$ is rather similar. We will later indicate how the following analysis of the representing function changes if $k = 2$. By Proposition 9.3 there are functions $F : \mathbb{R}^5 \to \mathbb{R}$ and $B_1, B_2 : \mathbb{R}^3 \to \mathbb{R}$ such that for all $f \in C^3(\mathbb{R})$ and $x \in \mathbb{R}$

$$Tf(x) = F\big(x, f(x), f'(x), f''(x), f'''(x)\big),$$
$$A_i f(x) = B_i\big(f(x), f'(x), f''(x)\big), \quad i = 1, 2. \tag{9.5}$$

Then $T(\mathrm{Id}) = 0$, $A_i(\mathrm{Id}) = \mathbf{1}$ translates into

$$F(x, x, 1, 0, 0) = 0, \quad B_i(x, 1, 0) = 1, \quad i = 1, 2,$$

equations which we will use in the following without further mention. Given arbitrary values $x, y, z, \alpha_1, \alpha_2, \alpha_3, \beta_1, \beta_2, \beta_3 \in \mathbb{R}$, choose functions $f, g \in C^3(\mathbb{R})$ with $g(x) = y$, $f(y) = z$, $g^{(j)}(x) = \beta_j$, $f^{(j)}(y) = \alpha_j$ for $j = 1, 2, 3$. Then the second-order chain rule operator equation (9.1) is equivalent to the functional equation for the three functions F, B_1, B_2 given by

$$F(x, z, \alpha_1\beta_1, \alpha_2\beta_1^2 + \alpha_1\beta_2, \alpha_3\beta_1^3 + 3\alpha_2\beta_1\beta_2 + \alpha_1\beta_3)$$
$$= F(y, z, \alpha_1, \alpha_2, \alpha_3)B_1(y, \beta_1, \beta_2) + F(x, y, \beta_1, \beta_2, \beta_3)B_2(z, \alpha_1, \alpha_2), \tag{9.6}$$

since

$$(f \circ g)'' = f'' \circ g \cdot (g')^2 + f' \circ g \cdot g'',$$
$$(f \circ g)''' = f''' \circ g \cdot (g')^3 + 3f'' \circ g \cdot g' \cdot g'' + f' \circ g \cdot g'''.$$

(ii) We choose particular values for the α_i's and β_i's in (9.6) to identify the structure of F, B_1 and B_2. Our first aim is to show that $F(x, z, \alpha_1, \alpha_2, \alpha_3)$ is an affine function of α_3, with coefficients depending on $(x, z, \alpha_1, \alpha_2)$ and that B_1 and B_2 are related by $B_1(z, \alpha_1, \alpha_2) = \alpha_1^2 B_2(z, \alpha_1, \alpha_2)$.

Choosing $\alpha_1 = \beta_1 = 1$, $\alpha_2 = \beta_2 = 0$ in (9.6), we get

$$F(x, z, 1, 0, \alpha_3 + \beta_3) = F(y, z, 1, 0, \alpha_3) + F(x, y, 1, 0, \beta_3), \qquad (9.7)$$

which yields for $\alpha_3 = \beta_3 = y = 0$ that $F(x, z, 1, 0, 0) = F(0, z, 1, 0, 0) + F(x, 0, 1, 0, 0)$.
For $x = z$ this implies $0 = F(x, x, 1, 0, 0) = F(0, x, 1, 0, 0) + F(x, 0, 1, 0, 0)$. Put
$G(x) := F(0, x, 1, 0, 0) = T(\mathrm{Id} + x)(0)$. Then $G(x) = -F(x, 0, 1, 0, 0)$ and hence
$F(x, z, 1, 0, 0) = G(z) - G(x)$. Put $x = z$ in (9.7) and interchange α_3 and β_3. Then

$$F(x, x, 1, \alpha_3 + \beta_3) = F(y, x, 1, 0, \beta_3) + F(x, y, 1, 0, \alpha_3) = F(y, y, 1, \alpha_3 + \beta_3),$$

so that $\widetilde{F}(\alpha_3) := F(x, x, 1, \alpha_3)$ is independent of $x \in \mathbb{R}$, with $\widetilde{F} : \mathbb{R} \to \mathbb{R}$ being
additive by (9.7) for $x = y = z$, $\widetilde{F}(\alpha_3 + \beta_3) = \widetilde{F}(\alpha_3) + \widetilde{F}(\beta_3)$. Choose $\beta_3 = 0$ and
$y = z$ in (9.7). Then

$$F(x, z, 1, 0, \alpha_3) = \widetilde{F}(\alpha_3) + G(z) - G(x). \qquad (9.8)$$

Next, put $\alpha_1 = 1$, $\alpha_2 = \beta_3 = 0$ in (9.6). Then

$$F(x, z, \beta_1, \beta_2, \alpha_3'\beta_1^3) = F(y, z, 1, 0, \alpha_3')B_1(y, \beta_1, \beta_2) + F(x, y, \beta_1, \beta_2, 0).$$

For $\beta_1 \neq 0$, $\alpha_3'\beta_1^3$ may attain arbitrary values, varying α_3'. Renaming the variables
$(\beta_1, \beta_2, \alpha_3'\beta_1^3)$ by $(\alpha_1, \alpha_2, \alpha_3)$, we get for $\alpha_1 \neq 0$, using (9.8)

$$F(x, z, \alpha_1, \alpha_2, \alpha_3) = F\left(y, z, 1, 0, \frac{\alpha_3}{\alpha_1^3}\right) B_1(y, \alpha_1, \alpha_2) + F(x, y, \alpha_1, \alpha_2, 0)$$

$$= F(x, y, \alpha_1, \alpha_2, 0) + \left(\widetilde{F}\left(\frac{\alpha_3}{\alpha_1^3}\right) + G(z) - G(y)\right) B_1(y, \alpha_1, \alpha_2). \quad (9.9)$$

Similarly, choosing $\beta_1 = 1$, $\beta_2 = \alpha_3 = 0$ in (9.6) and replacing $\alpha_1\beta_3$ by α_3, we get
for $\alpha_1 \neq 0$

$$F(x, z, \alpha_1, \alpha_2, \alpha_3)$$
$$= F(y, z, \alpha_1, \alpha_2, 0) + \left(\widetilde{F}\left(\frac{\alpha_3}{\alpha_1}\right) + G(y) - G(x)\right) B_2(z, \alpha_1, \alpha_2). \quad (9.10)$$

Take $y = z$ in (9.9) and $y = x$ in (9.10) and compare the results to conclude that
for any $\alpha_1, \alpha_2, \alpha_3 \in \mathbb{R}$ with $\alpha_1 \neq 0$

$$\widetilde{F}\left(\frac{\alpha_3}{\alpha_1^3}\right) B_1(z, \alpha_1, \alpha_2) = \widetilde{F}\left(\frac{\alpha_3}{\alpha_1}\right) B_2(z, \alpha_1, \alpha_2).$$

If $\widetilde{F}$ were identically zero, by equation (9.9) the function F would be independent
of α_3 and Tf would be independent of f''' at any point $x \in \mathbb{R}$, contrary to our

assumption. Hence, there is $c \neq 0$ with $\widetilde{F}(c) \neq 0$. Choose $\alpha_3 = c\alpha_1^3$. Then with $\overline{F}(t) := \frac{\widetilde{F}(ct)}{\widetilde{F}(c)}$,

$$B_1(z, \alpha_1, \alpha_2) = \overline{F}(\alpha_1^2) B_2(z, \alpha_1, \alpha_2).$$

For $f(x) := x + \frac{1}{3}x^3$, $A_2 f(x) = B_2(f(x), 1 + x^2, 2x)$ and $A_2 f(0) = B_2(0, 1, 0) = 1 \neq 0$. Since $A_2 f$ is continuous, there is $\epsilon > 0$ such that $A_2 f(x) \neq 0$ for all $x \in \mathbb{R}$ with $|x| \leq \epsilon$. Then for all $\alpha \in [1, 1 + \epsilon^2]$,

$$A_2 f(\sqrt{\alpha - 1}) = B_2(f(\sqrt{\alpha - 1}), \alpha, 2\sqrt{\alpha - 1}) \neq 0$$

and

$$\frac{A_1 f(\sqrt{\alpha - 1})}{A_2 f(\sqrt{\alpha - 1})} = \frac{B_1(f(\sqrt{\alpha - 1}), \alpha, 2\sqrt{\alpha - 1})}{B_2(f(\sqrt{\alpha - 1}), \alpha, 2\sqrt{\alpha - 1})} = \overline{F}(\alpha^2)$$

defines a continuous function for $\alpha \in [1, 1 + \epsilon^2]$. Since $\overline{F}$ is additive, Proposition 2.2 yields that $\overline{F}$ is linear. Since $\overline{F}(1) = 1$, we have $\overline{F}(t) = t$. Put $d := \widetilde{F}(1)$. Then $\widetilde{F}(t) = dt$ and for all α_1, α_2 with $\alpha_1 \neq 0$,

$$B_1(z, \alpha_1, \alpha_2) = \alpha_1^2 B_2(z, \alpha_1, \alpha_2).$$

Actually, B_1 and B_2 are also independent of z: Put $y = z$ in (9.10) to find

$$F(x, z, \alpha_1, \alpha_2, \alpha_3) = F(z, z, \alpha_1, \alpha_2, 0) + \left(d\frac{\alpha_3}{\alpha_1} + G(z) - G(x) \right) B_2(z, \alpha_1, \alpha_2).$$

Taking $y = x$ in (9.9), we get, using in addition that $B_1 = \alpha_1^2 B_2$,

$$F(x, z, \alpha_1, \alpha_2, \alpha_3) = F(x, x, \alpha_1, \alpha_2, 0) + \left(d\frac{\alpha_3}{\alpha_1^3} + G(z) - G(x) \right) \alpha_1^2 B_2(x, \alpha_1, \alpha_2).$$

Since the left-hand sides of the previous two equations are identical, so are the right-hand sides. Isolating the terms involving α_3 on one side, we conclude

$$d\frac{\alpha_3}{\alpha_1} \big(B_2(z, \alpha_1, \alpha_2) - B_2(x, \alpha_1, \alpha_2) \big) = \big[F(x, x, \alpha_1, \alpha_2, 0) - F(z, z, \alpha_1, \alpha_2, 0) \big]$$

$$+ \big[G(z) - G(x) \big] \big[\alpha_1^2 B_2(x, \alpha_1, \alpha_2) - B_2(z, \alpha_1, \alpha_2) \big]. \quad (9.11)$$

The right-hand side is independent of α_3 and hence the left-hand side, too, requiring that $B_2(z, \alpha_1, \alpha_2) = B_2(x, \alpha_1, \alpha_2)$. This means that B_2 and $B_1 = \alpha_1^2 B_2$ are independent of x and z. Put

$$B(\alpha_1, \alpha_2) := B_2(x, \alpha_1, \alpha_2) = B_2(z, \alpha_1, \alpha_2),$$

so that $\alpha_1^2 B(\alpha_1, \alpha_2) = B_1(x, \alpha_1, \alpha_2) = B_1(z, \alpha_1, \alpha_2)$. Since now the left-hand side of (9.11) is zero, so is the right-hand side. This yields for $x = 0$ with $G(0) = T(\mathrm{Id})(0) = 0$,

$$F(z, z, \alpha_1, \alpha_2, 0) = F(0, 0, \alpha_1, \alpha_2, 0) + G(z)(\alpha_1^2 - 1)B(\alpha_1, \alpha_2).$$

Let $F_0(\alpha_1, \alpha_2) := F(0, 0, \alpha_1, \alpha_2, 0)$. Then, using (9.10) with $y = z$,

$$
\begin{aligned}
F(x, z, \alpha_1, \alpha_2, \alpha_3) &= \left[F_0(\alpha_1, \alpha_2) + G(z)(\alpha_1^2 - 1)B(\alpha_1, \alpha_2) \right] \\
&\quad + \left[d\frac{\alpha_3}{\alpha_1} + G(z) - G(x) \right] B(\alpha_1, \alpha_2) \\
&= F_0(\alpha_1, \alpha_2) + \left[d\frac{\alpha_3}{\alpha_1} + \alpha_1^2 G(z) - G(x) \right] B(\alpha_1, \alpha_2). \quad (9.12)
\end{aligned}
$$

(iii) By (9.12), it suffices to determine the functions F_0 and B. We claim that B is independent of α_2 and multiplicative in α_1.

Insert formula (9.12) for F into (9.6) with $x = y = z = 0$, $\alpha_1 \neq 0 \neq \beta_1$ and isolate terms involving α_3 and β_3 on the left-hand side. This yields after some calculation

$$
\begin{aligned}
d\left(\frac{\alpha_3 \beta_1^2}{\alpha_1} + \frac{\beta_3}{\beta_1} \right) &\left[B(\alpha_1 \beta_1, \alpha_2 \beta_1^2 + \alpha_1 \beta_2) - B(\alpha_1, \alpha_2)B(\beta_1, \beta_2) \right] \\
&= \beta_1^2 F_0(\alpha_1, \alpha_2) B(\beta_1, \beta_2) + F_0(\beta_1, \beta_2) B(\alpha_1, \alpha_2) - F_0(\alpha_1 \beta_1, \alpha_2 \beta_1^2 + \alpha_1 \beta_2) \\
&\quad - 3d\frac{\alpha_2 \beta_2}{\alpha_1} B(\alpha_1 \beta_1, \alpha_2 \beta_1^2 + \alpha_1 \beta_2). \quad (9.13)
\end{aligned}
$$

Since the right-hand side is independent of α_3 and β_3, the same is true for the left-hand side. Using that $d \neq 0$, we get

$$
B(\alpha_1 \beta_1, \alpha_2 \beta_1^2 + \alpha_1 \beta_2) = B(\alpha_1, \alpha_2)B(\beta_1, \beta_2) = B(\alpha_1 \beta_1, \alpha_2 \beta_1 + \alpha_1^2 \beta_2), \quad (9.14)
$$

where the last equality is a consequence of the symmetry of the product in the middle in α and β. Given any $t, s \in \mathbb{R}$ and fixed values α_1, β_1 with $\alpha_1 \beta_1 \notin \{0, 1\}$, the linear equations

$$
\beta_1^2 \alpha_2 + \alpha_1 \beta_2 = t, \quad \beta_1 \alpha_2 + \alpha_1^2 \beta_2 = s,
$$

may be solved for α_2, β_2, since $\det \begin{pmatrix} \beta_1^2 & \alpha_1 \\ \beta_1 & \alpha_1^2 \end{pmatrix} = \alpha_1 \beta_1 (\alpha_1 \beta_1 - 1) \neq 0$. Hence for $\gamma_1 \notin \{0, 1\}$, $B(\gamma_1, t)$ is independent of t. For $f(x) = \frac{1}{2}x^2$, $A_2 f(\alpha_1) = B_2\left(\frac{\alpha_1^2}{2}, \alpha_1, 1\right) = B(\alpha_1, 1) = B(\alpha_1, 0)$, if $\alpha_1 \notin \{0, 1\}$. Since $A_2 f$ is continuous, $B(\alpha_1, 0)$ is continuous in α_1 and by (9.14) multiplicative,

$$
B(\alpha_1 \beta_1, 0) = B(\alpha_1, 0) \, B(\beta_1, 0).
$$

By Proposition 2.3, there is $p \in \mathbb{R}$ such that

$$
B(\alpha_1, 0) = |\alpha_1|^p \quad \text{or} \quad B(\alpha_1, 0) = |\alpha_1|^p \operatorname{sgn} \alpha_1.
$$

Since $B(\,\cdot\,, 0)$ is continuous, we need $p \geq 0$ in the first case and $p > 0$ in the second case. Thus we have for all α_1, α_2, also for $\alpha_1 = 1$ and $\alpha_1 = 0$,

$$
B(\alpha_1, \alpha_2) = B(\alpha_1, 0) = |\alpha_1|^p \{\operatorname{sgn} \alpha_1\} =: \widetilde{B}(\alpha_1). \quad (9.15)
$$

(iv) We finally determine the form of F_0. Since the left-hand side of (9.13) is zero, so is the right-hand side, and using also (9.15), we get for $\alpha_1 \neq 0 \neq \beta_1$

$$F_0(\alpha_1\beta_1, \alpha_2\beta_1^2 + \alpha_1\beta_2) + 3d\frac{\alpha_2\beta_2}{\alpha_1}\widetilde{B}(\alpha_1\beta_1)$$
$$= F_0(\alpha_1, \alpha_2)\beta_1^2\,\widetilde{B}(\beta_1) + F_0(\beta_1, \beta_2)\,\widetilde{B}(\alpha_1). \quad (9.16)$$

Let $\beta_1 = 1$, $\alpha_2 = 0$. After renaming $\alpha_1\beta_2$ as α_2, we get

$$F_0(\alpha_1, \alpha_2) = F(\alpha_1, 0) + F_0\left(1, \frac{\alpha_2}{\alpha_1}\right)\widetilde{B}(\alpha_1). \quad (9.17)$$

Similarly, for $\alpha_1 = 1$, $\beta_2 = 0$ we find after renaming variables

$$F_0(\alpha_1, \alpha_2) = F_0(\alpha_1, 0) + F_0\left(1, \frac{\alpha_2}{\alpha_1^2}\right)\alpha_1^2\,\widetilde{B}(\alpha_1).$$

Comparing the results, we conclude that $F\left(1, \frac{\alpha_2}{\alpha_1}\right) = F\left(1, \frac{\alpha_2}{\alpha_1^2}\right)\alpha_1^2$ which for $\alpha_2 = \alpha_1^2 =: \alpha^2$ given $F_0(1, \alpha) = F_0(1, 1)\alpha^2 =: b\alpha^2$. To identify also $F_0(\alpha_1, 0)$, put $\alpha_2 = \beta_2 = 0$ in (9.16). By symmetry in α_1 and β_1,

$$F_0(\alpha_1\beta_1, 0) = F_0(\alpha_1, 0)\beta_1^2\,\widetilde{B}(\beta_1) + F_0(\beta_1, 0)\widetilde{B}(\alpha_1)$$
$$= F_0(\beta_1, 0)\alpha_1^2\,\widetilde{B}(\alpha_1) + F_0(\alpha_1, 0)\widetilde{B}(\beta_1).$$

Take the difference of both right-hand sides and choose a fixed $\beta \notin \{0, 1, -1\}$. Let $c := \frac{F_0(\beta_1, 0)}{(\beta_1^2 - 1)\widetilde{B}(\beta_1)}$. Then $F_0(\alpha_1, 0) = c(\alpha_1^2 - 1)\widetilde{B}(\alpha_1)$. By (9.17),

$$F_0(\alpha_1, \alpha_2) = \left[c(\alpha_1^2 - 1) + b\frac{\alpha_2^2}{\alpha_1^2}\right]\widetilde{B}(\alpha_1).$$

Put $\alpha_1 = \beta_1 = 1$ and $\alpha_2 = \beta_2 = \frac{1}{2}$, to find using (9.16) and $\widetilde{B}(1) = 1$,

$$b + \frac{3}{4}d = F_0(1, 1) + \frac{3}{4}d = 2F_0\left(1, \frac{1}{2}\right) = \frac{b}{2},$$

so that $b = -\frac{3}{2}d$ is necessary for (9.16) to be satisfied. Formula (9.12) finally yields, using the formulas for F_0 and B

$$F(x, z, \alpha_1, \alpha_2, \alpha_3) = \left(\left[c(\alpha_1^2 - 1) - \frac{3}{2}d\frac{\alpha_2^2}{\alpha_1^2}\right] + \left[d\frac{\alpha_3}{\alpha_1} + \alpha_1^2 G(z) - G(x)\right]\right)\widetilde{B}(\alpha_1)$$
$$= \left[d\left(\frac{\alpha_3}{\alpha_1} - \frac{3}{2}\frac{\alpha_2^2}{\alpha_1^2}\right) + \alpha_1^2 H(z) - H(x)\right]\widetilde{B}(\alpha_1), \quad (9.18)$$

where we put $H(z) := G(z) + c$. Inserting formulas (9.18) and (9.15) for F and B and those for B_1 and B_2 in terms of B into (9.6), calculation shows that, conversely, these functions satisfy (9.6). Therefore, (9.18) gives the general solution of the functional equation (9.6).

(v) Hence by (9.5)

$$Tf = \left[d(f')^2 Sf + ((f')^2 H \circ f - H)(f')^2\right]|f'|^{p-2}\{\operatorname{sgn} f'\},$$
$$A_1 f = (f')^2 A_2 f, \quad A_2 f = |f'|^p \{\operatorname{sgn} f'\}, \tag{9.19}$$

for all $f \in C^3(\mathbb{R})$, where S denotes the Schwarzian. We need $p \geq 2$ to guarantee that Tf is continuous for all $f \in C^3(\mathbb{R})$. Since Tf, f, f', f'' and f''' are continuous for all $f \in C^3(\mathbb{R})$, it follows from the formula for T that $(f')^2 H \circ f - H$ is continuous in $x_0 \in \mathbb{R}$ for all $f \in C^3(\mathbb{R})$ and $x_0 \in \mathbb{R}$ with $f'(x_0) \neq 0$. A similar argument as in part (iii) of the proof of Theorem 4.1 then shows that H is continuous. For a detailed proof, cf. [KM3, p. 889].

For $p \geq 2$ or $p > 2$, formula (9.18) also holds for $\alpha_1 = 0$, i.e., when $f'(x) = 0$. To see this, take arbitrary $x_0, z, \alpha_2, \alpha_3 \in \mathbb{R}$ with α_2, α_3 not both zero. Choose $f(x) = z + \frac{\alpha_2}{2}(x - x_0)^2 + \frac{\alpha_3}{6}(x - x_0)^3$. Then for x close to x_0, $f'(x) \neq 0$. We consider the case $f'(x) > 0$. Then by (9.19)

$$Tf(x) = d\left[\alpha_3 \left(\alpha_2(x - x_0) + \frac{\alpha_3}{2}(x - x_0)^2\right)^{p-1}\right.$$
$$\left. - \frac{3}{2}(\alpha_2 + \alpha_3(x - x_0))^2 \left(\alpha_2(x - x_0) + \frac{\alpha_3}{2}(x - x_0)^2\right)^{p-2}\right]$$
$$+ \left(\alpha_2(x - x_0) + \frac{\alpha_3}{2}(x - x_0)^2\right)^{p+2} H(f(x))$$
$$- \left(\alpha_2(x - x_0) + \frac{\alpha_3}{2}(x - x_0)^2\right)^p H(x).$$

For $x \to x_0$, the left-hand side converges to $(Tf)(x_0) = F(x_0, z, 0, \alpha_2, \alpha_3)$ by the continuity of Tf and the right-hand side converges to 0 if $p > 2$ and to $-\frac{3}{2}d\alpha_2^2$ if $p = 2$ since H and f are continuous. Hence (9.18) and (9.19) also hold for $\alpha_1 = 0$, assuming $p \geq 2$.

(vi) We quickly mention how the analysis of F and B changes when $k = 2$, i.e., when $T : C^2(\mathbb{R}) \to C(\mathbb{R})$ and $A_1, A_2 : C^1(\mathbb{R}) \to C(\mathbb{R})$. The representing functions F, B_1 and B_2 depend on one argument less and (9.6) is replaced by

$$F(x, z, \alpha_1\beta_1, \alpha_2\beta_1^2 + \alpha_1\beta_2) = F(y, z, \alpha_1, \alpha_2)B_1(y, \beta_1) + F(x, y, \beta_1, \beta_2)B_2(z, \alpha_1).$$

In this case $\widetilde{F}(\alpha_2) := F(x, x, 1, \alpha_2)$ is independent of x and additive. Putting $G(x) = T(\operatorname{Id} + x)(0) = F(0, x, 1, 0)$, equations (9.9) and (9.10) are replaced by

$$F(x, z, \alpha_1, \alpha_2) = F(x, y, \alpha_1, 0) + \left[\widetilde{F}\left(\frac{\alpha_2}{\alpha_1^2}\right) + G(z) - G(y)\right] B_1(y, \alpha_1)$$
$$= F(y, z, \alpha_1, 0) + \left[\widetilde{F}\left(\frac{\alpha_2}{\alpha_1}\right) + G(y) - G(x)\right] B_2(z, \alpha_1).$$

This then yields that $B_1(z, \alpha_1) = \alpha_1 B_2(z, \alpha_1)$ and (9.11) has the analogue

$$d\frac{\alpha_2}{\alpha_1}\big(B_2(z, \alpha_1) - B_2(x, \alpha_1)\big) = \big[F(x, x, \alpha_1, 0) - F(z, z, \alpha_1, 0)\big]$$
$$+ \big[G(z) - G(x)\big]\big[\alpha_1 B_2(x, \alpha_1) - B_2(z, \alpha_1)\big],$$

with $d := \widetilde{F}(1) \neq 0$, since T is assumed to depend on the second derivative. This shows that $B(\alpha_1) := B_2(z, \alpha_1)$ is independent of $z \in \mathbb{R}$, and (9.12) is replaced by

$$F(x, z, \alpha_1, \alpha_2) = F_0(\alpha_1) + \Big[d\frac{\alpha_2}{\alpha_1} + \alpha_1 G(z) - G(x)\Big]B(\alpha_1),$$

with $F_0(\alpha_1) := F(0, 0, \alpha_1, 0)$. The analogue of (9.13) is

$$d\left(\frac{\alpha_2\beta_1}{\alpha_1} + \frac{\beta_2}{\beta_1}\right)\big[B(\alpha_1\beta_1) - B(\alpha_1)B(\beta_1)\big]$$
$$= \beta_1 F_0(\alpha_1) B(\beta_1) + F_0(\beta_1) B(\alpha_1) - F_0(\alpha_1\beta_1).$$

The right-hand side is independent of α_2 and β_2, implying that both sides are zero. Hence B is multiplicative, $B(\alpha_1\beta_1) = B(\alpha_1) B(\beta_1)$ and

$$F_0(\alpha_1\beta_1) = \beta_1 F_0(\alpha_1) B(\beta_1) + F_0(\beta_1) B(\alpha_1),$$

which yields that $F_0(\alpha_1) = c(\alpha_1 - 1)B(\alpha_1)$, and finally with $H(z) = G(z) + c$,

$$F(x, z, \alpha_1, \alpha_2) = \Big(d\frac{\alpha_2}{\alpha_1} + \alpha_1 H(z) - H(x)\Big)B(\alpha_1),$$

as the analogue of equation (9.18) in the case $k = 2$.

(vii) We now turn to the case $k = 1$, when $T, A_1, A_2 : C^1(\mathbb{R}) \to C(\mathbb{R})$ satisfy (9.1). If $F : \mathbb{R}^3 \to \mathbb{R}$, $B_1, B_2 : \mathbb{R}^2 \to \mathbb{R}$ represent T, A_1, A_2 according to Proposition 9.3,

$$Tf(x) = F\big(x, f(x), f'(x)\big), \quad A_i f(x) = B_i\big(f(x), f'(x)\big), \quad i = 1, 2,$$

for all $x \in \mathbb{R}$, $f \in C^1(\mathbb{R})$, we have as a replacement of (9.6)

$$F(x, z, \alpha_1\beta_1) = F(y, z, \alpha_1) B_1(y, \beta_1) + F(x, y, \beta_1) B_2(z, \alpha_1), \tag{9.20}$$

for all $x, y, z, \alpha_1, \beta_1 \in \mathbb{R}$. We again put $G(x) := F(0, x, 1) = T(\mathrm{Id}+x)(0)$, with $G(0) = T(\mathrm{Id})(0) = 0$. Let $\widetilde{B}_i(\alpha_1) := B_i(0, \alpha_1)$ and $E(\alpha_1) := F(0, 0, \alpha_1)$. Then, putting $y = 0$ in (9.20) and using the symmetry in (α_1, β_1), we have

$$F(x, z, \alpha_1\beta_1) = F(0, z, \alpha_1) \widetilde{B}_1(\beta_1) + F(x, 0, \beta_1) B_2(z, \alpha_1)$$
$$= F(0, z, \beta_1) \widetilde{B}_1(\alpha_1) + F(x, 0, \alpha_1) B_2(z, \beta_1).$$

Choosing $\beta_1 = 1$ in both equations, and also $z = 0$ in the first equation, we find, using also $B_i(z, 1) = 1$,

$$F(x, z, \alpha_1) = G(z)\, \widetilde{B}_1(\alpha_1) + F(x, 0, \alpha_1)$$
$$= G(z)\, \widetilde{B}_1(\alpha_1) + \big(E(\alpha_1) - G(x)\, \widetilde{B}_2(\alpha_1)\big). \qquad (9.21)$$

Inserting this into (9.20) with $y = 0$, we get after reordering terms

$$G(z)\big[\widetilde{B}_1(\alpha_1\beta_1) - \widetilde{B}_1(\alpha_1)\widetilde{B}_1(\beta_1)\big] - G(x)\big[\widetilde{B}_2(\alpha_1\beta_1) - B_2(z, \alpha_1)\, \widetilde{B}_2(\beta_1)\big]$$
$$= E(\alpha_1)\, \widetilde{B}_1(\beta_1) + E(\beta_1)\, B_2(z, \alpha_1) - E(\alpha_1\beta_1). \qquad (9.22)$$

The functions G and E cannot both be zero, since then by (9.21), $F = 0$ and $T = 0$.

(viii) Assume first that $G \equiv 0$ and $E \neq 0$. Then the left and hence also the right-hand side of (9.22) is zero, and hence $B_2(z, \alpha_1)$ cannot depend on z since the other terms do not depend on z. Hence

$$E(\alpha_1\beta_1) = E(\alpha_1)\, B(\beta_1) + E(\beta_1)\, B(\alpha_1) \qquad (9.23)$$

with $B := \frac{1}{2}(\widetilde{B}_1 + \widetilde{B}_2)$. By (9.21)

$$Tf(x) = F\big(x, f(x), f'(x)\big) = E(f'(x))$$

is continuous for all $f \in C^1(\mathbb{R})$. Choosing $f(x) = \frac{1}{2}x^2$ shows that E is continuous. Also $\frac{1}{2}(A_1 f(x) + A_2 f(x)) = B(f'(x))$, which implies that B is continuous too. By Proposition 2.13, the solutions of (9.23) are given by either

$$E(\alpha_1) = d\, \ln|\alpha_1|\, |\alpha_1|^p\, \{\operatorname{sgn}\alpha_1\}, \quad B(\alpha_1) = |\alpha_1|^p\, \{\operatorname{sgn}\alpha_1\},$$

or

$$E(\alpha_1) = d\, \sin(c\ln|\alpha_1|)\, |\alpha_1|^p\, \{\operatorname{sgn}\alpha_1\}, \quad B(\alpha_1) = \cos(c\ln|\alpha_1|)\, |\alpha_1|^p\, \{\operatorname{sgn}\alpha_1\},$$

or

$$E(\alpha_1) = \frac{d}{2}\big(|\alpha_1|^q\, \{\operatorname{sgn}\alpha_1\} - |\alpha_1|^r\, [\operatorname{sgn}\alpha_1]\big),$$
$$B(\alpha_1) = \frac{1}{2}\big(|\alpha_1|^q\, \{\operatorname{sgn}\alpha_1\} + |\alpha_1|^r\, [\operatorname{sgn}\alpha_1]\big),$$

for suitable constants d, c, p, q, r. These solutions are of the form given in Theorem 9.1 (b1), $k = 1$ and (b2); B in the last solution may be replaced for (9.22) by $B_1(\alpha_1) = |\alpha_1|^q\, \{\operatorname{sgn}\alpha_1\}$, $B_2(\alpha) = |\alpha_1|^r\, [\operatorname{sgn}\alpha_1]$, since only $\frac{1}{2}(B_1 + B_2)$ is uniquely determined. The last solution (d) of Proposition 2.13 does not apply here since $B(1)$ needs to be 1.

(ix) Now assume that $G \not\equiv 0$. Since the right-hand side of (9.22) is independent of x, so is the left-hand side. Choose $x \in \mathbb{R}$ with $G(x) \neq 0$. We also know that $G(0) = 0$. It follows from (9.22) that

$$\widetilde{B}_2(\alpha_1\beta_1) = B_2(z, \alpha_1) \widetilde{B}_2(\beta_1),$$

which again implies that $B_2(z, \alpha_1)$ is independent of z and that $\widetilde{B}_2$ is multiplicative, i.e., $\widetilde{B}_2(\alpha_1) = |\alpha_1|^p\{\operatorname{sgn}\alpha_1\}$ for some $p \geq 0$.

Now the right-hand and hence also the left-hand side of (9.22) are independent of z, yielding $\widetilde{B}_1(\alpha_1\beta_1) = \widetilde{B}_1(\alpha_1)\widetilde{B}_1(\beta_1)$. Therefore, $\widetilde{B}_1(\alpha_1) = |\alpha_1|^q[\operatorname{sgn}\alpha_1]$ for some $q \geq 0$. Since the left-hand side of (9.22) is zero, so is the right-hand side, and by symmetry

$$E(\alpha_1\beta_1) = E(\alpha_1) \widetilde{B}_1(\beta_1) + E(\beta_1) \widetilde{B}_2(\alpha_1)$$
$$= E(\alpha_1) \widetilde{B}_2(\beta_1) + E(\beta_1) \widetilde{B}_1(\alpha_1), \qquad (9.24)$$

$$E(\alpha_1)\big(\widetilde{B}_1(\beta_1) - \widetilde{B}_2(\beta_1)\big) = E(\beta_1)\big(\widetilde{B}_1(\alpha_1) - \widetilde{B}_2(\alpha_1)\big). \qquad (9.25)$$

Equation (9.24) implies for $\alpha_1 = \beta_1 = 1$ that $E(1) = 2E(1)$, $E(1) = 0$. By (9.21)

$$Tf = E(f') + G(f) \cdot \widetilde{B}_1(f') - G \cdot \widetilde{B}_2(f')$$

is continuous for all $f \in C^1(\mathbb{R})$. Hence, for functions f and x with $f'(x) = 1$ we have that $G(f) - G$ is continuous at x. This implies that G is continuous, similarly as in the proof of Theorem 4.1. Choose $f(x) = \frac{x^2}{2}$ to conclude that also E is continuous.

If $E \equiv 0$, $F(x, z, \alpha_1) = G(z) |\alpha_1|^q [\operatorname{sgn}\alpha_1] - G(x) |\alpha_1|^p \{\operatorname{sgn}\alpha_1\}$ by (9.21), yielding one of the solutions of (b2) of Theorem 9.1.

If $E \not\equiv 0$, choose $\beta_1 \in \mathbb{R}$ with $E(\beta_1) \neq 0$. Then (9.25) implies

$$\widetilde{B}_1(\alpha_1) = \widetilde{B}_2(\alpha_1) + c\,E(\alpha_1), \quad c := \frac{\widetilde{B}_1(\beta_1) - \widetilde{B}_2(\beta_1)}{E(\beta_1)}.$$

If $c = 0$, $\widetilde{B}_1(\alpha_1) = \widetilde{B}_2(\alpha_1) = |\alpha_1|^p \{\operatorname{sgn}\alpha_1\}$ and (9.24) gives

$$\frac{E(\alpha_1\beta_1)}{\widetilde{B}_2(\alpha_1\beta_1)} = \frac{E(\alpha_1)}{\widetilde{B}_2(\alpha_1)} + \frac{E(\beta_1)}{\widetilde{B}_2(\beta_1)}, \quad \alpha_1 \neq 0 \neq \beta_1.$$

Hence, $\psi(s) := \frac{E(\exp s)}{\widetilde{B}_2(\exp s)}$ is additive and continuous. Thus there is $d \neq 0$ such that $\psi(s) = ds$, $s \in \mathbb{R}$. Therefore

$$E(\alpha_1) = d \ln|\alpha_1| \widetilde{B}_2(\alpha_1),$$

and equation (9.21) gives

$$F(x, z, \alpha_1) = \big(d \, \ln|\alpha_1| + G(z) - G(x)\big)\widetilde{B}_2(\alpha_1),$$
$$\widetilde{B}_1(\alpha_1) = \widetilde{B}_2(\alpha_1) = |\alpha_1|^p\{\mathrm{sgn}\,\alpha_1\},$$

which is the solution in (b1) for $k = 1$.

If $c \neq 0$, with $d = \frac{1}{c}$, $E(\alpha_1) = d(\widetilde{B}_1(\alpha_1) - \widetilde{B}_2(\alpha_1))$, and with $H(x) := G(x) + d$, using again (9.21)

$$F(x, z, \alpha_1) = H(z)\,\widetilde{B}_1(\alpha_1) - H(x)\,\widetilde{B}_2(\alpha_1)$$
$$= H(z)\,|\alpha_1|^q\,[\mathrm{sgn}\,\alpha_1] - H(x)\,|\alpha_1|^p\,\{\mathrm{sgn}\,\alpha_1\},$$

which gives the last solution in part (b1) of Theorem 9.1.

This ends the proof of part (b) of Theorem 9.1, finding all solutions of (9.1) if $k \in \{1, 2, 3\}$. $\qquad\square$

Remark. As mentioned in Remark (b) after Theorem 9.1, the assumption of pointwise C^{k-1}-continuity of A_1 and A_2 for $k \geq 2$ is of a technical nature. Without this assumption, but keeping the isotropy condition, e.g., equation (9.7), would have to be replaced by

$$F(x, z, 1, 0, \alpha_3 + \beta_3)$$
$$= F(y, z, 1, 0, \alpha_3)B_1(y, 1, 0, \beta_3) + F(x, y, 1, 0, \beta_3)B_2(z, 1, 0, \alpha_3),$$

which admits solutions which do not satisfy $B_1(y, 1, 0, \beta_3) = B_2(z, 1, 0, \alpha_3)$ $= 1$ for all $y, z, \alpha_3, \beta_3 \in \mathbb{R}$. The continuous solutions of this equation for $x = y = z$ are given in Corollary 2.12. They involve additional exponential terms, e.g., in solution (a) of Corollary 2.12

$$F(x, x, 1, 0, \alpha_3) = c(x)\alpha_3 \exp(p(x)\alpha_3), \quad B_i(x, 1, 0, \alpha_3) = \exp(p(x)\alpha_3).$$

Equations corresponding to (9.9) and (9.10) should, however, yield a contradiction if $p(x)$ were non-zero, due to different orders of growth of $F(x, z, \alpha_1, \alpha_2, \alpha_3)$ in α_3 in both formulas.

9.3   The case $k \geq 4$

It still remains to prove part (a) of Theorem 9.1, namely, that the second-order chain rule equation (9.1) does not admit any solutions which depend non-trivially on the k-th derivative when $k \geq 4$. We prove this without the isotropy assumption.

Proposition 9.4. *Let $k \in \mathbb{N}$, $k \geq 4$. Suppose that $T : C^k(\mathbb{R}) \to C(\mathbb{R})$ and $A_1, A_2 :$ $C^{k-1}(\mathbb{R}) \to C(\mathbb{R})$ satisfy (9.1), that (T, A_1) is C^k-non-degenerate and that A_1 and A_2 are C^{k-1}-pointwise continuous. Then T does not depend on the k-th derivative.*

The basic reason that no operators T exist so that Tf would depend non-trivially on $f^{(k)}$ for some $k \geq 4$ is that the Faà di Bruno formula for $(f \circ g)^{(k)}$ contains too many terms, so that that functional equation for the representing functions F, A_1, and A_2 would require too many equations to hold for relatively few variables. This is true, in particular, since the operators A_1, A_2 would necessarily have the simple form $A_1 f = (f')^{k-1} A_2 f$, $A_2 f = |f'|^p \{\operatorname{sgn} f'\}$ with $p \geq 0$, as we will see in the proof.

Proof. (i) By Proposition 9.3 there are functions $F : \mathbb{R}^{k+2} \to \mathbb{R}$ and $B_1, B_2 : \mathbb{R}^{k+1} \to \mathbb{R}$ such that for all $f \in C^k(\mathbb{R})$ and $x \in \mathbb{R}$

$$Tf(x) = F\big(x, f(x), \ldots, f^{(k)}(x)\big),$$
$$A_i f(x) = B_i\big(x, f(x), \ldots, f^{(k-1)}(x)\big), \quad i = 1, 2.$$

The fact that A_1 and A_2 are isotropic was used only at the end of the proof of Proposition 9.3 to avoid the dependence of the functions B_i of the independent variable x which we may keep. We will apply the operator equation only for $x = y = z = 0$ and functions $f, g \in C^k(\mathbb{R})$ with $f(0) = g(0) = 0$. Then the functions F and B_i are independent of $x = 0$, $f(x) = 0$, and we may consider them as functions $F : \mathbb{R}^k \to \mathbb{R}$ and $B_i : \mathbb{R}^{k-1} \to \mathbb{R}$ for $i = 1, 2$. We have $T(\operatorname{Id}) = 0$, $A_i(\operatorname{Id}) = \mathbf{1}$. Isotropy is not needed to prove this. Therefore

$$F\underbrace{(1, 0, \ldots, 0)}_{k} = 0, \quad B_i\underbrace{(1, 0, \ldots, 0)}_{k-1} = 1.$$

The operator equation (9.1) then turns into a functional equation for F, B_1 and B_2,

$$F\big((f \circ g)'(0), \ldots, (f \circ g)^{(k)}(0)\big)$$
$$= F\big(f'(0), \ldots, f^{(k)}(0)\big) \cdot B_1\big(g'(0), \ldots, g^{(k-1)}(0)\big)$$
$$+ F\big(g'(0), \ldots, g^{(k)}(0)\big) \cdot B_2\big(f'(0), \ldots, f^{(k-1)}(0)\big).$$

For any $\alpha_1, \ldots, \alpha_k, \beta_1, \ldots, \beta_k \in \mathbb{R}$ choose $f, g \in C^k(\mathbb{R})$ with $f(0) = g(0) = 0$ and $f^{(j)}(0) = \alpha_j$, $g^{(j)}(0) = \beta_j$, $j \in \{1, \ldots, k\}$. Then $(f \circ g)''(0) = \alpha_2 \beta_1^2 + \alpha_1 \beta_2$. By the Faà di Bruno formula for $(f \circ g)^{(j)}$ for $j \geq 3$, we have

$$p_j(\alpha, \beta) := (f \circ g)^{(j)}(0)$$
$$= \alpha_j \beta_1^j + \binom{j}{2} \alpha_{j-1} \beta_1^{j-1} \beta_2 + q_j(\alpha, \beta) + j \alpha_2 \beta_1 \beta_{j-1} + \alpha_1 \beta_j$$
$$=: \alpha_j \beta_1^j + r_j(\alpha, \beta) + \alpha_1 \beta_j \tag{9.26}$$

where $q_3(\alpha, \beta) = 0$ and $q_j(\alpha, \beta)$ for $j \geq 4$ is the sum of monomials in the variables $(\alpha_1, \ldots, \alpha_{j-2}, \beta_1, \ldots, \beta_{j-2})$, each of which contains at least one α_ℓ and one β_m as a factor for some $2 \leq \ell, m \leq j-2$, and where $r_j(\alpha, \beta)$ depends only on $\alpha_1, \ldots, \alpha_{j-1}$ and $\beta_1, \ldots, \beta_{j-1}$. This is shown by induction on j. For a simple proof of the Faà di Bruno formula, we refer to Spindler [Sp]. Note that $p_j(\alpha, \beta)$ is *not* symmetric in (α, β). The operator equation (9.1) for (T, A_1, A_2) is then equivalent to the functional equation for (F, B_1, B_2) given by

$$
\begin{aligned}
F\big(\alpha_1\beta_1, \alpha_2\beta_1^2 + \alpha_1\beta_2, &p_3(\alpha, \beta), \ldots, p_k(\alpha, \beta)\big) \\
&= F(\alpha_1, \ldots, \alpha_k)B_1(\beta_1, \ldots, \beta_{k-1}) + F(\beta_1, \ldots, \beta_k)B_2(\alpha_1, \ldots, \alpha_{k-1}). \quad (9.27)
\end{aligned}
$$

(ii) We show that $F(\alpha_1, \ldots, \alpha_{k-1}, \alpha_k)$ is an affine function of α_k; more precisely,

$$
F(\alpha_1, \ldots, \alpha_k) = F(\alpha_1, \ldots, \alpha_{k-1}, 0) + d\frac{\alpha_k}{\alpha_1}B_2(\alpha_1, \ldots, \alpha_{k-1}), \quad (9.28)
$$

and that B_1 and B_2 are related by

$$
B_1(\alpha_1, \ldots, \alpha_{k-1}) = \alpha_1^{k-1}B_2(\alpha_1, \ldots, \alpha_{k-1}). \quad (9.29)
$$

This is similar to part (ii) of the proof of part (b) of Theorem 9.1:

Define $\widetilde{F}(\alpha_k) := F(1, 0, \ldots, 0, \alpha_k)$. Choosing $\alpha_1 = \beta_1 = 1$, $\alpha_2 = \cdots = \alpha_{k-1} = \beta_1 = \cdots = \beta_{k-1} = 0$ in (9.27), we get using (9.26) that $\widetilde{F}$ is additive, $\widetilde{F}(\alpha_k + \beta_k) = \widetilde{F}(\alpha_k) + \widetilde{F}(\beta_k)$. Next take $\alpha_1 = 1$, $\alpha_2 = \cdots = \alpha_{k-1} = \beta_k = 0$. Then by (9.26) and (9.27)

$$
F(\beta_1, \ldots, \beta_{k-1}, \beta_1^k\alpha_k) = \widetilde{F}(\alpha_k)B_1(\beta_1, \ldots, \beta_{k-1}) + F(\beta_1, \ldots, \beta_{k-1}, 0),
$$

using $B_2(1, 0, \ldots, 0) = 1$. Renaming variables yields for $\alpha_1 \neq 0$

$$
F(\alpha_1, \ldots, \alpha_{k-1}, \alpha_k) = \widetilde{F}\left(\frac{\alpha_k}{\alpha_1^k}\right)B_1(\alpha_2, \ldots, \alpha_{k-1}) + F(\alpha_1, \ldots, \alpha_{k-1}, 0).
$$

Similarly, starting with $\beta_1 = 1$, $\beta_2 = \cdots = \beta_{k-1} = \alpha_k = 0$, we find for $\alpha_1 \neq 0$,

$$
F(\alpha_1, \ldots, \alpha_{k-1}, \alpha_k) = \widetilde{F}\left(\frac{\alpha_k}{\alpha_1}\right)B_2(\alpha_1, \ldots, \alpha_{k-1}) + F(\alpha_1, \ldots, \alpha_{k-1}, 0). \quad (9.30)
$$

Comparing both equations, we conclude that

$$
\widetilde{F}\left(\frac{\alpha_k}{\alpha_1^k}\right)B_1(\alpha_1, \ldots, \alpha_{k-1}) = \widetilde{F}\left(\frac{\alpha_k}{\alpha_1}\right)B_2(\alpha_1, \ldots, \alpha_{k-1}).
$$

We now assume that Tf depends non-trivially on $f^{(k)}$. Then F has to depend on α_k. The previous formulas then show that $\widetilde{F}$ cannot be identically zero. Thus there is $c \neq 0$ such that $\widetilde{F}(c) \neq 0$. Let $\overline{F}(t) := \frac{\widetilde{F}(ct)}{\widetilde{F}(c)}$. Choose $\alpha_k := c\alpha_1^k$. Then

$$
B_1(\alpha_1, \ldots, \alpha_{k-1}) = \overline{F}(\alpha_1^{k-1})B_2(\alpha_1, \ldots, \alpha_{k-1}). \quad (9.31)
$$

For $f(x) = x + \frac{1}{k}x^k$, $A_2 f(x) = B_2(1 + x^{k-1}, (k-1)x^{k-2}, \ldots, (k-1)!x)$. Hence, $A_2 f(0) = B_2(1, 0, \ldots, 0) = A_2(\text{Id})(0) = 1$. Since $A_2 f$ is continuous, there is $\epsilon > 0$ such that $A_2 f(x) \neq 0$ for all $|x| \leq \epsilon$. Define $\varphi : [1, 1 + \epsilon^{k-1}] \to [0, \epsilon]$ by $\varphi(\alpha) = (\alpha - 1)^{\frac{1}{k-1}}$. Then $A_2 f(\varphi(x))$ is continuous and non-zero in $[1, 1 + \epsilon^{k-1}]$. Therefore

$$\frac{A_1 f(\varphi(\alpha))}{A_2 f(\varphi(\alpha))} = \frac{B_1\big(\alpha, (k-1)\,\varphi(\alpha)^{k-1}, \ldots, (k-1)!\varphi(\alpha)\big)}{B_2\big(\alpha, (k-1)\,\varphi(\alpha)^{k-2}, \cdots, (k-1)!\varphi(\alpha)\big)} = \overline{F}(\alpha^{k-1})$$

is continuous in $[1, 1 + \epsilon^{k-1}]$. By Proposition 2.2, the additive function $\overline{F}$ is linear. Using $\overline{F}(1) = 1$, we conclude that $\overline{F}(t) = t$ and hence $\widetilde{F}(t) = dt$ with $d := \widetilde{F}(1) \neq 0$. Now (9.31) implies (9.29), and (9.30) yields (9.28). By (9.26)

$$\frac{p_k(\alpha, \beta)}{\alpha_1 \beta_1} = \frac{\alpha_k}{\alpha_1}\beta_1^{k-1} + \frac{\beta_k}{\beta_1} + \frac{r_k(\alpha, \beta)}{\alpha_1 \beta_1},$$

where $\frac{r_k(\alpha, \beta)}{\alpha_1 \beta_1}$ depends neither on α_k nor on β_k. We now insert (9.28) into (9.27), use the last equation and isolate the terms having a factor α_k or β_k on the left-hand side. Calculation shows that

$$d\left(\frac{\alpha_k}{\alpha_1}\beta_1^{k-1} + \frac{\beta_k}{\beta_1}\right)\big[B_2(\alpha_1\beta_1, \alpha_2\beta_1^2 + \alpha_1\beta_2, \ldots, p_{k-1}(\alpha, \beta))$$
$$- B_2(\alpha_1, \ldots, \alpha_{k-1})B_2(\beta_1, \ldots, \beta_{k-1})\big]$$
$$= \beta_1^{k-1}F(\alpha_1, \ldots, \alpha_{k-1}, 0)B_2(\beta_1, \ldots, \beta_{k-1})$$
$$+ F(\beta_1, \ldots, \beta_{k-1}, 0)B_2(\alpha_1, \ldots, \alpha_{k-1})$$
$$- F\big(\alpha_1\beta_1, \ldots, p_{k-1}(\alpha, \beta), 0\big) - d\frac{r_k(\alpha, \beta)}{\alpha_1\beta_1}B_2\big(\alpha_1\beta_1, \ldots, p_{k-1}(\alpha, \beta)\big). \quad (9.32)$$

Since the right-hand side of (9.32) neither depends on α_k nor on β_k, and since α_k and β_k may be chosen arbitrarily, the factor of the term involving α_k and β_k on the left has to be zero, and we get, using the symmetry of the product,

$$B_2(\alpha_1\beta_1, \alpha_2\beta_1^2 + \alpha_1\beta_2, \ldots, p_{k-1}(\alpha, \beta)) = B_2(\alpha_1, \ldots, \alpha_{k-1})\,B_2(\beta_1, \ldots, \beta_{k-1})$$
$$= B_2\big(\alpha_1\beta_1, \beta_2\alpha_1^2 + \beta_1\alpha_2, \ldots, p_{k-1}(\beta, \alpha)\big). \quad (9.33)$$

The same argument as in the proof of Theorem 4.1, part (i), on the chain rule equation of first order now shows that B_2 is independent of the variables $\alpha_2, \ldots, \alpha_{k-1}$: For α_1, β_1 with $\alpha_1\beta_1 \notin \{0, 1, -1\}$ and arbitrary values of $t_2, \ldots, t_{k-1}, s_2, \ldots, s_{k-1}$ we may successively solve the equations

$$\left\{\begin{matrix} p_2(\alpha, \beta) = t_2 \\ p_2(\beta, \alpha) = s_2 \end{matrix}\right\}, \ldots, \left\{\begin{matrix} p_{k-1}(\alpha, \beta) = t_{k-1} \\ p_{k-1}(\beta, \alpha) = s_{k-1} \end{matrix}\right\},$$

since they may be reformulated in terms of the successive linear equations for $(\alpha_2, \beta_2), \ldots, (\alpha_{k-1}, \beta_{k-1})$, using (9.26),

$$\left\{\begin{matrix} \beta_1^j\alpha_j + \alpha_1\beta_j = t_j - r_j(\alpha, \beta) \\ \beta_1\alpha_j + \alpha_1^j\beta_j = s_j - r_j(\alpha, \beta) \end{matrix}\right\}, \quad j = 2, \ldots, k-1,$$

where $r_2 = 0$ and $r_j(\alpha, \beta)$ only depends on the variables $(\alpha_1, \ldots, \alpha_{j-1}, \beta_1, \ldots, \beta_{j-1})$ chosen before. The equations are uniquely solvable since $\det \begin{pmatrix} \beta_1^j & \alpha_1 \\ \beta_1 & \alpha_1^j \end{pmatrix} = \alpha_1 \beta_1((\alpha_1\beta_1)^{j-1} - 1) \neq 0$. Therefore, for all $\alpha \notin \{0, 1, -1\}$, by (9.33)

$$B_2(\alpha, t_2, \ldots, t_{k-1}) = B_2(\alpha, s_2, \ldots, s_{k-1}) =: B(\alpha),$$

and B_2 does not depend on $\alpha_2, \ldots, \alpha_{k-1}$ for $\alpha \notin \{0, 1, -1\}$. This also holds in the limit for $\alpha \in \{0, 1, -1\}$. Equation (9.33) then yields that B is multiplicative, $B(\alpha_1\beta_1) = B(\alpha_1) B(\beta_1)$. For $f(x) = \frac{x^2}{2}$, $A_2 f(x) = B_2(x, 0, \ldots, 0) = B(x)$ is continuous. By Proposition 2.3, there is $p \geq 0$ such that

$$B_2(\alpha_1, \ldots, \alpha_{k-1}) = B(\alpha_1) = |\alpha_1|^p \{\operatorname{sgn} \alpha_1\}.$$

Since the left-hand side of (9.32) is zero, so is the right-hand side. We use this only for $\alpha_1 = \beta_1 = 1$ and conclude, using $B(1) = 1$, that

$$F\big(1, \alpha_2 + \beta_2, \ldots, p_{k-1}(\alpha, \beta), 0\big) + d\, r_k(\alpha, \beta)$$
$$= F(1, \alpha_2, \ldots, \alpha_{k-1}, 0) + F(1, \beta_2, \ldots, \beta_{k-1}, 0). \quad (9.34)$$

Taking here, for $k \geq 4$, $\alpha_2 = \cdots = \alpha_{k-2} = \beta_{k-1} = 0$, and $\alpha_1 = \beta_1 = 1$ as before, we find from (9.26) and the explanation of the term $q_k(\alpha, \beta)$ in

$$r_k(\alpha, \beta) = \binom{k}{2}\alpha_{k-1}\beta_1^{k-1}\beta_2 + k\,\alpha_2\,\beta_1\,\beta_{k-1} + q_k(\alpha, \beta)$$

given there that

$$F(1, \beta_2, \ldots, \beta_{k-2}, \alpha_{k-1}, 0) + d\binom{k}{2}\alpha_{k-1}\beta_2$$
$$= F(1, 0, \ldots, 0, \alpha_{k-1}, 0) + F(1, \beta_2, \ldots, \beta_{k-2}, 0, 0)$$

Renaming variables, this means

$$F(1, \alpha_2, \ldots, \alpha_{k-2}, \alpha_{k-1}, 0) + d\binom{k}{2}\alpha_{k-1}\alpha_2$$
$$= F(1, 0, \ldots, 0, \alpha_{k-1}, 0) + F(1, \alpha_2, \ldots, \alpha_{k-2}, 0, 0).$$

Next, we choose $\beta_2 = \cdots = \beta_{k-2} = \alpha_{k-1} = 0$ in (9.34). Using (9.26), we then get

$$F(1, \alpha_2, \ldots, \alpha_{k-2}, \beta_{k-1}, 0) + d\,k\,\alpha_2\beta_{k-1}$$
$$= F(1, \alpha_2, \ldots, \alpha_{k-2}, 0, 0) + F(1, 0, \ldots, 0, \beta_{k-1}, 0).$$

Renaming β_{k-1} as α_{k-1}, both equations are identical except for the term involving $\alpha_2\,\alpha_{k-1}$. We conclude that necessarily

$$0 = d\left[\binom{k}{2} - k\right]\alpha_2\,\alpha_{k-1} = d\,\frac{k(k-3)}{2}\,\alpha_2\,\alpha_{k-1}. \quad (9.35)$$

Since α_{k-1} and α_2 can be chosen arbitrarily and $k - 3 \geq 1$, it follows that $d = 0$, hence $\widetilde{F} = 0$ and by (9.28)

$$F(\alpha_1, \ldots, \alpha_k) = F(\alpha_1, \ldots, \alpha_{k-1}, 0).$$

The assumption that Tf depended non-trivially on $f^{(k)}$ let to the conclusion that it does not depend on $f^{(k)}$, a contradiction. Note that equation (9.35) does not give a contradiction for $k = 3$. There we had the Schwarzian solution. This ends the proof of Proposition 9.4. □

9.4 Notes and References

Theorem 9.1 is due to König and Milman in [KM3] ($k \in \{0, 1, 2\}$) and [KM6] ($k \geq 3$).

A somewhat similar answer as in Theorem 9.1 appears in the study of the first additive function of the group of diffeomorphisms on the projective line, with coefficients in the λ-densities, cf. Ovsienko, Tabachnikov [OT]. This is essentially the study of *continuous* operators $T : \text{Diff}(\mathbb{RP}^1) \to C^\infty(\mathbb{RP}^1)$ satisfying the functional equation

$$T(f \circ g) = (g')^\lambda Tf \circ g + Tg,$$

for various values $\lambda \in \mathbb{R}$, i.e., for operators of the specific form $A_1 g = (g')^\lambda$, $A_2 f = \mathbf{1}$ on different function spaces. This equation generalizes the chain rule in a different fashion than the one studied here. As it turns out, the cohomology groups are non-trivial only for the values $\lambda \in \{0, 1, 2\}$, and they are represented by cocycles constituting derivations of orders $1, 2, 3$, respectively. In the last case, the non-trivial cocycle corresponds to the Schwarzian derivative, [OT, p. 20]. This should be compared to Theorem 9.1 with (formally) $p = 0$ and $\lambda = 2$ when T is the Schwarzian and $A_1 g = (g')^2$, $A_2 f = 1$. Note here that we do neither assume the continuity of T nor a specific form of the operators A_1 and A_2. For $q = 0$ and $\lambda = 1, 0$, respectively, the non-trivial cocycles are represented by f'' and $\ln |f'|$, respectively, [OT]. Clearly, for diffeomorphisms f, $f' \neq 0$. We thank D. Faifman for bringing [OT] to our attention.

Bibliography

[A]   J. Aczél; *Lectures on functional equations and their applications*, Academic Press, 1966.

[AD]   J. Aczél, J. Dhombres; *Functional equations in several variables*, Cambridge Univ. Press, 1989.

[AAM]   S. Alesker S. Artstein-Avidan, V. Milman; A characterization of the Fourier transform and related topics, in: A. Alexandrov et al. (ed), *Linear and complex analysis, Dedicated to V.P. Havin*, Amer. Math. Soc. Transl. **226**, *Advances in the Math. Sciences* **63** (2009), 11–26.

[AAFM]   S. Alesker, S. Artstein-Avidan, D. Faifman, V. Milman; A characterization of product preserving maps with applications to a characterization of the Fourier transform, *Illinois J. Math.* **54** (2010), 1115–1132.

[AO]   A. Alexiewicz, W. Orlicz; Remarque sur l'équation fonctionelle $f(x+y) = f(x) + f(y)$, *Fundamenta Math.* **33** (1945), 314–315.

[AFM]   S. Artstein-Avidan, D. Faifman, V. Milman; On multiplicative maps of continuous and smooth functions, Geometric aspects of functional analysis, *Lecture Notes in Math.* **2050**, 35–59, Springer, Heidelberg, 2012.

[AKM]   S. Artstein-Avidan, H. König, V. Milman; The chain rule as a functional equation, *J. Funct. Anal.* **259** (2010), 2999–3024.

[AM]   S. Artstein-Avidan, V. Milman; The concept of duality in convex analysis, and the characterization of the Legendre transform, *Annals of Math.* **169** (2009), 661–674.

[B]   S. Banach; Sur l'équation fonctionelle $f(x+y) = f(x)+f(y)$, *Fundamenta Math.* **1** (1920), 123–124.

[BS]   K. Böröczky, R. Schneider; A characterization of the duality mapping for convex bodies, *Geom. Funct. Anal.* **18** (2008), 657–667.

[D]   J. Dieudonné; On the automorphisms of classical groups, *Mem. AMS* **2** (1951), 1–95.

© Springer Nature Switzerland AG 2018
H. König, V. Milman, *Operator Relations Characterizing Derivatives*,
https://doi.org/10.1007/978-3-030-00241-1

[FKM]   D. Faifman, H. König, V. Milman; Submultiplicative operators on C^k-spaces, preprint, July 2016.

[F1]   D. Faifman; A characterization of the Fourier transform by the Poisson summation formula, *C. R. Math. Acad. Sci. Paris* **348** (2010), 407–410.

[F2]   D. Faifman; A family of unitary operators satisfying a Poisson-type summation formula, *Lecture Notes in Math.* **2050**, 191–204, Springer, 2012.

[F3]   D. Faifman; Personal communication.

[Fe]   M. Fekete, Über die Verteilung der Wurzeln gewisser algebraischer Gleichungen mit ganzzahligen Koeffizienten, *Math. Zeitschr.* **17** (1923), 228–249.

[Fr]   M. Fréchet; Pri la funkcia ekvacio $f(x + y) = f(x) + f(y)$, *Enseignement Mathématique* **15** (1913), 390–393.

[GS]   H. Goldmann, P. Šemrl; Multiplicative derivations on $C(X)$, *Monatsh. Math.* **121** (1996), 189–197.

[G]   Th. Gronwall; Note on the derivatives with respect to a parameter of the solutions of a system of differential equations, *Ann. Math.* **20** (1919), 292–296.

[Gr]   P. Gruber; The endomorphisms of the lattice of norms in finite dimensions, *Abh. Math. Sem. Univ. Hamburg* **62** (1992), 179–189.

[GMP]   J. Gustavsson, L. Maligranda, J. Peetre; A submultiplicative function, *Indag. Math.* **51** (1989), 435–442.

[H]   Ph. Hartman; *Ordinary Differential Equations*, 2nd ed., Birkhäuser, 1982.

[HP]   E. Hille, R. Phillips; *Functional Analysis and Semi-Groups*, Amer. Math. Soc., Providence, 1968.

[Ho]   L. Hörmander; *The Analysis of Linear Partial Differential Operators I*, Springer, 1983.

[H]   L. Hua; Supplement to the paper of Dieudonné on the automorphisms of classical groups, *Mem. AMS* **2** (1951), 96–122.

[J]   A. Járai; *Regularity Properties of Functional Equations in Several Variables*, Springer, 2005.

[Ke]   H. Kestelman; Sur l'équation fonctionelle $f(x + y) = f(x) + f(y)$, *Fundamenta Math.* **34** (1947), 144–147.

[KM1]   H. König, V. Milman; Characterizing the derivative and the entropy function by the Leibniz rule, with an appendix by D. Faifman, *J. Funct. Anal.* **261** (2011), 1325–1344.

[KM2]   H. König, V. Milman; The chain rule as a functional equation on $\mathbb{R}^n$, *J. Funct. Anal.* **261** (2011), 861–875.

[KM3]   H. König, V. Milman; A functional equation characerizing the second derivative, *J. Funct. Anal.* **261** (2011), 876–896.

[KM4]   H. König, V. Milman; An operator equation generalizing the Leibniz rule for the second derivative, *Lecture Notes in Math.* **2050**, 279–299, Springer, 2012.

[KM5]   H. König, V. Milman; An operator equation characterizing the Laplacian, *St. Petersburg Math. Journal (Algebra and Analysis)* **24** (2013), 631–644.

[KM6]   H. König, V. Milman; Operator equations and domain dependence, the case of the Schwarzian derivative, *J. Funct. Anal.* **266** (2014), 2546–2569.

[KM7]   H. König, V. Milman; Rigidity and stability of the Leibniz and the chain rule, *Proc. Steklov Inst.* **280** (2013), 191–207.

[KM8]   H. König, V. Milman; The derivation equation for C^k-functions: stability and localization, *Israel Journ. Math.* **203** (2014), 405–427.

[KM9]   H. König, V. Milman; Submultiplicative functions and operator inequalities, *Studia Math.* **223** (2014), 217–231.

[KM10] H. König, V. Milman; Rigidity of the chain rule and nearly submultiplicative functions, *Lecture Notes in Math.* **2169**, 235–264, Springer, 2017.

[KM11] H. König, V. Milman; The chain rule operator equation for polynomials and entire functions, in: Convexity and Concentration, *IMA Proc. in Math. and its Appl.*, 613–623, Springer, 2017.

[KM12] H. König, V. Milman; A note on operator equations describing the integral, *J. Math. Physics, Analysis, Geometry* **9** (2013), 51–58.

[KM13] H. König, V. Milman; The extended Leibniz rule and related equations in the space of rapidly decreasing functions, *J. Math. Physics, Analysis, Geometry* **14** (2018).

[LS]   G. Lešnjak, P. Šemrl; Continuous multiplicative mappings on $C(X)$, *Proc. Amer. Math. Soc.* **126** (1998), 127–133.

[M]   A.N. Milgram; Multiplicative semigroups of continuous functions, *Duke Math. J.* **16** (1949), 377–383.

[Mr]   J. Mrčun; On isomorphisms of algebras of smooth functions, *Proc. Amer. Math. Soc.* **133** (2005), 3109–3113.

[MS]   J. Mrčun, P. Šemrl; Multplicative bijections between algebras of differentiable functions, *Ann. Acad. Sci. Fenn. Math.* **32** (2007), 471–480.

[OT]   V. Ovsienko, S. Tabachnikov; *Projective Differential Geometry, old and new. From the Schwarzian Derivative to the Cohomology of Diffeomorphism Groups*, Cambridge Univ. Press, 2005.

[PS]   G. Pólya, G. Szegö; *Aufgaben und Lehrsätze aus der Analysis I*, Springer, 1970.

[RTRS]   J. M. Rassias, E. Thandapani, K. Ravi, B. V. Senthil Kumar; *Functional Equations and Inequalities*, World Scientific, 2017.

[S]   W. Sierpinski; Sur l'équation fonctionelle $f(x + y) = f(x) + f(y)$, *Fundamenta Math.* **1** (1920), 116–122.

[Sz]   L. Székelyhidi; *Convolution Type Functional Equations on Topological Abelian Groups*, World Scientific, 1991.

[Sp]   K. Spindler; A short proof of the formula of Faà di Bruno, *Elem. Math.* **60** (2005), 33–35.

Subject Index

additive function, 10, 13, 14, 86, 93, 181

annihilation of affine functions, 7, 114, 128

axiom of choice, 7, 11, 27

Cauchy equation, 10

chain rule, 2, 4–7, 9, 19, 29, 53–58, 62–66, 70–72, 75, 77–79, 86, 87, 91, 92, 104, 107, 111, 113, 131, 161, 163, 166, 183

  equation, 7, 181

  inequality, 91, 92, 97, 98, 101, 105, 107, 110

  operator equation, 169

  perturbed, 91, 104, 107, 110

  relaxed, 77, 79

  second-order, 5, 19, 114, 161, 163, 164, 166, 169, 178

C^k-spaces, 2

coboundaries, 71

cocycles, 71, 183

cohomology, 71, 183

compound product, 4, 71, 166

compound sum, 5

convolution, 1, 2, 89

differential operators, 9, 135

directional derivative, 115, 117, 118, 129, 130

Faà di Bruno formula, 58, 59, 179, 180

Faifman example, 51

Fourier transform, 1–3, 89

functional equation, 2–4, 6, 9, 10, 18, 19, 23, 24, 26, 27, 30, 45, 48, 58, 87, 113, 114, 120, 128, 137, 169, 174, 179, 180, 183

Gronwall's inequality, 91

half-bounded function, 6, 53, 56, 72, 161

Hamel basis, 8, 11

Hessian, 114

homogeneous solution, 113, 115, 134, 147

homothetic, 6, 39, 151, 156

initial conditions, 2, 4, 6, 9, 30, 54, 70, 73, 164, 165

inner automorphism, 65, 71

involution, 3

isotropic, 118, 130, 163, 165–167, 169, 179

Laplacian, 4, 7, 9, 113, 114, 128, 129, 136

Legendre transform, 3

Leibniz rule, 2, 4, 6, 7, 9, 15, 19, 29, 30, 32–36, 38, 39, 41, 42, 49–52, 55, 75–77, 79, 80, 113, 115, 117, 134, 143–146, 148, 150

  extended, 38–41, 44, 47, 52, 55, 56, 72, 117, 150, 152, 158, 166

  higher-order, 7, 84

  perturbed, 7, 79–81, 84, 86, 136

  relaxed, 76, 77

  second-order, 19, 113–115, 117, 128, 129, 134, 141, 144, 162, 166

localization, 6, 7, 10, 19, 30, 33, 35, 36, 38, 39, 65, 85, 97, 104, 107, 108, 110, 112, 114, 137, 142, 143, 147, 151, 156, 159

  on intervals, 31, 99

locally bounded function, 16, 80, 83

locally non-degenerate, 86, 87

locally surjective, 64

lower-order operator, 7, 161

© Springer Nature Switzerland AG 2018
H. König, V. Milman, *Operator Relations Characterizing Derivatives*,
https://doi.org/10.1007/978-3-030-00241-1

Author Index

© Springer Nature Switzerland AG 2018
H. König, V. Milman, *Operator Relations Characterizing Derivatives*,
https://doi.org/10.1007/978-3-030-00241-1

Printed in the United States
By Bookmasters